W0115707

# Extinct

# Extinct: A Compendium of Obsolete Objects

Edited by Barbara Penner,
Adrian Forty, Olivia Horsfall Turner
and Miranda Critchley

**REAKTION BOOKS**

Published by Reaktion Books Ltd
Unit 32, Waterside
44–48 Wharf Road
London N1 7UX, UK
www.reaktionbooks.co.uk

First published 2021

Copyright © The Authors 2021

All rights reserved

No part of this publication may be reproduced, stored in a retrieval
system, or transmitted, in any form or by any means, electronic,
mechanical, photocopying, recording or otherwise, without the prior
permission of the publishers

Published with support from HERA (Humanities in the European Research Area)

Humanities in the European Research Area

Printed and bound in India by Replika Press Pvt. Ltd

A catalogue record for this book is available from the British Library

ISBN 978 1 78914 452 9

# Contents

**Introduction**
Barbara Penner, Adrian Forty,
Olivia Horsfall Turner
and Miranda Critchley 09

**Acoustic Location Device**
Bryony Quinn 21

**Action Office Acoustic Area
Conditioner: The 'Maskitball'**
Kristen Gallerneaux 25

**Air-curtain Roof**
Laurent Stalder 29

**All-plastic House**
Carola Hein 33

**Arsenic Wallpaper**
Lucinda Hawksley 37

**Arundel Print**
Tanya Harrod 41

**Asbestos-cement Rondavel**
Hannah le Roux 45

**Ashtray**
Catherine Slessor 49

**Atmospheric Railway**
Niall McLaughlin 53

**Cab-fare Map**
Paul Dobraszczyk 57

**Central Heating**
Mario Carpo 61

**Chaparral 2J: The 'Sucker Car'**
Eirik A. G. Bøhn 65

**Chatelaine**
Iris Moon 69

**The Clapper**
Charles Rice 73

**Close-constraint Key**
Ben Vandenput 77

**Concorde**
Thomas McQuillan 81

**ConvAirCar**
Emily M. Orr 85

**Cybernetic Anthropomorphic Machines**
Lydia Kallipoliti 89

**Cybersyn**
Hugo Palmarola and Pedro Ignacio Alonso 93

**Cyclegraph**
Barbara Penner 97

*Cyclops 1*
Thandi Loewenson 101

**Dougong**
Guang Yu Ren and Edward Denison 105

**Dymaxion House**
Barry Bergdoll 109

**Edison's Anti-gravitation Under-clothing**
Bob Nicholson 113

**Electrotype Pattern**
Angus Patterson 117

**Fisher-Price Peg Figures**
Mark Morris 121

**Flashcube**
Harriet Harriss 125

**Flying Boat**
David Edgerton 129

**Glass Lantern Slide**
Daniel M. Abramson 133

**Globe of Mars**
Lucy Garrett 137

**High-pressure Water Mains**
Adrian Forty 141

**House Environment**
Eszter Steierhoffer 145

**'Hummingbird' Taxi**
Lucinda Hawksley 149

**Incandescent Light Bulb**
Mari Hvattum 153

**Integrated Radio/TV Cabinet**
Anders V. Munch 157

**Invacar: The 'Invalid Carriage'**
Elizabeth Guffey 161

**Kodachrome**
Tacita Dean 165

**Letraset**
Robin Kinross 169

**Leucotome**
Carsten Timmermann 173

**Manchester Pail System: 'Dolly Vardens'**
Barbara Penner 177

**Mechanical Polygraph**
Danielle S. Willkens 181

**Medical Wax Model**
Thomas Kador 185

**Memo**
Adrian Forty 189

**Milk Spoon**
Hugo Palmarola 193

**MiniDisc**
Priya Khanchandani 197

**Minitel**
Shahed Saleem 201

**Moon Towers**
Bryony Quinn 205

**Nikini**
Rachel Siobhán Tyler 209

**'No Nonsense' Fountain Pen**
Pippo Ciorra 213

**North Bucks Monorail City**
Gillian Darley 217

*Notgeld*
Tom Wilkinson 221

**Oil from Coal**
David Edgerton 225

**Optical Telegraph**
David Trotter 229

**Paper Aeroplane Ticket**
Gökçe Günel 233

**Paper Dress**
Olivia Horsfall Turner 237

**Pasilalinic-sympathetic Compass**
Richard Taws 241

**Phase-change Chemical Heat-storage Barrel**
Daniel Barber 245

**Player Piano**
Hal Foster 249

**Pneumatic Postal System**
Jacob Paskins 253

**Polaroid SX-70**
Deyan Sudjic 257

**Public Standards of Length**
David Rooney 261

**Pyrophone**
Tim Boon 265

**Realistic Wax Mannequin**
Maude Bass-Krueger 269

**Rotring, Letratone, MiniCAD**
Tony Fretton 273

**Scaphander: 'Man-boat'**
Steven Connor 277

**Scarificator**
Thomas Kador 281

**Serving Hatch**
Tim Ainsworth Anstey 285

**Sinclair C5**
Simon Sadler 289

**Skirt Grip**
Amy de la Haye 293

**Slide Rule**
Adrian Forty 297

**Slotted Screwdriver**
Richard Wentworth 301

**Space Frame**
Catherine Slessor 305

**Stanley 55 Combination Plane**
Nikos Magouliotis 309

**Telephone Table**
Edwin Heathcote 313

**Teletype**
James Purdon 317

*Théâtrophone*
Carlotta Darò 321

**Think City Electric Vehicle**
Kjetil Fallan 325

**Trombe Wall**
Paul Bouet 329

**T-shirt Plastic Bag**
Johanna Agerman Ross 333

**Ultratemp® Roasting Rack**
Christian Parreno 337

**UV-radiated Artificial Beach**
Maarten Liefooghe 341

**Vertical Filing Cabinet**
Zeynep Çelik Alexander 345

**Water Bag**
Sarah Bell 349

**Writing Case**
Barry Curtis 353

**Zeppelin**
Jeremy Myerson 357

**Biographies** 361

**References and Further Reading** 368

**Acknowledgements** 387

**Photo Acknowledgements** 388

# Introduction

Barbara Penner, Adrian Forty,
Olivia Horsfall Turner and Miranda Critchley

*And as natural selection works solely by and for the good of each being, all corporeal and mental endowments will tend to progress towards perfection.*
Charles Darwin, *On the Origin of Species* (1859)

This book is a collection of objects that once populated the world, but do so no longer. Some of the artefacts and technology it contains were once ubiquitous; others barely made it into existence, not much more than an idea or a prototype. We are interested not simply in why these things – some of them once very familiar – disappeared, but in what their disappearance tells us about the world we have created for ourselves.

The process of the disappearance of objects and technology is sometimes referred to as obsolescence, and sometimes – and this is the description we have chosen to focus on – extinction. Both terms contain certain assumptions about how and why things disappear, while neglecting other, no less pertinent, possibilities. 'Extinction' is explicitly a borrowing from theories of natural selection and evolution, and, like all analogies, makes certain things clearer, while obscuring others. The economist Amartya Sen warns, 'Darwin's general idea of progress . . . can have the effect of misdirecting our attention, in ways that are crucial in the contemporary world.'[1]

One particular obfuscation that arises when Darwin's ideas of evolution are applied to artefacts is the assumption that it is only the fittest, the best or the most appropriate objects and technology that survive. In this model, design, like nature, is thought to be an optimization machine always pushing forward – progressing towards perfection. When things disappear, they do so, it is implied, because of their own inadequacy or their unsuitedness to their conditions. Part of the purpose of this book is to probe and question this seeming inevitability.

Its other purpose is to use extinct objects to recall other ways and possibilities of engaging with the world. Why are extinct objects suited for this task? We suggest that, at the moment of their invention, technology and products must all project forward in some way. The act of design and manufacturing is anticipatory; to be conceived of and made, a thing is necessarily imprinted with an idea of future needs, demands or ways of living which it may then help to bring about. As the architectural theorists Beatriz Colomina and Mark Wigley also observe, 'Design is a form of projection, to shape something rather than find it, to invent something and think about the possible outcomes of that invention.'[2] This projected vision of the future may not be heroic or utopian; indeed, it is more often mundane and humble. But the result is that even the most insignificant design's extinction speaks of a road not taken, a future rerouted or unrealized.

As we will discover through the 85 examples gathered in this book, there are countless practical explanations for and reasons why things become extinct. But in considering their purpose and rationale, we encounter the ghosts of futures that never came to pass, their projections having proved unfounded, short-lived, misguided – or, as in the case of William Gaddis's account of the player piano, all too prescient of the worst of what was to come. Extinct though they may be, these objects retain the imprint of possible futures, some of which we may be glad to have left behind and others whose relevance is recovered today.

We believe that a study of extinct objects has much to offer here and now. Narratives of technology tend to be innovation-focused and do not pay much attention to cast-offs or dead ends; they emphasize novelty and vision and are infused with a sense of destiny. But this book argues that the history of objects becomes far richer when we also consider the underside of progress: the conflicts, obsolescence, accidents, destruction and failures that are an integral part of modernization. Considering these can open up fresh perspectives on modernization's modes of operation, which is our particular concern.

Darwin's *On the Origin of Species* was published in 1859, eight years after the Great Exhibition in London. More than any other single event, the Great Exhibition serves as an index to the material transformations that accompanied industrialization, the shift to factory production and the harnessing of new power sources. Not only did the exhibition show-case the technological advances and goods of the previous two decades (for an example, see Angus Patterson's entry on Electrotypes), but it also presaged developments to come, most memorably through the Crystal Palace's own construction, a revolutionary demonstration of the potential of iron, glass and prefabrication.

Many of the contradictions and paradoxes of industrial capitalism were fully on display at the Great Exhibition as well. With its list of international exhibitors, it promoted a liberal ideology of free trade and open markets; yet, with its strong colonial presence, it signalled its dependence on commodities, captive markets and cheap labour. From the start, it was obvious that the fruits of prosperity at the Great Exhibition would never be equally distributed. And, for those who cared to see it, the terrible human and environmental cost of the new methods of manufacture and urbanization were already evident, if not in the Palace itself, then in its immediate environs, the streets of London.

In the light of these contradictions, we begin to understand that evolutionary theory and narratives of progress had a crucial role to play in modernization: they were required to naturalize the impact of capitalism and to ensure its continued spread. This was certainly the view of the cultural historian Lewis Mumford, who, in his monumental *Technics and Civilization* (1934), argued that the function of evolutionary theory in industrial society was not to explain technical change, but to normalize the inequities produced by capitalism. In the Darwinian model, the enrichment of the bourgeoisie became proof of their strength and their right to exploit the labour of those supposedly weaker than themselves. Observing that the phrase 'the survival of the fittest' was a tautology – 'for survival was taken as the proof of fitness' – Mumford notes sardonically, 'that did not decrease its usefulness.'[3]

But, for the most part, narratives of progress were able to sweep such concerns aside. Against what Mumford called 'tooth and claw' accounts of Victorian social order, a more benign account of capitalism emerged that held – and largely continues to hold – that it lifts up those places where it settles, rippling out to bring jobs and improve basic living conditions for all. In particular, technical innovations and infrastructural improvements are positioned as the mechanism by which capitalism's benefits are delivered; as they bring about greater ease of movement and more rapid communication, so the theory goes, they help to create a better-informed, more equal and less restive populace.

When set against such advantages, resisting progress can easily be positioned as dangerous and perverse. One of those who believed absolutely that mechanical invention would underwrite the general advance of humankind, the *philosophe* Denis Diderot, wrote in bewilderment at those who stood in progress's way: 'How bizarre is the working of the human mind! . . . The mind distrusts its powers. It stumbles in self-created difficulties.' Diderot justified his own epic *Encyclopédie*, which from 1752 onwards compiled hundreds of engravings of manufacturing technology, with the claim that 'our descendants, by becoming better instructed, may as a consequence be more virtuous and happier.'[4] Even though most of the trades and industries that Diderot captured so painstakingly would be transformed or rendered obsolete with the coming of the steam age, the faith in progress that he articulated helped to lay the ground for it.

We find faith in progress everywhere by the mid-nineteenth century. It was a staple of boosterish mass-market publications such as the *Illustrated London News*, which were firmly pro-development and represented vast metropolitan improvements with unwavering enthusiasm. (Its images, latterly, have helped to fix the idea of the visionary Victorian engineering age in our minds.) When applied to the field of what Diderot called 'the mechanical arts' – a broad category that included everything from agriculture to iron founding – the emergent idea of extinction, and the corresponding belief in perfectibility, portrayed technological development as a kind of internally propelled, irresistible and positive force.

We see the idea at work in the writing of the influential modernist design historian Sigfried Giedion. In *Mechanization Takes Command* (1948), one of the few accounts of the industrial arts that approaches the *Encyclopédie* in its ambition, Giedion discusses the formal and stylistic evolution of everyday designs in implicitly Darwinian terms. In one typical passage, he uses heroic language to describe the washbasin's struggle to achieve its right form: 'Like a kernel emerging from its shell, the washbasin through the decades breaks loose from its envelope of furniture.' Giedion sees abandonment of the Victorian 'weakness for adornment' as proper and inevitable given the advance of industrial improvements. 'Only with the advent of mass-produced enamel and earthenware', he explains, 'could natural forms truly pierce through.'

The model of change in this passage is striking. The designed products of industrial processes are described as if they emerge naturally from mechanical developments; no human agents are involved (hence why Giedion claims his is an 'anonymous' history). Natural – that is, modern – forms are established through an impersonal evolutionary process, which is linear and one-way. Old forms are to be shed and left behind; later in the same passage, Giedion speaks despairingly of trends to re-encase bathroom equipment in furniture as 'backsliding', retrograde and unnatural.[5] The question of the washbasin's adornment becomes charged with moralism: progress in design is equated to social progressiveness; the unadorned washbasin confirms the forward march of civilization.

The ideas underpinning the evolutionary model of technological innovation did not pass unchallenged. In the wake of the Second World War, a flood of innovations was released, from injection-moulded plastics to Polaroid cameras, which drove consumerism to new heights, spurred on by the corporate embrace of 'planned obsolescence' or, we might say, planned extinction, whereby one product model was deliberately phased out to encourage the purchase of another. By the 1960s committed opposition to such strategies had emerged. Counter-cultural movements particularly rejected the loss of traditional skills, occupations and social relations and the waste of natural resources that accompanied capitalist production.

Other than refusing development entirely, designers also embraced the idea of appropriate technology and sought to create products that were less resource-hungry and more responsive to local communities. In his book *Small Is Beautiful* (1973) the economist E. F. Schumacher, the guru of the appropriate technology movement, identified the tendency of Western onlookers to impose Western criteria of success – such as Western standards of consumption, value systems and behaviour patterns – on to countries operating under very different conditions, constraints and cultures. The net result was that, even as colonized countries achieved independence from colonial rule, the technological fixes promoted by development agencies tended to reassert colonial-era power relations. Emphasizing local solutions and expertise, appropriate technology became a way to disrupt the naturalized flow of progress from centre to periphery.

Historians of design and technology also questioned the evolutionary model. In the 1980s Adrian Forty debunked Giedion's biological account of mechanization, stating firmly, 'The design of manufactured goods is determined not by some internal genetic structure but by the people and the industries that make them.'[6] Feminist scholars insisted that design histories were remiss in emphasizing the technical side of production and ignoring the responses of consumers. Historians of science and technology associated with Actor–Network Theory traced the interconnected and diffuse human and non-human actors who usher in – or fail to usher in – technological innovation. (And, in the context of *Extinct*, it is worth acknowledging Bruno Latour's *Aramis, or the Love of Technology*, 1993, as an influential case study of one such failure.) Most recently, the way in which Social Darwinism and eugenics have shaped industrial design, architecture and urban planning has been exposed and fiercely denounced by those who study race and the exclusionary nature of the built environment.[7]

But, despite the principled resistance to them, evolutionary models retain their allure in contemporary culture. In fact, the rise of computing, automation and artificial intelligence has only further enshrined the belief in progress and the worship of technological innovation. We live in an age of product drops and continual upgrades. One reason for the endurance of an evolutionary model may lie in the way patents are registered in the first place; the requirement that each invention cite 'prior art', that is, precedents from which it has drawn, reinforces the idea of innovation as a genetic chain. But more generally, as the historian Jill Lepore observes, narratives of continual innovation reflect the vested interests of those who tell/sell them: 'People who are in the business of selling predictions need to present the past as predictable . . . and in histories written by futurists the machines just keep coming.'[8]

This book's scepticism towards accounts of uninterrupted innovation has been significantly informed by the historian – and *Extinct* contributor – David Edgerton's *Shock of the Old* (2006). In this seminal work, Edgerton demolishes typical narratives of technological progress mainly by tracing which technologies are in actual widespread use around the world. As he argues, moving away from innovation-centric history towards a use-centred one not only interrupts notions of progress, but changes our list of what inventions have had the most impact on modernity. Edgerton's own list of significant technology includes many objects that do not usually make the top ten of 'inventions that changed the world', since it includes 'the rickshaw, the condom, the horse, the sewing machine, the spinning wheel, the Haber-Bosch process, the hydrogenation of coal, cemented-carbide tools, bicycles, corrugated iron, cement, asbestos, DDT, the chain saw and the refrigerator'.[9]

While we have approached our subject from a different angle from that of Edgerton, focusing on interrupted developments rather than enduring ones, this book largely upholds his most novel insights. Edgerton observes that older technology quietly survives almost everywhere because it is readily available and cheap to operate, or very well suited to local conditions. This was confirmed for us in the early days of our project, when we had to confront the fact that it is difficult to find things that are truly extinct. To enter the world of extinct objects is to enter the world of the undead, where few things expire completely. Many become dormant, waiting to be revived in another form or another place as circumstances change; Edgerton's own entry on oil-from-coal processes makes just this point. And many leave residual traces, in the form of design features (skeuomorphs), language or practices that persist after the object itself is gone.

Surveying the objects contained in this book also affirms Edgerton's other key insight: the fictitiousness of the linearity that underpins narratives of progress. Rather than being a story of smooth and universal progress, technology and design advance unevenly, in starts and stops, often looping back or leaping tracks to pick up a strand of development abandoned long ago. Now that we are in the Age of the Anthropocene, with its emphasis on deep geological time, the concept of extinction becomes even more complex: climate change has accelerated extinctions of all kinds and yet we know that our traces will long endure. Extinction is not obliteration. As the environmental writer David Farrier puts it, 'The entire atmosphere now bears the marks of our passage, like a vast geochemical trace fossil of the journeys we have taken and the energy we have consumed.'[10]

For all these reasons, we realized that we could not be overly dogmatic about what this book would include. We began by inviting a broad range of writers – curators, critics, artists, architects and academics – to identify and write about a case of extinction at any scale. (For the sake of simplicity, we refer to these case studies collectively as 'objects', but they include tools, equipment, structures and infrastructures relating to key vectors of modern life, from bodies and homes to workplaces and cities.) As proposals came in, we abandoned the idea that extinction meant objects would suddenly and totally disappear. In many cases, objects continue in use even after production ceases. Others are kept alive by hobbyists and collectors who prize them for their nostalgia value; even as this book was being written, several extinct objects, notably the Zeppelin, were resuscitated thanks to crowdfunding.[11] More formally, many objects are preserved by heritage bodies or in museums of science and design.

So long as we know how something works, it is always, in theory, possible to revive it or to implant it elsewhere. As this project advanced, we became less concerned with the completeness of a proposed object's extinction, which is almost impossible to verify anyway, and more focused on the question of what set it on the road to extinction in the first place. Rather than seeing extinction as something that happens naturally, we encouraged authors to pin down precisely the varied processes and forces bringing about that extinction – often much less obvious than one might think at first. The case studies of extinction here thus tend to defy easy categorization or pat explanation. To begin to make sense of them, we have proposed six categories of extinction – Failed, Superseded, Enforced, Defunct, Aestivated and Visionary.

We initially thought 'failed' objects, which are removed from circulation because they don't work, would be a large category. Yet only a few case studies in these pages – most famously Concorde – died because of catastrophic and spectacular technological failure. Instead, a far more typical category proved to be 'superseded', in which an object is overtaken by a more 'advanced' model that supposedly performs the same function more efficiently. Crucially, however, new models often perform this function quite differently, emphasizing that, when something is rendered extinct, it is not only objects but a web of related skills, habits and associations that disappear. This is one reason why innovations are sometimes resisted or adopted only slowly – and why some industrial designers insist that, in order to be accepted, new products must be implanted with elements of the old. With another nod to Darwin, the industrial designer Henry Dreyfuss dubbed this 'survival' form.[12]

Extinction is most clear-cut and easily traced in the 'enforced' category: cases where extinction is brought about by central shifts in government policy or by regulatory bodies, whose intervention proves decisive either in terms of eradicating an object or practice, or in promoting one type of technology or infrastructure over another. The stated reasons for these extinctions might be economic or environmental, either in the sense of conserving resources or of protecting public safety, or they could be the result of political decisions. Most of these objects remain viable technologically, and might remain in limited use, but the risks or costs associated with them are no longer deemed acceptable and they are clearly on their way out. Prime exhibits here are the ashtray and the T-shirt plastic bag.

Less dramatic and singular than 'enforced', 'defunct' extinctions occur most frequently with commercial products. Sometimes defunct objects never take off owing to a misreading of the market or a lack of consumer buy-in. Some cannot be mass-produced or continue to operate at reasonable cost; others require larger infrastructure that is never built. Some are found to pose dangers to users, or their original *raison d'être* evaporates. A large number of objects simply fall victim to shifting fashions and aesthetics, although they may continue to circulate as curios. And sometimes practices associated with the use of certain objects survive even after the objects themselves become defunct. So, in the case of the optical telegraph, which was all but abandoned after the Napoleonic wars, David Trotter argues that the form of communication it established – the public performance of private messaging – remains ubiquitous today.

In those cases where extinctions are planned – for instance, when objects are legislated out of existence – this is invariably done in the name of the public good and itself becomes proof of progress, tautologically reconfirming the rightness and modernity of the evolutionary model. Sometimes extinction does seem progressive (it is hard to argue for bringing back arsenic wallpapers or leucotomes, for example), but we should be wary of always accepting this justification. Extinction in many of our examples was demonstrably arbitrary and ideological; the Chilean milk spoon, for instance, represents a national health initiative to improve child nutrition that was abandoned following the country's right-wing military coup. And the Chinese dougong became defunct in an era of British imperialism, in which traditional wooden structures gave way to supposedly superior iron- and steel-frame constructions.

But the dougong also demonstrates how many extinct objects have complex afterlives. Soon after its extinction, it was brought back by a mix of Chinese and foreign architects as a formal icon of 'Chineseness'. Now rendered in stone or concrete, and appreciated for its symbolic rather than its structural function, it stands as one of our 'aestivated' objects. In the natural world, aestivation refers to a state of dormancy, in which organisms wait out hostile conditions before reviving. To qualify as 'aestivated' in this collection, we have stipulated that a whole object must be revived (rather than, say, a component of it), but it need not appear in exactly the same form; it can reappear with some kind of material adaptation that makes it suitable for current conditions or needs, which themselves may have changed in the interim.

Taken together, the case studies in these pages present a picture of changeability and contingency, and emphasize the sheer range of forces that must align for technology to succeed. This point is driven home through the example of electric vehicles. The earliest model discussed here, the 'Hummingbird' Taxi, appeared in 1897, more than 120 years ago, and the other, the Think City Electric Vehicle, was launched in 1998 with a final iteration in 2008. Both were workable and appealing because they promised to improve the environment (the Hummingbird was meant to counter the effects of the 'insanitary horse'). Yet neither took. Today, of course, Tesla is sweeping all before it; and however tempting it is to attribute the company's success to a single factor – such as an extraordinary leader – it is clearly down to a confluence of economic, technical, environmental, political and social factors. Considering some of the more admirable aspects of earlier designs (Think City's later models were made of 95 per cent recyclable materials), it becomes hard to regard Tesla's product as *the* optimal or predestined one.

In compiling charge sheets of deferred opportunities and losses, however, we risk missing the larger value of extinct objects. Inevitably, to study extinction is to run up against limits: the constraints of cost, the lack of political will, the inherent conservatism of markets and the collective failure of imagination. But extinct objects can operate equally as containers of potential and of provocation – and arguably they are most compelling when seen in this light. This is most explicit in the propositions or prototypes in the 'Visionary' category: some are experimental, playfully exploring technical possibilities; others set out to articulate different, more liberated visions of future design or of society. Occasionally they serve as a teasing commentary on progress itself: 'Edison's Anti-gravitation Under-clothing' mocks the reverence for heroic inventors in a way that still sticks, given our ongoing faith in corporate techno-optimists such as Bill Gates or Elon Musk to solve the world's problems.

And, finally, many of the extinct objects here act as stores or repositories, offering alternative visions of how we might deal with problems in ways large and small: how we might still address the problem of cities, sustainably store and carry water, or even mask workplace noise. While they may be suggestive, it is important to stress that this book is not meant as a 'how to' manual for future technological fixes. For what is most compelling about these entries is their reminder that extinct objects represent not only technology but other ways of thinking, making and interacting with the world, other attitudes towards the body, crafts, copies, beauty, art, communications, movement, leisure, love, class, cultural identity, nature and artificial intelligence. Ultimately, every extinct object embodies a vision of the future, a vision that, even if the object itself has been superseded, is still in some way available to us.

# Acoustic Location Device
Bryony Quinn

Acoustic location devices were originally developed as part of an early-warning system against aerial attack during the First World War, but, almost immediately, their primary use was as technology for the guidance of ordnance. At their most basic, the designs resembled huge trumpets or shells that amplified an operator's hearing so as to be able to locate and track a target by sound alone.

In the tumultuous early decades of the twentieth century, many countries developed their own strangely sculptural variations that defy general description – although one particular Japanese model was aptly known as a 'war tuba'. The model pictured here, a German Goerz Sound Locator, is similar to one purchased by the British War Office in 1929 from C. P. Goerz of Vienna for comparison with the current British patterns. It could be dismantled quickly and loaded efficiently on to vehicles for relocation. The main function of the original Goerz Sound Locator was to guide the spotlights that would pick out aircraft for the guns on the ground, but, conversely, operators demanded complete silence in order to work. Any local disturbance – conversation or gunfire, for example – was intolerable to the user of a machine.

One of the earliest and clearest accounts of the use of the acoustic location devices occurred during the defence of London from Zeppelin attacks, which began in 1915. Under the command of Admiral Sir Percy Scott, the safeguarding of the city from above involved the careful observation and armament of coastline and high land points across southeastern England. Despite the improvement in design of anti-aircraft guns and cannon during the early days of the First World War, by September 1915, as the frequency and terrible success of aerial bombardment of London escalated, these weapons were found to be an inadequate means of protection for the capital. And so Alfred Rawlinson, who had worked in both the Army and the Navy and assisted in the organization of the Paris anti-aircraft defences the previous winter, was invited by Scott to share his considerable experience.

In his book recounting this period, *Defence of London, 1915–1918* (1923), Rawlinson describes how evident it was that enemies were seeking safety for their attack, 'principally by rendering their airships as far as possible "invisible" by taking advantage of the protection of clouds when within range of our guns'. The guns could, of course, spray bullets and missiles in a generally informed direction, but the strategy proved to be no real deterrent to the

Goerz Sound Locator
in use in the UK, 1937.

targets. The main problem, Rawlinson identified, presented itself in a series of fundamental questions: 'How were we to effectively *hit* an object which we could not *see*?' He quickly followed this by a deeper proposition: 'If we cannot see it, how do we know it is there?'

Rawlinson had been trained within a tradition of conflict that relied on the ability to visually distinguish friend from foe, live target from defenceless tree. The modern state of warfare changed that by allowing the bombing of towns and cities from above, cloaked by darkness or cloud and conducted at a distance that was far less discriminating of its targets than any conflict before it. 'The answer', Rawlinson continued, 'is childishly simple . . . as we cannot aim using our *eyes*, we shall be obliged to shoot – if we are to shoot at all – by using our *ears* to aim with.'

The design of the first British acoustic location devices, which are widely recorded to be the earliest in effective use, resembled two gramophone trumpets pointing in the same direction and fixed to a pole that pivots freely on the horizontal plane; a compass set at the point around which the trumpets and pole revolve, and where an operator would be positioned; and, finally, two pipes attached to two stethoscopes that lead up to the pivot and fit into the two trumpets. 'It only then remained to fix a man's head to the pole,' Rawlinson deadpanned, 'insert the ends of the stethoscope pipes in his ears, and to tell him to "listen", and to turn his head *and the pole with it* in the direction of the sound which he heard.' Bearings taken by several 'listening machines' (as Rawlinson called the operators and their devices) were then telephoned to headquarters and the enemy's track plotted until it entered the range of the expectant anti-aircraft guns.

Acoustic location devices sit comfortably in a collection of military technology that extends the human senses: telescopes that lengthen sight to determine the approach of distant enemies, and night-vision goggles that allow the wearer to see beyond daylight. With acoustic location came a sonic reach that could cover whole landscapes and coastlines. What made the devices new and distinct from a spyglass or telescope was their ability to detect movement that occurred out of sight. By taking advantage of the feature of sound to reflect around or through a fixed or opaque object, acoustic location could effectively 'see' around or through environments that would otherwise obscure the advancing enemy: hills or mountains, large buildings, clouds and so on.

The earliest acoustic location devices were operated by at least one controller, and were portable. For this reason they should not be confused with the monolithic (and more handsome) concrete 'sound mirrors', a number of which still dot the English coastline. Although these permanent and static structures were a slightly later development in acoustic location technology, both portable and static systems came to the same end with the advent of

Radio Detection and Ranging in the early 1930s. So-called radar took the same principle of identifying a target by sound, but instead detected the live site of the enemy by radio signals.

The intention of both is the same – to locate an invisible target – but whereas radar efficiently converted a radio signal into a visual image for several viewers to observe (once a screen was introduced into the design), acoustic location devices had the odd effect of projecting an image of the target using sound, rather than light, directly into the mind of an operator. The use of such an instrument presented a psychological problem, that of translation, for every user. Information will alter as it shifts between the senses, and, despite the initial success of the 'listening machines', there were discrepancies in the preciseness with which different individuals located the same sound.

The accuracy of all the devices relied on the acuity of the listener in distinguishing the sounds of enemy movement, and determining its direction, altitude and range. This dependency, and the resultant subjective variances, was a problem that was identified at the beginning of the use of acoustic location devices, and a striking (but futile) solution was proposed. 'Who will be the man upon whose powers of *hearing* we may most confidently rely and by what means shall we recognize him?' asked Rawlinson. 'In the realm of "hearing",' he declared, 'the *blind* must always reign supreme.' And so it was that blind men were some of the most successful early operators of this new technology, best able to see what was invisible to everyone else. Ultimately, though, it was the failure to overcome the discrepancies between individuals' interpretation of the sounds, as well as interference from other sound sources, that encouraged the development of a technology less sensitive to operator subjectivity – radar.

Acoustic location devices and radar operated simultaneously throughout the 1930s, until the accuracy and reliability of radio signals surpassed that of the 'sound mirrors' and 'war tubas'. In the meantime, many variants were developed by different countries to reduce auditor subjectivity; some affixed large shells directly to the head of the operator as 'personal parabolas', while others experimented with honeycomb structures or strange and technical curves the likes of which would not be seen again until science fiction hit mainstream cinema decades later. All to no avail: by the early 1940s the manual devices for acoustic location such as Goerz had produced and Rawlinson had championed were obsolete.

# Action Office Acoustic Area Conditioner: The 'Maskitball'

## Kristen Gallerneaux

The bubbles and baubles of the 1960s and 1970s. The sensory-depriving wombs of Italian design dreams: egg-shaped chairs, hovering over tulip pedestals. In *The Prisoner*, 'Rover' bounce-chases Patrick McGoohan down the beach, emitting its bass roar. An inflatable plastic suit by Archigram, a personal dome office in the discomfiting age of the new bureaucratic-everywhere. Radomes lurk on the horizon. The great blue marble, seen from space. Spheres – white, clear, sinister, omniscient – run the range from security cameras on the ceiling to peculiar plastic seating. All carry some form of implied sentience: to hold and to calm, or to observe and report.

A different kind of sphere perched on the edge of an office cubicle is known as the 'acoustic area conditioner' (AAC), and informally as 'the maskitball'. First conceptualized by the designers Robert Propst and Jack Kelley at the Herman Miller Research Corporation (HMRC) in Ann Arbor, Michigan, as part of the Action Office system, the AAC was marketed from 1975 as a noise-masking solution and privacy device. The amusing moniker arose because 'it was basketball season', according to Kelley, who was then acting as HMRC's Director of Systems Development. 'And phonetically, it fit as a kind of joke.'[1] Over its short lifespan, the maskitball was the most stylish ICBM of inter-office sound pollution on the market.

Widely hailed as a game-changer in twentieth-century office design, the great innovation of Action Office was its holistic approach. Offices, in Propst's mind, had become chaotic wastelands, filled with siloed-off workers, ad hoc furniture arrangements and piles of clutter. Action Office was meant to act as a modular system of freestanding panels that could be arranged into mutable, porous workspaces with built-in organization systems to keep clutter out of sight. 'People shouldn't be planted like onions in pots to sit somewhere,' Propst said.[2] Propst's vision gave rise to the modern office cubicle, and as it was adopted from 1964, new sonic situations in the workplace emerged that themselves required design attention.

What was the best way to deal with the noise of an open office? One option was expensive ad hoc acoustic systems, cobbled together from existing speakers and sound generators, requiring the skill of acoustic engineers to install and maintain them. An alternative was the so-called Shushing system, which used background noise to cancel other noise, a phenomenon Propst observed in 1963. To a fellow engineer, he described sitting in a restaurant with his back turned to a loud talker, seated nearby. Ambient sound blurred

Acoustic Area Conditioner, Herman Miller Research Corporation, USA, 1975.

the person's words. 'What interested me especially', wrote Propst, 'was that this was not a very noisy restaurant but somehow in this case, the background noise was a very effective interference.'[3]

A prototype sound device was made to act on Propst's insights, followed by the first successful installation of six AAC units at the HMRC office. In 1975 the first 25 AACs appeared in Herman Miller sales data, sandwiched among other Action Office noise-dampening accessories – ceiling tiles, carpet, drapery, screens and plants. The panels that made up the cubicles themselves were supposedly able to 'tune' a room through hushing layers of perforated metal, fibreglass and cloth. Ideally, the maskitballs would punctuate the office, perched atop 203-centimetre (80 in.) wall panels every 3–3.5 m (10–12 ft). Adding units would increase the zonal cloud of privacy, making it difficult to aurally pinpoint the source of the masking effects.

The AAC plugged into existing electrical wall outlets and was low maintenance and low cost. It also gave individuals the ability to adjust bass and treble to their taste, depending on privacy needs or to account for noisier times of day. High frequencies emitted from the top of the device, bouncing off the ceiling and dispersing to blend with ambient chatter. Mid and low frequencies diffused from the conditioner's equator band, free to dissipate.

To avoid the irritating, insistent sounds of conventional masking systems, the AAC's electronic circuits included a modulator that continuously randomized the pitch of sounds. As explained in *The Action Office Acoustic Handbook* (1972), when the AACs were used in groups, their sound waves interlocked and were perceived as pleasant, natural and comfortable. White noise worked, within preset limits; amid worries that the conditioner might be used as 'a weapon in office conflicts', it was programmed not to exceed 50 decibels, the level at which noise is thought to lead to intellectual fatigue and communication breakdown.[4]

The AAC was not the first white-noise machine. In the world of consumer electronics, that honour goes to Jim Buckwalter's Marpac SleepMate of 1962 (rebranded as the 'Dohm' in 2010). SleepMate, however, resembles nothing so much as a flipped-over dog bowl. By contrast, Propst was always interested in considering how a noise-masking machine should *look*, and decided that it should be 'unobtrusive in appearance and at the same time unembarrassed at being a masking sound source'.[5]

In consultation with Kelley, the device took the form of an ultra-modern white sphere. Kelley believed a sphere was 'the least insulting shape to stick up in the air'; significantly, it was also the best one to hide the electronics.[6] The maskitball's two hemispheres were banded at the middle with an inset speaker void that emphasized its rotundity. Encased in its opaque exoskeleton were solid-state transistorized electronics. A latecomer to the space age, the maskitball was in good company as it gobbled up noise in offices filled with rounded cubicle corners. This was the masking of sound, aestheticized. A brochure advertising the AAC went all out: 'You've got to see it to hear it. And you've got to hear it to believe it.'[7] The genius of the maskitball's designers was to recognize the value in allowing some of the bustling sounds of work to remain. These sounds are the acoustic signature of an organization, communicating 'this is who we are, what we are doing', and as such are preferable to the awkward vacuum of overly deadened sound. An acoustic conditioner relies on the fact that a certain amount of background sound must be present for its effect to close the gap between its own emitted frequencies and irritating office noise. You need noise to cancel noise.

The maskitball worked well, although production ceased in the early 1980s. Why Herman Miller discontinued it is not clear, especially since the need for it did not go away; not only did Action Office continue to be produced, but its model of open-plan and collaborative working had, by this point, spread widely. Today flexible co-working offices are the norm. But the industrial chic aesthetic many of them adopt, with polished concrete, raised ceilings and exposed metal HVAC systems and open tables, are an acoustical nightmare. The twentieth-century clatter of typewriters and ringing telephones has been replaced by the click of twenty-first-century laptop keys and the buzz of smartphones set to vibrate.

Even if the problem the maskitball was designed to treat is persistent and pervasive, the remedies have changed. Instead of a systems-based approach, responses have become more targeted and personalized. People use a new generation of devices to create acoustic silos. To counter the slap-back echo, workers isolate themselves with noise-cancelling headphones, signalling 'I am trying to concentrate.' The most basic function of an acoustic conditioner – to provide a stream of background frequencies to mask noise – has found new life in a seemingly infinite menu of apps and streaming music services, white noise and ambient playlists.

White noise is maternal, the first sounds we hear in the womb. Imagine yourself inside a seashell, with waves of ambient noise shushing in. Today white noise is a prescription. As a chronic insomniac, I doze off nightly to a blend of rain and static that broadcasts on the slightly squashed spherical body of the Google Home Mini beside me – a close cousin of Propst and Kelley's device. Lately, as I drift off, I've been troubled by the fact that while approximately 1,800 AAC units were sold, they seem to have all but vanished. It's as though the maskitball never existed, except for one lonely prototype in the Robert Propst archive and in Herman Miller promotional materials. There is not a single sphere to be found on eBay or on the mid-century modern antiques market, or spotted randomly at a swap meet. Are they bouncing around in a secret office-product landfill somewhere?

'OK Google: can you tell me where all the maskitballs went?'

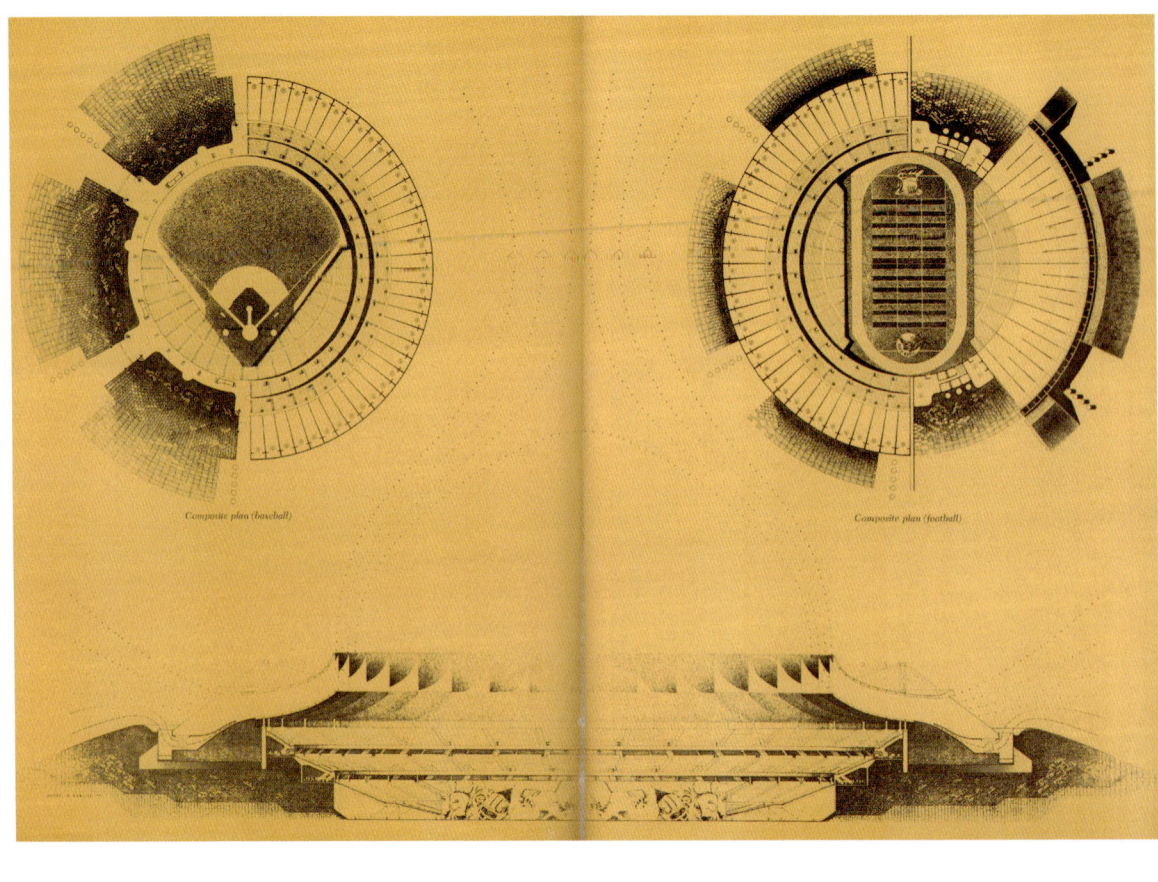

Composite plan (baseball)

Composite plan (football)

# Air-curtain Roof
## Laurent Stalder

Of the many aspirations of modern architecture, one of the most persistent was for the dematerialization of its built substance. Anticipated in nineteenth-century discussions of glass architecture, where 'walls and ceilings invisible to the eye' made it seem as if buildings might evaporate into 'solidly shaped' air, the possibility of architecture without substance was brought even closer in the early twentieth century. Describing the qualities of Le Corbusier's buildings of the 1920s, the Swiss critic Sigfried Giedion wrote: 'air flows through them! Air becomes a constituent factor!'

Yet, as long as buildings were still enclosed with glass, the dissolution of architecture into air remained a metaphorical rather than an actual postulate. Only after the Second World War did the idea of a wholly insubstantial architecture start to look feasible. In *Évolution générale de l'art actuel vers l'immatérialisation (et non pas dématérialisation)* of 1958, the German architect Werner Ruhnau and the French artist Yves Klein collaborated to devise various proposals for warm-air roofs, initially for the covering of a café in front of the theatre in Gelsenkirchen, northwestern Germany, and a year later for an amphitheatre in the ruins of the monastery in Bad Hersfeld in central Germany, where it would have both mechanically deflected rain and also caused it to evaporate. Ruhnau took his cue from the air curtains of contemporary department stores, and from air-conditioned open-plan offices, hoping that at some point in the future entire areas of the outdoor environment could be similarly climate-controlled. Whereas Richard Buckminster Fuller envisaged cities roofed by glass domes, Ruhnau saw them being covered by 'compressed-air canopies'.[1] Thus Ruhnau's projects are characterized not as an architecture of transparent walls and ceilings shielding an artificially generated comfortable interior atmosphere, but rather as a far more radical dissolution of the physical building itself into a series of airflows with the corresponding air-blowing apparatus. Just how doggedly this quest carried on being pursued, right up to the last decades of the twentieth century, is demonstrated in the British architect Cedric Price's 'Air Roof' project, conceived in 1980 with the Canadian architect Peter L. E. Goering. Price proposed an installation to keep rain and snow out of stadiums and other open public spaces by means of an airstream generated by turbines.

Yet the possibilities for an air roof envisaged by the post-Second World War avant-garde rested on inventions made at the opening of the twentieth

Plans and section for open-air sports stadium with air-curtain roof, Peter Goering, *Canadian Architect*, November 1971.

century. On 28 February 1903 the American machine engineer Theophilus
Van Kannel, inventor of the revolving door, registered a patent for an air
curtain, and was given legal protection on 8 November 1904. The aim of
the installation was 'to furnish a means of excluding wind, snow, rain, and
dust from open doorways so as to furnish the protection afforded by closed
doors without obstruction to the free movement of passers which is caused
by doors adapted to exclude the elements.' The system consisted of a series of
'nozzles' that could be mounted in the floor, in the ceiling or at the sides, and
which pointed outwards, protecting the entrance without interfering with
the circulation of people. In defining this goal, but likewise in its separation
of the entrance and exit gates, the air curtain explicitly drew on the revolving
door, with the advantage of entirely barrier-free access.

Van Kannel's invention provided the basis for numerous prototype air
curtains for walls and for roofs, distinguished either by their increasingly
powerful performance or by their additional shielding functions: the doubled
and adjustable Air Screen (1911) for shop entrances, ideal to protect against
sudden gusts of wind; the Insect-proof Portal (1923) with its five vertically
directed columns of air, which promised to 'entrap and destroy' pests; an
Insect and Dust Protection Unit (1924), explicitly equipped with an air roof,
consisting of a table with shelves above it protected on top and at the sides by
an air curtain; and a Gaseous Curtain (1927), thought up by the artist László
Moholy-Nagy for cinema projection. In addition were two patents using
an air curtain or air roof to divide off entire rooms, already in the interwar
period anticipating Klein and Ruhnau: the 'Technique for the Closing Off
of Rooms', patented in Germany in 1932, using horizontal or vertical
'air walls' actuated by 'kinetic energy'; and the 'Method and Means for
Delimiting, without Solid Walls, Confined Zones in a Field of Atmosphere'
patented in France in 1939, the prototype of which was illustrated by the
example of a stadium where the air curtain isolated the climate of the public
stands from that of the playing field without obstructing vision or sound.

From the 1950s onwards the air curtain became a standard device in
department stores and other public buildings, reaching a developmental high
point in 1960 with the 27-metre-wide (89 ft) opening in the entrance area
of the Pan Am Terminal in New York, but the air-curtain roof, on the other
hand, remained for the most part an object of scientific research. Exploration
into its feasibility was led from 1968 onwards by the Institute for Aerospace
Studies in Toronto. Accompanying the research reports are various proposed
models for concrete architectural applications: an interior covered by a lateral
airstream, a provisional shelter covered with a conical airstream, a pavement
protected by a horizontal airstream and, most prominently, a stadium shielded
by a vertically tapering Annular Air-curtain Roof.[2] Along with reducing the
noise generated by the air movement that would have limited the general

application of the system, the deflection of water represented a particular challenge that the Toronto engineers continued to dedicate themselves to solving well into the mid-1980s, be it for air-curtain roofs or air-ring roofs.[3]

Research into air-curtain roofs never advanced beyond a small-scale prototype of the Annular Air-curtain Roof and the construction of an air canopy above the air-extraction system above a parking garage in Toronto. With a few exceptions, such as air curtains in electric arc furnaces, it rarely found its way into practical use. Although the reasons for the abandonment of the idea were never categorically stated, the air-curtain roof would appear to have been condemned to failure for its impracticability not only technically but above all on account of the energy requirements, estimated at 10,000–36,000 kW to cover a stadium 100 m (328 ft) in diameter. Yet precisely herein lies the paradigmatic nature of the air-curtain roof for twentieth-century architecture: its energy dependency, which can be best encapsulated in the catchphrase 'on and off', meant that while it never came into being, as an idea it remained very much alive, standing for a radical reformulation of the discipline of architecture no longer as the art of construction, but as the regulation of the various performances a building fulfils. Technology of one kind or another may become extinct, yet it is outlived by concepts latent within it.

Although architecture abandoned projects with air roofs in the 1980s, contemporary philosophy, in the wake of a growing awareness of the potential of air and of the dangers arising from its contamination or insufficiency, became increasingly preoccupied with air. Luce Irigaray's fundamental essay *L'oubli de l'air* (1983), and more recently Peter Sloterdijk's *Luftleben* (2002) and Bruno Latour's 'Air-Condition' (2005), all encouraged a reconceptualization of this vital non-substance, but the radical correlate, an architecture of air, has received very little attention. This is all the more surprising because the air roof, whether implemented or not, is possibly a more effective tool of cognition than any philosophical apparatus in making visible the simultaneous dimensions of air as milieu, substance, material and energy.

# All-plastic House
## Carola Hein

Ever since the invention of synthetic polymer-based plastics at the beginning of the twentieth century, designers and architects were tantalized by their possibilities for building. Initially, the uses they envisioned ranged from the small-scale – light switches and furniture – to whole elements – windows and walls. But after the Second World War, architects, in alliance with chemical companies, began to explore the application of plastics to the entire house. For architects, the complete mass-produced all-plastic house seemed to offer a solution to the housing shortage and a chance to experiment formally. For chemical companies, it appeared to be a profitable use for their products. Their collaborations resulted in a number of all-plastic houses, memorable symbols of hopeful futurism. One of the most successful was the Futuro house, developed in 1968 by the Finnish architect Matti Suuronen. Nonetheless, the dream of the all-plastic house soon died, and, although plastic did end up reshaping domestic environments, the forms in which it did so were very different from what had been first envisaged.

The height of experimentation in all-plastic houses was in the late 1950s and early 1960s. Probably the most famous of these was the 'House of the Future' designed by the Massachusetts Institute of Technology (MIT) architects Marvin Goody and Richard Hamilton and sponsored by the chemical company Monsanto. Its location in Disneyland, California, guaranteed that it was seen by millions of visitors. Opened in 1957, the 119-square-metre (1,280 sq. ft) pod-like structure with curved plastic walls cantilevered over a concrete block. The cross-shaped floor plan, with the kitchen at its centre, was designed for a typical family of the 1950s: two parents and two children. Modular and flexible, in line with the designs of Richard Buckminster Fuller, the attraction offered a view of future living – in the year 1986 – with new technology including ultrasonic-wave dishwashers, picture phones and atomic food preservation to make the life of the typical stay-at-home mother easier. The display included small household items, such as microwave ovens, that would much later be constructed of plastic. In introducing new forms and technology, the house was meant to exemplify a modern lifestyle: clean, functional and fun.

However wacky some of its innovations may now seem, the House of the Future was the outcome of a serious and ambitious research collaboration over several years between key players in design, production and education. It was also highly strategic, as Monsanto – in common with many other

Monsanto 'House of the Future', designed by Marvin Goody and Richard Hamilton, Disneyland, Anaheim, California, 1957.

manufacturers that had grown big on wartime contracts – looked for ways to diversify its products for a peacetime market. Housing was one of the arenas to which it turned its attention. Monsanto worked closely with MIT, funding studios to 'forecast possibilities that can be achieved when, in the future, we may take maximum advantage of the inherent properties of plastics as applied to house fabrication'. By 1956 plans had advanced to the point that MIT could announce that, with the company's sponsorship, its Division of Building Engineering and Construction, together with the Department of Architecture, was designing and constructing a plastic 'House of Tomorrow', in which 'Structural shape and architectural design have gone hand in hand, and a great deal of pioneering in structural design has been made necessary by the relatively new and untried structural properties of the materials.'[1]

For the professors at MIT, the goal of the collaboration was nothing less than to develop an integrated, structural approach to plastics in construction. Robert K. Mueller of Monsanto's Plastics Division pointed to the diverse possibilities of plastic materials in architecture and design: 'the future of plastics in building is limited only by our imaginations and the public acceptance of new concepts in living.'[2] And where better to win over the public to this integrated approach than at Disneyland? The number of visitors guaranteed the House of the Future high exposure. In 1957, the first year of its opening, an average of 60,000 people every week entered the building. By the time it was demolished, in 1967, it had been visited by around 20 million people. Few designs better capture the technological utopianism of the prosperous post-war years than the House of the Future.

In the 1960s European architects from Ionel Schein to Jean-Benjamin Maneval also designed all-plastic houses. As for their American colleagues, the new material and a widespread belief in technology seemed to provide them with the perfect answer to political, economic and societal challenges of the time. The most widely distributed prefabricated dwelling made of plastics was the Futuro, a ski cabin that could be easily heated and erected on difficult terrain. Composed of fibreglass-reinforced polyester, it looked like a flying saucer on stilts. Its designer assumed that by the 1980s entire blocks would be built from plastic homes, yet fewer than one hundred were built and sold (and even fewer survive, although those that remain are cherished by collectors worldwide).[3] The prospect of living in freestanding prefabricated capsules neither attracted the kind of purchaser able to afford plots large enough for single-family homes, nor fulfilled the requirements of state-funded dense, multi-storey housing programmes.

There were additional reasons for the failure of the all-plastic house. The oil crises of the 1970s and the increased cost of plastic construction may have played a role, but the end of the plastic houses had started earlier. As early as 1965 the American architect R. D. Gay was lamenting that the 'completely built plastic home seems to have disappeared'. He blamed the failure on a lack of coordination among builders, all of whom insisted on using their own industrialized systems, despite the fact that they were mostly 'ill-conceived, unsuitable, based on ignorance of basic research and national requirement, and generally ugly'. The academic sector came in for criticism, too, for not showing more leadership in the area: 'There are no "codes of practice"; no text books are available for the design of building structures.'[4]

The main reason cooperation stalled, however, was that plastic manufacturers became disillusioned, having overestimated the influence architects carried in the world of housing and construction. Quite simply, architects lacked the necessary leverage to get houses into production at scale. Hence, plastic manufacturers tried a new approach. Instead of following comprehensive architect-led design for entire prefabricated and repeatable structures, the petrochemical and building industries turned to smaller, more discrete elements of the home: products such as plastic bathroom units, insulation, windows, furniture, tiles, Lego toys and dolls' houses. This emphasis on domestic objects minimized their risk: if products failed, the research and development cost and capital investment were much smaller, but if they succeeded, the potential market was much greater.

This direct approach required that plastics manufacturers pay more attention to the contracting sector and consumers. Indeed, advertisements for these plastic elements spoke directly to consumers, especially women and children, who were now seen as an essential segment of the domestic market. Just such a product was the Formica Vanitory unit advertised under the slogan 'Too Bad Dad' – a must-have for women and girls, with a counter space around the sink, and drawers for towels, laundry and medicines. Formica, manufactured by De La Rue, was an easy-to-clean, long-lasting, heat-resistant laminate. This particular product targeted young female teenagers, who, in league with their mothers, were supposed to plead with their fathers for a place to wash, do their make-up and get dressed. A local Formica fabricator, found in the 'Plastics' section of the phone directory, could build the Vanitory in the colours and patterns selected by the women and girls in the household. And in the end, even dads might be able to enjoy a shave there.

As the Formica Vanitory shows, rather than creating a revolution in housing, plastic was ultimately deployed in the home in a much more everyday and piecemeal way than architects had initially wished, adapting and softening post-war interiors. While not achieving the comprehensive vision of the House of the Future, plastics nonetheless became the stuff and surface of future life, a background condition for the gendered landscape of post-war mass consumerism.

# Arsenic Wallpaper
## Lucinda Hawksley

When the Swedish-German chemist Carl Wilhelm Scheele was experimenting with copper arsenide, he made a discovery that would take the world
of interior decor by storm – and one that would lead, unwittingly, to many
deaths. In 1778 he invented a vivid green pigment that became known by a
variety of names, including Scheele's Green, emerald green, Paris green and
arsenic green.

In the early nineteenth century British interior designers were excited by
two important changes: the creation of a machine that could produce long
strips of wallpaper (instead of the previous small squares) and the repealing
of the paper tax. Suddenly, wallpaper was affordable. Following the Great
Exhibition in 1851, which showcased a number of wallpaper manufacturers,
the fashion for what most Victorians called 'paper hangings' spread across the
social classes. The popularity of Scheele's Green as a wallpaper pigment also
grew – even though the use of arsenical pigments had been outlawed
in many other European countries.

In Britain, the first problems associated with arsenic dyes came to light
in factories, where workers suffered mysterious health problems. The illnesses
varied from lung problems caused by breathing in arsenic-rich dust (used in
the creation of flock wallpapers), to skin problems caused by touching arsenic
dye, and from arsenical poisoning when pigment entered the bloodstream
through cuts or abrasions, to painful inflammation when arsenic irritated the
nasal passages and eyes. Because these conditions were so varied, and because
workers' rights were so routinely ignored, the problem was at first largely
dismissed by the establishment. The medical community grew increasingly
concerned, however, and articles about the dangers of arsenic-coloured
wallpaper appeared in the *British Medical Journal* and *The Lancet* as early as
the 1850s. But it was another two decades before the designers, manufacturers
and general public began to listen.

Even though arsenic was well known as a rat poison and was used in
homes throughout the country to control vermin, arsenical pigments were
employed to colour walls, textiles, furniture, clothing and children's toys.
Arsenic green was even used as a food colouring (often with fatal results).
In the world of wallpaper, Scheele's richly green pigment staunchly remained
one of the most sought-after colours.

The most famous wallpaper designer of the mid-Victorian age was
William Morris, who was known not only for his interior design work, but

Wallpaper with arabesque
pattern, printed with
Scheele's copper-arsenic
green, UK, 19th century.

Failed

for his connection with the Pre-Raphaelite movement, his founding of the Arts and Crafts movement and his campaigning philanthropy. It was his company, Morris, Marshall, Faulkner & Co. (later simply Morris & Co.), that made his name synonymous with wallpaper – and one of his favoured pigments was an arsenical green. Although history now associates arsenic solely with the colour green, it was actually used in the production of most wallpaper colours, meaning that almost every wallpaper design from the early and mid-Victorian period was liberally laced with poison.

In 1862 newspapers around Britain reported the death of a three-year-old girl named Ann Amelia Turner, who had lived in Limehouse, in the East End of London. Much was made of the tragedy of her death, because she was the last child to die in what had been a family of four children. Her bereft parents, and the local community, were told initially that the other children's deaths were caused by diphtheria. With Ann Amelia's illness, however, a local doctor began to question the earlier diagnosis; diphtheria was hugely contagious, yet none of their neighbours had become ill. Soon it became apparent that all four children died from arsenic poisoning, caused by green wallpaper in their home.

Many journalists were now on a crusade to expose the often fatal health problems experienced by workers in wallpaper factories – yet the problem continued to be ignored by businesses, and by the wallpaper-buying public. Customers wanted arsenical-green paper hangings, and designers were only too happy to oblige. The biggest stumbling block to making people understand the dangers of arsenic wallpaper was that not everyone living with it was affected, and even those who did become ill were not necessarily affected in the same way as other sufferers. It took a long time for scientists to discover that – aside from the risk of children licking the walls – the main danger lay in an invisible arsenious gas that emanated from wallpaper in damp, and especially mouldy, conditions. Another reason that many people refused to believe the rumours was simply because not everyone proved susceptible to arsenic poisoning. For example, if an entire family were living in one arsenic-papered room, sometimes only a few family members became ill from the arsenious gas. This convinced many people that the source of these mysterious illnesses could not be the wallpapered walls.

Both Morris and his business partner Edward Burne-Jones decorated their homes with arsenic wallpaper, and neither they nor their families suffered any obvious ill effects. Morris became furious about the ever-increasing clamour for wallpaper to be free of arsenic – although when the public began talking with their wallets, he was smart enough to listen.

The very first arsenic-free wallpaper to be produced in Britain had been made by the forward-thinking William Woollams & Co. in 1859, yet almost every other manufacturer ignored that firm's lead. In the 1870s, however, Morris, an astute businessman, understood that his customers no longer trusted arsenical colours, and he bowed to pressure. In 1875 Morris & Co., together with its manufacturing company, Jeffreys & Co., proudly proclaimed on its new wallpaper catalogues that all its papers were now free of arsenic.

No legislation was ever passed in the British Isles to prevent the use of arsenic in the manufacture of wallpaper, and the extinction of arsenic wallpaper in Britain was achieved entirely by campaigning doctors and journalists, and because of a change in public opinion. Yet, despite this public change of heart, Morris and many other designers mourned the loss of the brilliant colours that could be created only by using arsenic. In private, Morris compared the furore to the historic witch-hunts in Salem, Massachusetts; in 1885, already a decade after Morris & Co. had switched to the production of arsenic-free wallpaper, he wrote to his friend Thomas Wardle: 'As to the arsenic scare a greater folly it is hardly possible to imagine: the doctors were bitten as people were bitten by the witch fever.'

# Arundel Print
Tanya Harrod

When John Ruskin, the polymath art critic, saw the Arundel Society chromolithograph of Giorgione's Castelfranco Altarpiece (c. 1504) for the first time, he remarked that it 'announces itself clearly to you as a work of art'.[1] Although he had never seen the original, he had no hesitation in placing the colour print, itself based on a watercolour copy, in the category of 'art'. A copy of a copy, the print, which reached the Arundel Society's subscribers in 1879, was made without any kind of photographic intervention just at the time when photography was increasingly being used to record works of art.

We might expect that the chromolithographic reproductions the Arundel Society published from 1856 onwards would swiftly have been superseded by photography. But in fact the society continued its work 'to collect diligently and with discrimination the highest and best examples of Art and to bring them before hundreds of English minds' until 1897.

The Castelfranco Altarpiece print captured colour to a degree of accuracy not previously seen in printed reproductions. While from the fifteenth century onwards prints had reproduced the line and tone of paintings, and even attempted to suggest colour, they nonetheless remained firmly monochrome until well into the eighteenth century, when attempts were made to develop colour-engraving processes by figures such as Jakob Christoph Le Blon. Even after that, though, any colour was generally applied by hand.

The chromolithographs published by the Arundel Society were testament to a world of skill that evolved with the development of colour lithography from the 1830s. First, there was the ability of highly educated copyists such as the Viennese artist Eduard Kaiser, who was employed to make the watercolour copy of Giorgione's painting. Working as part of a team producing what can usefully be described as scholarly reproductive prints, he created a greatly reduced image of the same size as the resulting chromolithograph. He then sent his watercolour to London for examination by the society's Council, and, once approved, it went to the printing firm Storch & Kramer in Berlin, chosen for the society by Ludwig Grüner, Prince Albert's chief artistic advisor. A key-line drawing was created as a guide, but the judgement of the art director or 'visualizer' was crucial in breaking down the watercolour's constituent colours and working out the printing sequence. The watercolour was then reproduced using as many as thirty lithographic stones and a wide range of coloured inks.

Arundel Society chromolithograph of Giorgione's Castelfranco Altarpiece, or *The Virgin and Child Enthroned with Saints Francis and Nicasio or Liberale*, in original frame, 1879.

Arundel Society chromolithographs, chiefly copying early Renaissance Italian, German and Netherlandish paintings and frescoes, are among the most sophisticated non-photographic colour reproductions ever made. They entered museums and art colleges all over the world, while in the homes of their middle- and upper-class subscribers they were valued in their own right.

In William Morris's study at Kelmscott House in Hammersmith, west London, there hung the Arundel Society chromolithograph from 1888 of Botticelli's *Primavera* (c. 1477–82). At the auction of Oscar Wilde's effects in April 1895, his Arundel prints were listed on a pre-sale poster along with paintings, blue-and-white china and Thomas Carlyle's writing desk. In the 1880s the collector Charles Drury Edward Fortnum had framed Arundel Society chromolithographs of the *Primavera*, Melozzo da Forlì's *Sixtus IV Giving Audience* (1476) and the Castelfranco Altarpiece in his study at Hill House, Stanmore, Middlesex, along with objects of *vertu* such as his sixteenth-century Italian wood-and-plaster box, now in the Ashmolean Museum, Oxford.

We have, therefore, to think back to a world in which the possession of copies, as distinct from originals, conferred cultural capital; in which copies were seen as a vital educational tool; and in which copies were reviewed and discussed globally in national and regional newspapers and journals as seriously as original works of art, and were treated as objects of study by local antiquarian and literary societies.

The Arundel Society Castelfranco Altarpiece also reminds us that during the second half of the nineteenth century and well into the twentieth, the value of photography as a reproductive tool was not accepted unreservedly by all art historians. Some, such as Vincenzo Marchese, wrote as early as 1866 of 'le nobile conquiste della fotografia' ('the noble conquest of photography'), but there were dissenters, such as the American Richard Offner. 'Unhappily', Offner wrote in 1927, 'photography is largely an interpretative affair. It has this in common with general artistic practice, that the result is determined by the whim and genius of the operator, and the camera is only one of the determinants of the result.'[2] Although he created a photographic archive that remains at the Institute of Fine Arts in New York, he was keenly sensitive to the limitations of photography. Likewise, writing in 1944, the historian of medieval wall paintings Ernest William Tristram recommended a faithful drawing of a work of art as enabling 'patient, minute and persevering scrutiny', adding, 'It is precisely this which makes an intelligent drawing of so much more value as a record of a wall-painting than a photograph.'[3]

Today newspapers and journals do not regularly debate the quality of photographic images of works of art. Nor, generally, do serious collectors hang framed reproductions in their homes. Few would now claim that an intelligent transcription can teach us as much as an encounter with the original work of art, as Ruskin did after a weekend spent studying the Arundel Society chromolithographs of the Brancacci Chapel frescoes in Santa Maria del Carmine, Florence.

But perhaps we should heed the thoughtful verdict of the industrial designer Christian Barman writing in the *Penrose Annual* in 1949:

> An Arundel print is not even a direct translation; it is a translation from a translation . . . It would be too much to expect that such a print should fully transcribe the original as transcription is judged by the standard of modern photography; but as translations they had the same qualities of scholarship and understanding as our best translations of Homer and Dante.[4]

Sitting before a PowerPoint presentation or thumbing through an illustrated art book, we forget that analogue and digital photographs are also translations. Arundel Society prints, in their obsolete strangeness and beauty, deserve re-examination. By the early twentieth century they were regarded as remembrances of the Victorian age, as modernist sensibilities demanded an encounter with the artwork's singularity either face to face or through the apparently unmediated medium of photography.

For Walter Benjamin, writing in 1935–6, the photographic reproduction lacked the aura of the original and, indeed, worked to undermine it. Yet the reversal of values that he envisaged, brought about by the dominance of photography and cinema, has failed to materialize. Today the authentic work of art has never been more fetishized. Arundel prints, although no longer produced, are not themselves extinct, as their continuing presence in public and private collections affirms, yet they testify to the extinction of an aura once held by the finest copies of original works of art.

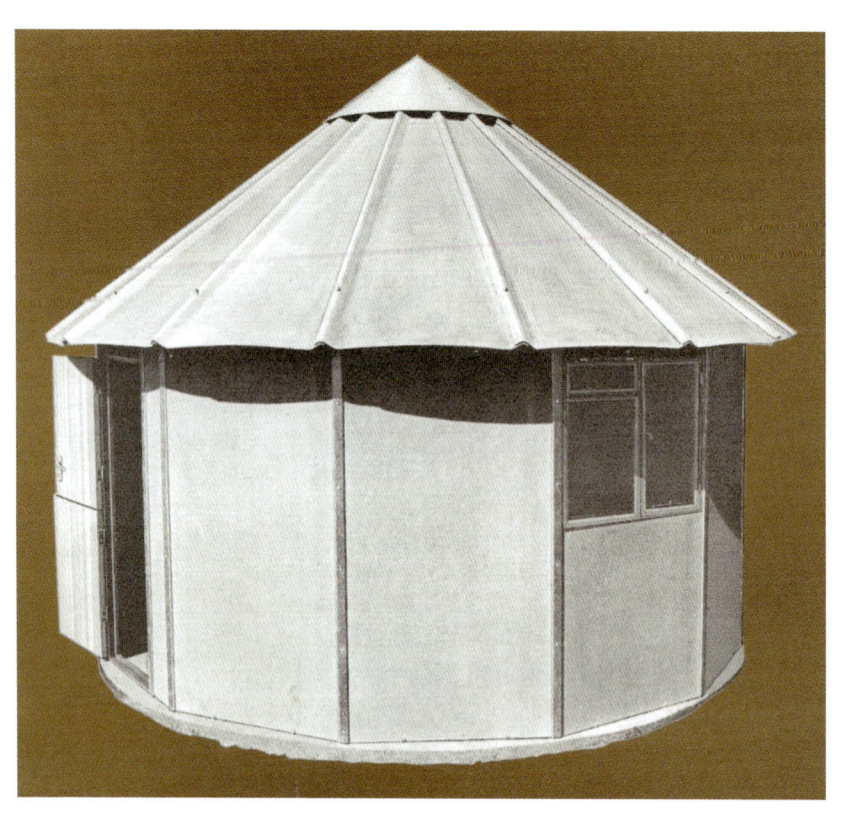

# Asbestos-cement Rondavel
## Hannah le Roux

The rondavel, or *rondawel* in Afrikaans – a round, single-roomed building, usually walled in earth and roofed in thatch – is not quite extinct, at least not in rural southern Africa. Its conical form uses material efficiently, long grass and thatchers are common, and it is comfortable. Its use confounds many of the constructed boundaries of language, class and race, and even the uncertain origin of the word, whether in *rodavallo*, *rundtafel* or *dewals*, suggests that it had diverse connections to Portuguese, Dutch or Malaysian builders in the Cape Colony. Despite its exotic name, it is very likely that the form is widely accepted because of its similarity to, if not its origin in, Bantu dwelling types from northern regions of the continent. Few other structures have such diverse associations with both colonial and colonized built cultures, and remain so affectively important to all.

Nineteenth-century missionaries such as John Smith Moffat found the round dwellings of indigenous Africans disquieting, in part because their clustered assemblages supported polygamous arrangements that were difficult to maintain in rectangular, multi-roomed houses. At the same time, settlers sometimes commissioned these structures, so that 'the mutual appropriation of aesthetic forms, as much by Europeans as by Tswana, led to a steady convergence in frontier housing styles . . . [that] entailed the way in which space was constructed and experienced; lived, so to speak, in the round. Or oblong.'[1]

Even after a wider range of building technology became available, the rondavel was still chosen by white architects at moments of reconnection with tradition, as in the Mamelodi campus for African teachers, built in the 1940s.[2] Today this hybrid of African and European colonial forms is commonly found in beach resorts and game lodges, sustaining fantasies of bush living. The cool, dark interiors of thatched rondavels offer respite from summer weather, and holiday resorts sometimes offer them as budget accommodation. They are difficult to divide into straight-walled areas and so tend to be used as bedrooms, scattered around rectangular bathroom and kitchen blocks. This leisure use, however, is usually a temporary one, and the permanent residents of rondavels on farms or in rural villages are almost always black Africans. Farm labourers could be delegated to build them, without plans, from local thatch grass and adobe, enabling farmers to avoid the obligation to construct more modern housing for workers.

Asbestos-cement rondavel from Everite catalogue, South Africa, 1974.

A modern and commercial version of the rondavel was also developed and marketed widely as a substitute for the home-grown originals, until, after a few decades, it was quietly taken out of production. The technology of the asbestos-cement rondavel was detailed comprehensively in the 1970s in a series of catalogues published by Everite, a subsidiary of the Swiss company Eternit owned by the Schmidheiny family; the company dominated the South African production of asbestos-cement building materials from the 1950s onwards. It is not clear when its roof-panel kit, or the kits for complete rondavels, first appeared, but according to the brochure from 1974, 'Everite asbestos-cement rondavel roofs have now been used for many years and their popularity is growing daily due to their aesthetically pleasing appearance, and functional design. They may be seen all over South Africa, on farms; at hotels and pleasure resorts; as shelters and stores; on change-rooms and site offices and on attractive cottages.'

The Everite asbestos-cement rondavel roof came in three sizes, with nominal interior diameters of 3.7, 4.9 and 6.1 m (12, 16 and 20 ft). The wedge-shaped roof sheets were stiffened by rolled edges that would overlap with the adjacent ones before being bolted down to a radial structure of steel purlins and rafters. The kit for a roof 4.9 m in diameter included 440 kg (970 lb) of asbestos-cement panels and 132 kg (291 lb) of steel supports, which met at a conical ring, finished with a 17-kilogram (37½ lb) roofing cap. Everite also offered a fully prefabricated octagonal structure, 2.4 m (5⅓ ft) high, with asbestos-cement boards, insulation and ceilings set between steel struts, and door and window frames.

What would have compelled farmers and resort owners to choose the kit, at a cost of R885.55 in 1972, over an organic and locally assembled building? The catalogue suggests that its appeal lay in the material qualities of asbestos-cement, which was 'incombustible, rot-proof, rodent- and termite-proof', along with its fast erection by two people in two days and its 'readily demountable' nature. The South African National Parks embraced the technology in areas where local thatchers had been displaced in the development of nature resorts. Distributed by rail, it also fitted into the logic of South African spatial development where mining and agricultural centres coincided with stops on the national lines. In semi-desert areas, such as the northern Cape mining towns of Kuruman and Prieska, where thatch grass didn't grow, cargo trains would depart filled with asbestos fibres and return with asbestos-cement building material.

By the mid-1970s the link between asbestos and fatal disease had become both a public concern and a commercial risk, and the Everite rondavel was doomed to a slow death. South African medical research had documented a connection between asbestos and cancer in the 1960s, and in 1974 bans on the material started coming into effect.[3] In the United States, the first

Remaining half of Everite
asbestos-cement rondavel,
Klipriviersberg, South
Africa, 2019.

lawsuits were launched that year against building-materials companies for selling asbestos-cement without warning and exposing workers in the industry to carcinogenic fibres, leading to the liquidation of major manufacturers in the face of class-action claims.

The elimination of asbestos-cement from Everite's South African products happened only after 2002, perhaps because 'financial constraints' had suspended the company's research into substitutes until 1996. This delayed extinction was most probably related to a relatively late extension of the international bans on the use of asbestos to include roofing materials, as well as the disposal of Eternit's holdings to local businesses that were less vulnerable to public concerns. In the wake of South Africa's elections in 1994, won resoundingly by a party with a mandate to provide both 'housing for all' and jobs, cheap and locally produced asbestos-cement roofing still had a role to play.

The asbestos-cement rondavel is a curiosity in the history of South African low-cost housing, which is more often associated with rows of rectangular matchbox dwellings. But the original model of the rondavel displaced by the asbestos-cement version is enjoying a resurgence. As a drive around Maseru or rural KwaZulu will reveal, African compound owners have resumed building earth and thatch rondavels within clusters of dwellings, sometimes reserving these spaces for communicating with their ancestors. Bringing the rondavel back from extinction, their return to this form deploys the building's organic and symbolic relations, without the toxic legacy of its asbestos-cement version.

# Ashtray
## Catherine Slessor

My mother possessed a superlative ashtray. Mounted on a waist-high stand, it took the form of a chrome-plated bowl with a gridded lid. Faintly reeking, it stood to attention in our 1960s suburban living room like some engorged trophy, ready to swing into action at the flick of a lighter. Beyond its role as passive receptacle, its heft hinted at a more nefarious purpose. You could probably have bludgeoned someone to death with it.

My mother smoked, as everyone did back then, her poison of choice being Benson & Hedges. The lustrous gold of the B&H packs neatly counterpointed the silver gleam of the chrome. Not only suggestively glamorous, the ashtray was also eminently functional, concealing unsightly butts in its bowl. Like many smokers' requisites (holders, lighters, cigarette cases), its design conspired to dignify the fugitive, messy act of smoking.

I still have that ashtray. Emblematic of a particular time and atmosphere, like a chrome-plated madeleine, it seemed too majestic to discard, although, in a measure of the socio-cultural distance travelled between then and now, it is never used. Instead, it forms part of my personal Museum of the Ashtray, joining a dozen or so specimens smuggled out of restaurants and bars over the years, that stands in mute testament to simpler times and less healthful pleasures.

While the history of tobacco and smoking is well documented, the ashtray is a more obscure object of art and desire. It started life as the quotidian 'ash pan' that first accompanied the introduction of tobacco from the New World to Europe in the late fifteenth century. From its early use in pipes and cigars, smoking tobacco in cigarette form became widespread during and after the Crimean War, when British soldiers emulated their Ottoman Turkish comrades, who had begun rolling tobacco in strips of old newspaper. It was an American, James Bonsack, who pioneered the mass production of cigarettes, with his eponymous Bonsack rolling machine, imported to England in 1883.

While rudimentary versions of ashtrays existed before the modern era, the idea of a consciously designed *objet de délice* started to gain ground only in the early twentieth century, as more women took up smoking. Form and ornamentation became more considered as the ashtray moved from the public realm into the private domestic sphere as part of a progressive, modern lifestyle. Art nouveau and art deco proved especially fertile stylistic terrain, and invariably artists and architects also got in on the act. In 1967 Salvador

Ashtray for Air India, designed by Salvador Dalí, decorated by Jules Teissonniere, porcelain, manufactured in Limoges, France, 1967.

Dalí devised a limited-edition ashtray in Limoges porcelain for Air India as a squiggle of a serpent and some contorted swans wrapped around a shell. (As payment, he requested a baby elephant, which Air India duly transported from Bangalore to Spain.) Less flamboyantly, the Danish polymath Arne Jacobsen designed a classic Mid-century Modernist version in the form of a laconic stainless-steel hemisphere that tipped up to disgorge and conceal the butts in the body of the ashtray, thus keeping everything Nordically neat. Jacobsen also designed a hearse, so he might be said to have astutely hedged his bets.

The ashtray became a coveted souvenir. When Quaglino's, the society restaurant in St James's, London, was taken over by the gourmand restaurateur Terence Conran in the 1990s, 25,000 of the establishment's distinctive Q-shaped metal ashtrays were appropriated by light-fingered diners over the course of a decade. That's an average of seven per day. This should come as no surprise, since a well-turned ashtray is not just a quotidian object, but often a thing of curious beauty, invested with sentimental significance. It is a memento: of that meal, that drink, that place, that time, that person, that occasion. Famously, people used to nick the ashtrays from Concorde, when you could still smoke on planes. During his tenure at Mirabelle restaurant in Mayfair, Marco Pierre White, once the *enfant terrible* of haute cuisine, printed the terse injunction 'Property of Marco Pierre White' on the bottom of his signature fish-shaped ashtrays. To no avail.

As White's experience showed, it does not pay to provide patrons with nice things. Better to go cheap and dirty, as generations of pub landlords instinctively grasped. Worlds away from the fripperies of hand-blown Murano, the heavy glass ashtray bulging with butts was a familiar yet gruesome staple of English hostelries. Usually the size of a manhole cover, it also came in handy in the event of a fight. Being cleanable and durable, glass makes an especially effective 'ash receiver'. Plastic, by contrast, is apt to burn and stain, as anyone who has attempted to stub out their cigarette in a yellow Ricard triangle, a fixture of French cafés, will testify.

From its heady days of pomp and plenitude, adorning every restaurant table, every bar counter and every living room sideboard, the ashtray has today atrophied to a grubby metal maw tacked outside offices, where shivering, diehard smokers congregate fitfully for conspiratorial drags during their allotted screen breaks. No longer standing sentinel at the nation's lunches and cocktail hours, or coming as standard in cars, trains and planes, the ashtray is en route to extinction – in the Anglo-American world, at least – collateral damage from the UK's ban from 2007 on smoking in enclosed indoor places. For aficionados, not being able to smoke in a bar seemed especially harsh, since that was 'the last public place you could go to be a dropout, a nonconformist, refusenik, a time-waster, a bohemian, a hider from reality, a bum, a rebel, a bore, a heathen', as Peggy Noonan wrote in the *Wall Street Journal* after New York's smoking ban was implemented in 2003.

Alternatives such as vaping have, to some extent, emerged to fill the void, spawning their own range of kit and associated paraphernalia. Cigars still endure among a certain type of tycoon, requiring specialized ashtrays to keep the glowing cigar butt perfectly horizontal so that it will burn evenly. Beyond humidor connoisseurship lies the more chaotic terrain of the pot smoker, who will always 'need a place to ash', as marijuana websites helpfully put it. Yet for all this, the direction of travel is clear: the ashtray has moved from status symbol to an object of secrecy, shame and subterfuge. Because, of course, the ashtray is not only an adjunct to social pleasure, but a memento mori, a reminder that you are dancing with death. The World Health Organization cites smoking tobacco as the world's single greatest preventable cause of death, and it is still the leading killer in the developed world, a smiling assassin invited to your table, insistently hovering.

Like delusional Norma Desmond frozen in a spotlight of her imagining in *Sunset Boulevard*, the ashtray is not destined for a miraculous comeback. It bestrode the twentieth century, as quotidian as a cruet set; today it skulks in the well-ordered cupboards of ironists and fetishists. The day may come when, if you produce one, people will be genuinely bemused as to its function. Perhaps it was never essential anyway. The American actor John Goodman recalls the advice he received from the legendary *bon fumeur* Peter O'Toole when the two men were filming together in 1991. During a break, Goodman asked to borrow an ashtray. O'Toole, with characteristic insouciance, flicked his ash on to the floor and declared: 'Make the world your ashtray, my boy.'

THE KINGSTOWN AND DALKEY ATMOSPHERIC RAILWAY—STARTING OF THE TRAIN.

# Atmospheric Railway
## Niall McLaughlin

From 1784 to 1825 innovative British engineers such as James Watt, Richard Trevithick and George and Robert Stephenson developed a combination of locomotive steam engines and metal tracks, setting the template for the development of railways around the world. The locomotive was the main drawback of these railways. The principle that its source of power had to be carried along with it was inefficient, since moving the engine and the fuel consumed 50 per cent of the total energy of the system. Steam locomotives were also so heavy that they broke the tracks, requiring the rails to become heavier and costlier. The cutting edge of innovation lay in optimizing the relationship between the weight of the engine, its purchase on the track and the ever-increasing strength of iron and steel rails.

Some engineers attempted to solve the problem at a conceptual level by thinking of how to get rid of the locomotive entirely. The inventor George Medhurst proposed a system of moving people and objects through an evacuated tube, based on the idea that the vehicle was like a piston. When a vacuum was created in front of the vehicle, it would be drawn along by the difference in atmospheric pressure. In the 1830s a collaboration between Joseph d'Aguilar Samuda, a shipbuilder, and Samuel Clegg, a gas engineer, resolved a proposal for an operative system. They illustrated a cast-iron pipe 23 cm (9 in.) in diameter lying between the two railway lines, parallel to them. A piston was pulled through the pipe by differential pressure, and was attached to travelling carriages by a rod, which drew them along. The engines that powered the system stood beside the track in buildings arrayed at intervals along the line.

The new system was first put to public use in the hamlet of Dalkey, on the outskirts of Dublin. An atmospheric railway drew passengers from the nearby town of Kingstown up the hill to the village, and the carriages rolled back down using gravity. Designed by Charles Blacker Vignoles using Samuda's patent, the railway propelled up to two hundred people along at a steady 48 kph (30 mph). This novel invention drew crowds of day trippers. The dirty soot and smuts associated with rail travel were entirely absent, and the swift, silent acceleration was an amazing new sensation. It must have felt like being whisked into the future. Frank Ebrington, an engineering student, briefly held the world land-speed record when he inadvertently released the brake of a detached carriage in Kingstown. He was sucked up the hill at terrifying speed, arriving in Dalkey a mere 75 seconds later.

The Kingstown and Dalkey Atmospheric Railway, engraving from the *Illustrated London News*, 6 January 1844.

The great achievement of this short length of track was to demonstrate that it was possible to traverse a significant incline with relative ease. Locomotives, with their great weight, tended to slip when running on tracks above a modest gradient. Samuda's system opened up the possibility of more efficient track planning, with fewer cuttings and lower construction costs. Engineers flocked to Dalkey to examine it in action. One visitor was Isambard Kingdom Brunel, who had been invited to design a railway in Devon over a prescribed route with difficult terrain. His optimistic enthusiasm for the system was entirely at odds with the view of his rival, George Stephenson, who doubted the reliability of the longitudinal valve and did not believe the promised efficiencies. He called it 'a great humbug'.

The South Devon Railway was designed by Brunel and Samuda on the basis of atmospheric propulsion, and the first section of the line was working by 1848. But the shortcomings of the system immediately became evident. The junction between the travelling piston and the rod drawing the carriages was a leather contraption sealed by beeswax and tallow. Tallow melted in summer, froze in winter and proved to be a delicacy for rats. The leather was desiccated by the vacuum in the pipe. Seals failed constantly. The pump houses had to power up for each passing train. If the train was late, inefficiencies in fuel and labour resulted. The cost per passenger mile was exorbitant. A telegraph was installed along the line to alert the power stations of impending arrivals, but it was not enough to solve the problems and Brunel advised a gathering of shareholders to abandon the system. It was a significant blow to his reputation, and the costliest failure of his career. Nevertheless, it is hard not to reflect that this monumental failure contained all the ingredients of a visionary system of integrated technology: a railway with carriages, an evacuated tube and an electrical cable all running along together.

The South Devon Railway lasted less than a year, and a few years later, in 1854, the Dalkey line was also closed. A leafy path ascends to the village now and is known as the Metals, arriving at the beautifully named Atmospheric Road. In Devon, where the project is remembered as the Atmospheric Caper, you find the remains of the old standing-engine houses. They were remarkable structures, designed by Brunel, with grand halls and chimneys as big as campaniles. They possessed a soaring optimism. No wonder the powerhouse in Starcross had a century-long afterlife as a chapel. Today one can sit across the road in the Atmospheric Railway Inn and drink a toast to hope and ingenuity.

The Dalkey Atmospheric Railway has an echo in fiction. Flann O'Brien, the twentieth-century Irish surrealist and satirist, had a well-documented fascination with railways. In his novel *The Dalkey Archive* (1964), the main character is a deranged visionary who designs a sealed container from

Remains of Brunel's railway pipe at Didcot Railway Centre, UK, 2011.

which the air is evacuated, allowing time to travel faster to aid the maturation of whiskey.

In common with many extinct objects, the atmospheric railway may simply have been born at the wrong time. Conceptually, it was a beautiful solution to the problems of rail transport, since it removed the unwieldy locomotive. It was let down by leather and tallow and an inability to learn from comparable contemporary solutions. In 1841 Robert Davidson's first electrical locomotive, *Galvani*, was abandoned because the batteries it carried were underpowered. It was a further forty years before electric trains became a public success in Berlin. Nowadays, the motive force of trains is carried by electrical cables with no moving engine. The Maglev system, recently built in Shanghai, uses magnetic force fields to hold the carriage just above the line. And the newest rail technology, being developed by Elon Musk in South Korea, places electrified carriages, propelled by magnetic levitation, into evacuated tubes to minimize air resistance. It is believed that this kind of system could produce speeds six times faster than air travel. Back in Dalkey, Ebrington, the one-time 'fastest man on earth', would raise a glass of whiskey to that.

MOGG'S POSTAL-DISTRICT AND CAB-FARE MAP.

MOGG'S
LONDON
AND ITS
ENVIRONS
The Latest Surveys.

E. S. Mogg

# Cab-fare Map
## Paul Dobraszczyk

It's easy to forget that taxicabs were horse-drawn vehicles longer than they have been motorized ones, and that passengers relied for a long time on printed information to work out their fares. The earliest vehicles for hire in London were the hackney coaches, established in the early seventeenth century and overseen from 1694 by the Hackney Coach Commissioners, who regulated prices according to both distance and time. Printed information for passengers began to appear in the early eighteenth century, initially as lists of fares and coach regulations included in engraved city maps, and later as books of fares in their own right that were later required by law to be carried in all cabs. Yet books of fares, no matter how well designed, were clearly problematic to use, whether carried in a pocket or consulted in a cab. Information in book format could never be found speedily; pages had to be turned, indexes consulted, and destinations and cab stands memorized.

The prominent London map-makers William and Edward Mogg attempted to address these problems with their mid-Victorian series of postal-district and cab-fare maps, the example illustrated here being published in 1859. Superimposed onto a conventional topographic map of London are grid squares at half-mile (0.8 km) intervals, labels of the postal districts, and the 4-mile (6.4 km) radius from Charing Cross (shown as a dark circle) that marked the transition from a sixpence to a shilling fare per mile. In addition, referencing aids are included around the edges of the map: letters along the top and bottom; numbers on the sides. In the 33-page index that accompanied the map and listed 3,000 places, readers were instructed on how to use the map. First, they were to locate their required destination in the index, and memorize the letter and figure of the square required. By consulting the map and matching the letter and figure to those given around its edges, the user could find the required place 'instantly'. Whether or not this process was any less laborious than flicking through the pages of a book is questionable. Nevertheless, cab-fare maps proliferated in the second half of the nineteenth century, and all manner of ingenious solutions were put forward to facilitate the easy calculation of fares in a rapidly expanding metropolis of increasing complexity. Some covered their maps with concentric circles or even triangles to help users work out their fares more quickly; others drew on the value of novelty to sell their products, with cab-fare maps printed on everyday objects from guidebooks to handkerchiefs.

London Postal-district and Cab-fare Map, Edward Mogg, 1859.

'The Cabman's Shelter. Enter Mrs. Giacometti Prodgers. Tableau!', from *Punch, or The London Charivari*, 6 March 1875.

Visitors to Victorian London were constantly warned of the danger of being ripped off by unscrupulous cabbies. The Metropolitan Police began regulating fares after 1853, mainly as a consequence of the widespread extortion of visitors by cabmen during the Great Exhibition of 1851. Cab-fare maps promised an easy solution to this perennial problem, but they were predicated on an essential paradox: even though fares by distance were fixed by the police, no journey was ever the same, whether in terms of time taken or route chosen, and this understandably resulted in countless disputes between passengers and drivers. One particular passenger, unusually a woman with the Dickensian title of Mrs Caroline Giacometti Prodgers, dominated the police reports in the early 1870s. Prodgers made a point of testing the honesty of cabmen by riding a mile to its exact limit and then asking the fare. After a time, she became so dreaded that the warning cry of 'Mother Prodgers' would send every cabman within hail dashing up side streets to escape; her notoriety was such that even the satirical journal *Punch* devoted a cartoon to her exploits.[1] Prodgers based her seemingly endless knowledge of London's distances on a book of fares she compiled herself, which she always carried in her pocket. By 1890, however, even she had become fed up with the constant arguments and was said to pay her fare 'without making an appearance in the police court necessary'.[2]

The invention of the modern taximeter by Friedrich Wilhelm Gustav Bruhn in Germany in 1891, first used in a motorized cab in 1897, signalled the demise of the era of printed information for cab passengers. By 1914 all new motorized cabs – and by then these already outnumbered horse-drawn ones – were required to carry an automated taximeter (the now commonly used 'taxi' or 'taxicab' are shortened versions of taximeter cab), and the need for extraneous printed information dwindled. But, as with all forms of automation, something valuable was lost, namely something of the human skills required to navigate the city on one's own terms (and also to negotiate a delicate social relationship), as well as more than a century's worth of the ingenuity and creativity of map-makers and letterpress printers. What became extinct, then, was a sophisticated and richly diverse form of graphic communication that had evolved to mediate both social relationships and those with the city.

Today, taximeters in cabs are invariably coupled with computerized satnav systems, the latter being the present-day equivalent of the printed maps of the Victorian period, albeit without their variety and visual appeal. The electronic maps show both driver and passenger the 'correct' route to be taken, the taximeter the corresponding fare. With online fare estimators now available to anyone with a smartphone, to ask the fare is no longer a necessary form of polite interaction between passenger and driver. Yet, even with all this technology, the uncertainty of the price of a taxi journey remains: the city outside can never be brought to order to ensure that each journey is conducted according to the same variables. Whenever we get into a taxicab, as the Victorians knew all too well, we are all still at the mercy of the city we can never control; and we still need the skills that their map-makers and map users valued so highly.

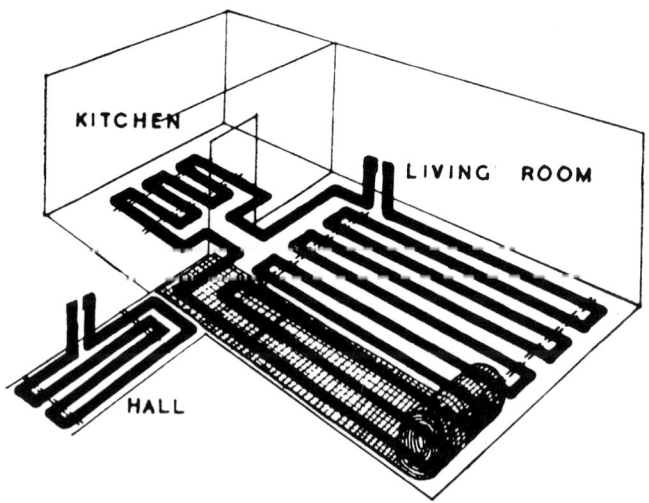

Electrical floor heating showing circuits of cables.

Section of floor showing heating cables in screed.

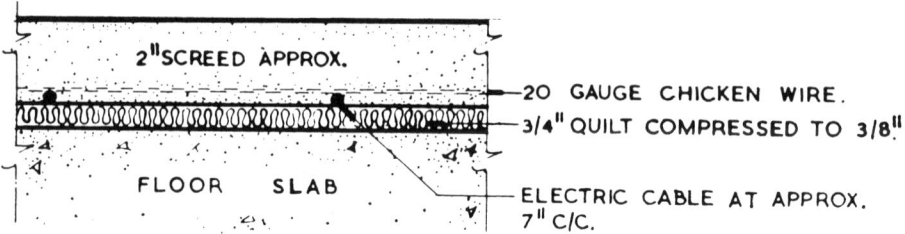

2" SCREED APPROX.

FLOOR SLAB

20 GAUGE CHICKEN WIRE.

3/4" QUILT COMPRESSED TO 3/8"

ELECTRIC CABLE AT APPROX. 7" C/C.

KITCHEN

LIVING ROOM

HALL

# Central Heating
## Mario Carpo

After the loss of Roman engineering, domestic heating in medieval Europe was for centuries a sorry ad hoc operation: rooms in some buildings had fireplaces, in which people lit fires when they could. Then came stoves, which, alongside the radiant heat they delivered, could also be used to generate hot air, hot water or steam; as Benjamin Franklin and a few others soon found out, these fluids could carry heat some distance, thus separating the production of heat from its localized consumption. The logical next step was a bigger stove, or furnace, tucked away in a service room, often in a basement; thence, heat and hot water would be carried to a whole house or multi-storey building – a system often called central heating. As the technical logic of the Industrial Revolution gained ground, it appeared that even bigger economies of scale could be reaped by centralizing and mass-producing domestic heat for entire town blocks, or even cities, as long as the cost of distribution and the related thermal loss would be less than the savings generated by more efficient and cleaner, industrial-grade heat-producing plants (district heating, *chauffage urbain*, *Fernwärme* and so on). The New York City Steam Company started operations in 1882 and its successor, Consolidated Edison, to this day delivers steam for heating (and increasingly other purposes) to most of Manhattan.

When heat is produced remotely, and delivered to residents by dint of some heavy infrastructure, domestic comfort is at the mercy of various technologies for local temperature control. Electric space heaters and gas burners can be switched on and off at will. Hot-water radiators and steam radiators typically have valves that can be opened and shut, or do so automatically. Vents can be opened and closed at will, too. But for other modes of heating local control may be more impervious, and for some centralized systems a personal, discretionary control of room temperature has often been seen as almost impossible. This did not happen for any overarching technical reason. In many parts of the world non-customizable, standardized heating was for a long time a social, political and ideological choice.

As with so many other myths of twentieth-century modernism, the idea of a standardized, uniform and universal thermal environment emerged in the course of the technophilic 1920s, and found one of its purest formulations in Le Corbusier's early writing.[1] Corb's plan of keeping all humans in perfectly sealed, purified environments at a steady temperature of 18°C (just over 64°F) – everywhere and at all times – may appear today like the

Diagrams of electric underfloor central heating system for Barbican Estate, London, 1959.

hallucination of a madman, yet it was for at least two generations of modernists a realistic and operative design inspiration, instantiated and materialized in many instances and experiences and cultural technologies of modern life, from the geodesic dome to pressurized air travel. In the mid-twentieth century underfloor heating, a relatively new mode of central heating, appeared best suited to fulfil and express this quintessentially modernist dream. When properly engineered, underfloor heating delivers gentle, isotropic, homogeneous heat at constant temperatures; it is as ubiquitous as it is invisible: due to thermal inertia (the delay whereby different temperatures between surfaces result in actual heat transfer) it is slow to react to changes in the supply of or demand for heat, and therefore works best in an 'always on' mode. Underfloor heating is ideally a cradle-to-grave provision, offered to all citizens, equally, by the invisible hand of some benevolent higher authority, under the exclusive control and jurisdiction of a remote, impersonal and often unreachable central administration.

Unlike ancient hypocausts, where heating came from flues built into the floor substructure – a technique largely maintained in many eastern countries until industrialization – modern underfloor heating uses small hot-water pipes embedded in the floor. A UK patent from 1907 describing a similar system, called 'panel warming', appears to have been exploited mostly in Switzerland, and there is anecdotal evidence of some underfloor heating in use in Europe and the United States before the Second World War. In London Oscar Faber, the noted structural engineer, retrofitted panel heating to some rooms of the Bank of England before the war, and in the rebuilding of the House of Commons immediately after.[2] Nobody knows why Chamberlin, Powell and Bon, the firm that designed the Barbican Estate in London from 1955 onwards, chose underfloor technology for heating all 2,014 flats of the estate, and by electric cables rather than hot-water pipes embedded in all floors – a choice unprecedented at that scale, and which must have seemed at the time daring, outlandish and quirky. It was approved without discussion, and carried out diligently from 1963, when building started, until 1976, when the last of the three residential high-rises of the estate was completed. The Barbican heating system was designed between 1956 and 1959:[3] 1956 is also the date of the UK's first Clean Air Act, of the Suez Canal energy crisis and of the inauguration of the first nuclear power station in the UK – all of which may explain the choice of electricity, instead of fossil-fired heating. In 1959 the architects went the extra mile to explain that ubiquitous underfloor heating in the estate was intended only to provide a uniform 'background', on top of which residents were free to add individual heating from their own electrical fires or space heaters, 'to their personal liking', but at their own expense.[4]

Honeywell T-86 round thermostat, designed by Henry Dreyfuss, USA, 1953.

To this day the 4,000 or so residents of the Barbican are also free to open (some) windows when the temperature in their flats goes up beyond their taste. That's about all the freedom they get. The humble bimetallic thermostat has been known since the 1830s, and as an electric switch since 1886. Thermostats are perfectly compatible with most central-heating technology, yet they are never mentioned in the Barbican technical documentation. It is a commonplace of media studies that the television remote control spelled the rise of Postmodernism. By introducing choice, differentiation, variation and granular customization to the control of the immediate environment of the human body, the room thermostat is the rival and nemesis of modernist centralized heating. The first thermostatic systems were developed in France at the end of the eighteenth century to provide steady temperatures for poultry breeding, but modern electromechanical thermostats (and today's electronic, networked thermostats) allow for and encourage controlled variations in the distribution of heat, and today's postmodern dwellers expect that temperatures – just like everything else in the digital economy – should be served on demand: where needed, when needed, as needed.[5] The political and ideological project of a standardized, machine-made thermal environment, where all receive the same amount of centrally regulated heat at all times, has been consigned to the dustbin of technical and social history.

# Chaparral 2J: The 'Sucker Car'
Eirik A. G. Bøhn

Over the course of less than a decade from the mid-1960s onwards, lax regulations and leaps in technology placed motor racing at the forefront of developments in engineering and thrilled spectators with exponential increases in speed and danger. Nowhere was this demonstrated more dramatically than in the nationally televised and widely popular Canadian-American Challenge Cup, or Can-Am. In contrast to the nimble machines that figured in the blue-blooded circus of European Grand Prix racing, Can-Am relied on the brute force of large displacement engines, attracting large crowds as the series toured North American raceways.

Nicknamed the 'Sucker Car' or the 'Vacuum Cleaner', the Chaparral 2J was unveiled by constructor team Jim Hall and Hap Sharp in 1970. It represented a watershed in the development of mechanical downforce, the mastery of which would enable a car to corner at greater speeds thanks to increased surface traction. In addition to the great forward thrust provided by its 650-horsepower Chevrolet ZL1 aluminium engine, the car featured an auxiliary two-stroke 45-horsepower snowmobile motor that powered two large-diameter fans, adopted from a mobile howitzer cannon. Reversing the principle of the hovercraft, the fans mechanically sucked the car to the ground, shifting 273 cu. m (9,650 cu. ft) of air per minute from underneath the car and producing 998 kg (2,200 lb) of downward thrust. The seal between the vehicle and the surface was provided by a curtain of the NASA-developed polycarbonate material Lexan, which operated in unison with the car's suspension through a system of pulleys and cables. It worked. After its debut at Watkins Glen, New York, in the hands of Jackie Stewart, bemused competitors reported how the sucker car sprayed oil, dust and debris from the track onto the cars in its wake as it darted through the corners, its twin fans screaming.

In the public imagination, the unprecedented loudness of the Chaparral represented a violent transgression of the frontiers of speed and a distinctly home-grown mechanical prowess. Fusing home-garage entrepreneurialism with aerospace technology, the car demonstrated to the American public the seemingly boundless engineering agency of determined individuals. Yet in the background of Hall and Sharp's project lurked the largely covert assistance of Chevrolet, an alliance that was downplayed in the press, with the notable exception of the Chevrolet newsletter *Corvette News* (August/September 1970), whose photo feature detailed the machine's suction apparatus and Lexan skirting and proclaimed: 'The car you see on the next four pages is

Chaparral-Chevrolet 2J,
Sports Prototype driver
Vic Elford at the wheel,
Can-Am Riverside,
11 January 1970.

perhaps the most incredible race car ever designed and built. It's the new Chaparral 2J. It promises to revolutionize everyone's ideas about race cars. Unless, of course, some spoil sports get together and ban it.'¹ To a society infatuated with numerical analysis, from the sublime numbers of the space programme to the Rand Corporation's quantification of the ongoing war in Vietnam, the piece spoke a familiar scientific language, informing the reader that the downforce had been calculated at 3,000 ft (915 m) above sea level in Midland, Texas, a somewhat misleading description of what had essentially been an approximate process of trial and error.

Celebrating this amiable dichotomy between the garden-shed mechanic and corporate might, the Chevrolet newsletter described the design process from inspired idea to digitally assisted construction:

> What would happen if the fans were reversed so that instead of hovering, it clung? Novel. New. Untried. Off to the computer to see what design parameters were involved. Programming. Reprogramming. An explosion of arithmetic disgorged from the computer to give Chaparral Cars what they wanted to know . . . In fact, a rivalry sprang up between the live, flesh-and-blood engineers and the inanimate computer. The design resulted in about two parts live engineer, one part computer.²

While the attraction of the automotive industry lay largely in technological advances, racing was intrinsically linked to individual, heroic protagonists. Reassuringly, for the immediate future of auto racing, greasy fingers still held sway over digital technology.

Precisely in its monstrous hybridity, the Chaparral 2J represented a marriage between the car and the hot-rod mechanic, both symbols of American individualism, and the greater forces of the nation's aerospace and military–industrial complex, incorporating military and NASA components. In keeping with this technological ambiguity of the high and the low, the 2J also came with a style wholly its own, contributing immeasurably to the inherent visual drama of racing. Although the front end retained the essential wedge shape associated with Can-Am racing, the car departed from the design of the competition with its prosaic boxlike posterior and covered rear wheels, recalling the shapes of heavy industry rather than the sleek aerodynamics associated with speed. In a grid dominated by a distinctive palette of yellow, orange, candy red and ocean blue machines, the Chaparral's white livery, jet-black details and bare aluminium centre panels evoked the familiar aesthetics of space travel. The styling made no attempt to hide the complementary thrusts of the machine, one propelling the car forward, the other pushing it to the ground.

Part conventional racing car and part suction chamber, the Chaparral 2J was visually distinct from, far louder than, and took the corners 50 per cent faster than the competition. But the merger of technology that made it possible was an uneasy one, causing the sucker car to be plagued by mechanical problems. As Hall later explained, it was precisely the vehicle's hybrid nature that posed the greatest challenge: 'I think the hardest part about it was that it was like having two cars, but in one space.'[3] Regardless of a series of losses caused by technical faults, the 2J was immensely popular with fans whenever it took to the grid, and epitomized a period when liberal regulations stimulated fantastic innovations that pointed to an even faster future. Voicing the spirit of a self-consciously pioneering period of motor racing, the celebrated driver Brian Redman explained in an interview a few months before the debut of the Chaparral: 'Every year we say it's impossible for cars to go any faster on such and such a circuit, but they do.'[4]

In the end, however, the spoil sports had their way. With the realization that the lack of regulatory restriction would spawn unfair advantage alongside novel expressions of automotive individualism, the elastic rules of the Can-Am racing series that had made the car possible in the first place soon shut around it. A year after its debut, the sucker car was banned by the Sports Car Club of America – a decision that marked the Chaparral 2J and suction-assisted prototype cars for extinction. Not long thereafter, in 1974, the Can-Am series itself folded, replaced by the more homogeneous – and, by comparison, lacklustre – Formula 5000. From this point onwards, homegrown engineering and the pioneering DIY individualism it represented would play a lesser part in drawing the American public to the racetrack.

# Chatelaine
Iris Moon

The chatelaine was an accessory with chains worn at the waist, to which users could attach items such as watches, keys and trinkets. Used by both men and women from the medieval period onwards, this precursor to the modern key ring became in the course of the eighteenth and nineteenth centuries a statement fashion accessory, growing in shape and changing in material depending on the fickle taste of each period. Ultimately a victim of fashion, the chatelaine fell prey in the twentieth century to uncinched waistlines, spacious, amorphous pockets that could contain anything, and the rising use of shoulder bags.

The word 'chatelaine' derives from the medieval French term for the mistress of a castle; it eventually came to encompass the collection of keys she carried at her waist, as the symbolic keeper of the household. Although the earliest usage of 'chatelaine' to describe the accessory was not until 1828, waist-hung appendages were already popular by the eighteenth century. Goldsmiths, watchmakers and toymen – makers and retailers of small metal accessories and objects of virtue – shaped an entire vocabulary around the chatelaine in the eighteenth century. They advertised chains, strings, stay hooks, etuis (cases) and equipages. The last of these had three basic components: the curved clasp, decorated on the front and plain on the back, which was hooked to the waist; chains linking the front of the clasp to containers; and a large etui, usually flanked by smaller containers for snuff, bonbons or thimbles. Elaborate rococo examples could be set with diamonds and precious stones. Polished and cut-steel chatelaines also became fashionable. Industrial entrepreneurs such as the English manufacturer Matthew Boulton designed steel watch chains, toys and accessories for chatelaines, exporting them in massive quantities to France from his Soho Manufactory in Birmingham.

Made of disparate components and materials, the chatelaine was a combinatory object unlike many of the guild-regulated and restricted luxury goods of the eighteenth century. Dangling accessories could be changed. Users could replace broken or obsolete components with newer ones. Around the 1750s, the Chelsea porcelain manufactory made tiny porcelain seals, trinkets and bonbonnières shaped as animals and cupids and fitted with rings at the top so that they could be put on a chain. A tiny cupid bearing the motto *Amor Vincet Omnia* (Love Conquers All) could be swapped for a small knife or tweezers. New and old components could exist together.

Chatelaine with calendar, steel, gold, blue glass, probably French, late 18th century.

Chatelaines became gendered expressions of identity: practical necessities for women, and a showy timepiece for men. At a time when pockets were separate attachments hidden deep in the folds of a woman's dress, the chatelaine made tools easily accessible and visible. A chased gold and agate etui containing a pair of gold scissors, a needle, a knife, a tiny spoon and tweezers would serve as a status symbol: stylish but practical. By contrast, William Hogarth's prints often pictured matrons carrying nothing more than a set of keys on a giant iron ring, symbolizing their no-nonsense attitude to household management. Men's chatelaines became controversial, tied up in the dandyish culture of the eighteenth century. Waistcoat pockets accrued a staggering number of seals and keys, and a pair of watches, one real and one fake. In Paris, men took classes in order to 'learn how to walk to make the maximum noise from the jangling of their trinkets'.[1] Around 1772 the macaronis, a group of fashionable London men who adopted extreme forms of dress after travelling to Italy on the Grand Tour, made accessories a key part of their collective identity. The chatelaine accompanied the requisite macaroni accessories of a dress sword, paper-thin leather shoes with buckles, a tiny tricorn hat and a humongous corsage. They introduced what was known as the 'macaroni hang'. Banishing the central clasp found on a typical chatelaine, the wearer slung a chain loosely at the front of the waist, forming a gentle curve that inadvertently mimicked the shape of macaroni pasta. Watches and charms dangled tantalizingly from the two loose ends.

The example shown here, probably French, is typical of the end of the eighteenth century, and demonstrates how the chatelaine bundle was a carrier of sentimentality, too, as the aggregator of keys and seals, both of which were loaded with symbolic meaning. Although goldsmiths marked chatelaines that were made of precious metal, this cheaper, unmarked example demonstrates the fashion for accessories made of steel and blue glass in the second half of the eighteenth century. The clasp at the back allows easy placement at the waist. The seven chains below are attached to a miniature corkscrew, a decorative bow and a small key. It is possible that the coarser steel chains were later additions.[2] Unusually, the central plate contains a calendar instead of an hourly watch. Abbreviations for the days of the week have been engraved in French next to a series of numerical dates on a circular enamel plate, which can be adjusted by turning the tiny, star-shaped dial at the centre. While the steel gives it a severe quality, the pair of doves and the overturned flower basket on the clasp suggest a sentimental aspect; perhaps the calendar's purpose was to hasten the passage of time between the meetings of two lovers.

The combinatory quality of the chatelaine and its adaptability to a succession of changing fashion regimes make its extinction from our wardrobes particularly surprising. Prominently visible on the waspish waists of women

Portrait of Hannah Maley (Mrs Johannes Cornelis) Cuyler wearing a chatelaine, unknown artist, *c.* 1790, oil on canvas.

until the end of the nineteenth century, it all but disappeared as a fashion accessory in the twentieth century, and is today reduced to nothing more than a carabiner key chain. What ultimately sealed the chatelaine's fate was a different handheld device: the purse or handbag. Yet few today realize that the 'it' fashion accessory of the late twentieth century had at one time dangled from the noisy chains of the chatelaine.

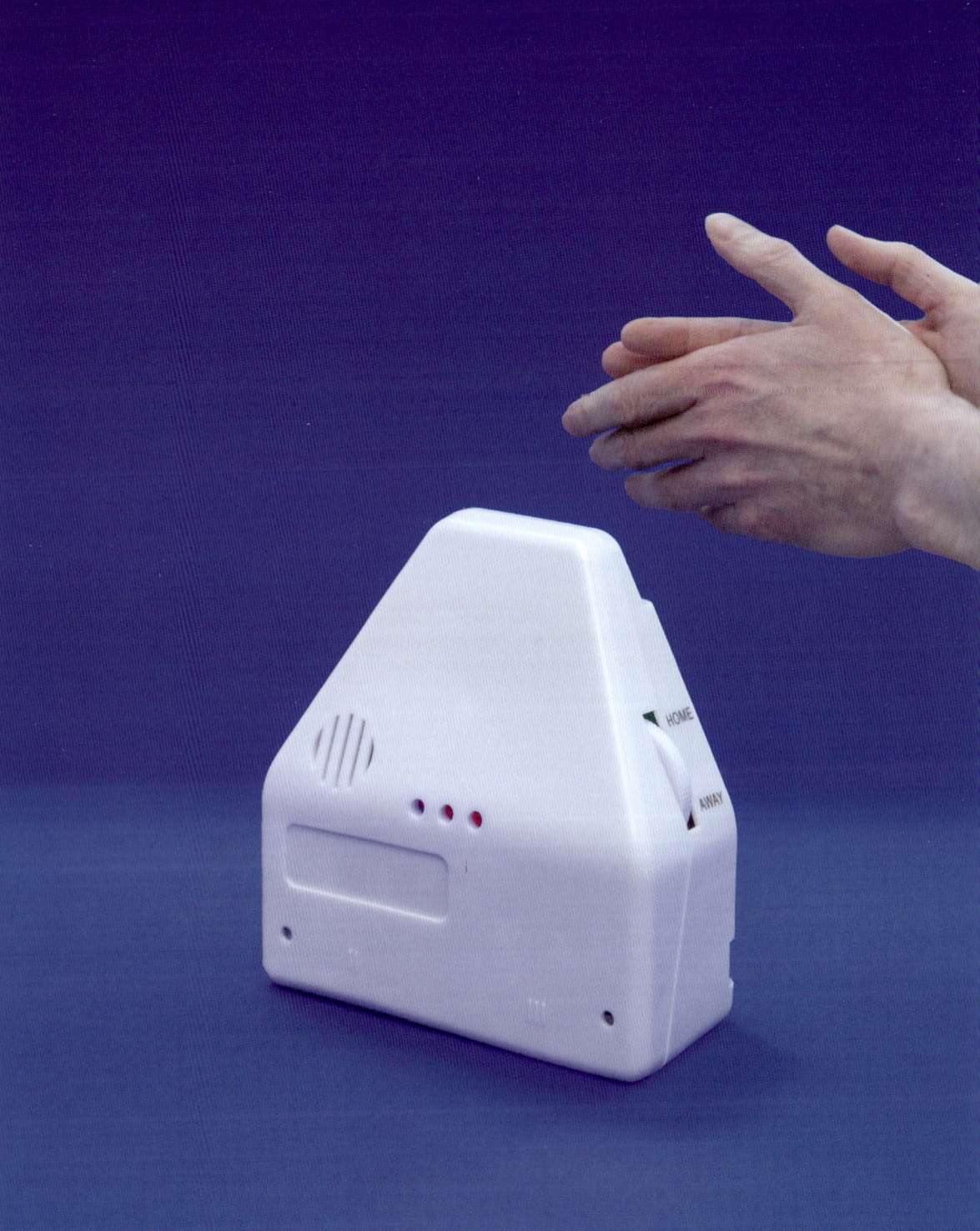

# The Clapper
## Charles Rice

'Clap on, clap off, it's The Clapper!' Rarely has an advertising catchphrase been more inane or yet more precise about an object's function. The Clapper, marketed from the mid-1980s by the American businessman Joe Pedott, is a sound-activated switch that turns appliances on and off with a clap of the hands. Plugging directly into a power socket, it looks like a double adaptor. Two appliances can be plugged into it, each controlled by a different sequence of claps that are programmed into the device. The Clapper's selling point was convenience. Without moving from your lounge chair or bed you could turn on the television or turn off the light; returning home in the dark, all you had to do was clap for illumination. The Clapper also came with a security feature. In 'away' mode, it would switch on a light at any sound, then switch it off and reset itself a few moments later. The Clapper reduced a domestic command to its pure, performative moment: the clap that demands immediate action. Such an act would have been cringeworthy if it hadn't been couched in the camp humour that pervaded The Clapper's marketing.

Now that we converse with Alexa, Google Assistant and Siri, and adjust just about anything in our homes remotely via a smartphone, The Clapper's claims to convenience seem quaint. Yet it shouldn't simply be seen as an outmoded novelty. As the patents registering its design reveal, it is part of a chain of innovation in automation and sensing technology that has culminated in today's smart home. Its first patent, filed in 1985, was for a 'Sound activated light switch', its claim being simply for 'the ornamental design', that is, the outward appearance of the switch.[1] This device was originally known as 'The Great American Turn On', its inventors having engaged Pedott to help market it. When he discovered that the switch didn't really work – it had a tendency to short-circuit television sets plugged into it – Pedott bought the rights to it and hired an engineer to redesign its inner workings.[2]

A second patent, filed in 1993, was for a 'Method and apparatus for activating switches in response to different acoustic signals', and includes a flow diagram that describes how The Clapper's sensing and switching technology works.[3] In investigating a patent's claim, assessors research the 'prior art', that is, how it relates to other patents filed for similar or related inventions. Eleven other patents are cited in connection to the second Clapper patent, including a 'Wheelchair-mounted control apparatus' filed by New York University in 1978, a 'Voice-controlled welding system' from 1979, and a 'Method and device for voice-controlled operation of a

The Clapper, Joseph Enterprises, Inc., USA, c. 2000.

'Clap On! Clap Off!', packaged The Clapper as sold today, Joseph Enterprises, Inc., USA.

telecommunications terminal' from 1986. Record is also kept of subsequent patents citing the patent in question. There are currently 89 patents citing The Clapper's second patent, including an 'Intrinsic console with position-able programmable multi-function multi-position controllers' filed by the Canadian Space Agency in 1998, a 'Sound-actuated system for encouraging good personal hygiene in toilet facilities' from 2001, 'Audible sound detection control circuits for toys and other amusement devices' filed by Mattel in 2003, a raft of patents filed by Skybell Technologies in relation to doorbells, and – perhaps most interesting of all – 'Forming computer system networks based on acoustic signals' filed by Apple in 2014.

Such a set of associated patents upends regular understanding of innovation and technological development. Automated wheelchairs link to industrial machines, mechanized toys, space technology, doorbells and Apple's advancement of the Internet of Things, all via a sound-activated switch marketed as a novelty. The ability to link this assortment of inventions is the genius of a patent system built on the double condition of disclosure and protection. Protection grants a licence for exclusive use; disclosure allows the invention to be made public, enabling the differentiation and protection of further inventions and applications. Particular objects may become extinct, but patterns of invention continue along branching paths.

In the particular anonymous history its patent traces, The Clapper marks a critical threshold. It is the point at which we can distinguish the branching trajectory of 'intelligence' in objects. In however primitive a way, The Clapper allowed us to feel as though we were communicating with objects. But the clap was a unidirectional command. Now Alexa and company talk back. They listen and remember, storing our requests as data from which an evermore sophisticated picture of preferences and behaviour can be built. Apple's acoustic network patent from 2014, which discloses 'the establishment of data communications between devices based through the use of acoustic signals', indicates how the tables have been turned: the clap has developed into the code of machine interaction.[4] What was once the language of our command is today the conduit for the distribution of our data.

Now an object of retro-kitsch fascination, The Clapper is still marketed by Pedott's company Joseph Enterprises, Inc., the latest being a *Star Wars* Darth Vader version that delivers portentous aphorisms about the use of the Force when switching things on and off. Despite this, it is clear the device is functionally extinct as a result of the subsequent inventions to which it is linked. Emerging in the heyday of television marketing, it now has virtually no presence on social media. It comes from a time before the data threshold. It is just an enhanced switch, not a device that gathers and shares information. It doesn't talk to other devices, except, now, in the voice of Darth Vader's faux villainy. Yet, up against Alexa's guile, perhaps such a camp portent should register as critique. Do we indeed underestimate the dark side of home automation?

# Close-constraint Key
## Ben Vandenput

The *Schliesszwangschlüssel* or *Durchsteckschlüssel*, German for 'close-constraint key' or 'run-through key', looks as though it is straight out of a surrealist work of art. It is definitely not, for the key was actually used to lock and unlock the outer doors of tenement houses and apartments in Berlin and its suburbs from the 1910s until the end of the 1960s. The Berlin locking system is in use today only as a vestigial technology. It has been replaced by intercoms, keypads and other digitally controlled mechanisms that do not require physical keys to secure communal doors.

The close-constraint key was about 10 cm (4 in.) long and inscribed with the name 'Kerfin & Co. Berlin', the company that in 1912 took out a patent on the design and called it the 'System Schweiger' after its inventor, Johann Schweiger. It was Schweiger's Wedding-based firm that Albert Kerfin had taken over in 1893. The key was a bizarre-looking object. Rather than a bow at one end of the shaft and a bit at the other, like a normal key, the run-through key had two bits, one at each end. Without a bow that could easily be attached to a key ring, the key came with a holder that could itself be put on a key ring to guard against loss. The right-hand and left-hand bits were identical, apart from the fact that they were grooved on opposite sides in order to activate different parts of the lock.

The lock itself had not one but two keyholes, one on each side of the door. One was positioned vertically, as usual, with a round aperture to receive the key's shaft above a triangular aperture to receive the bit. The second keyhole used the same round aperture to receive the key's shaft, but had another, triangular aperture for the bit positioned horizontally, at 90° to the first. To unlock the door, one had to insert the key into the vertical keyhole and turn it 270° anticlockwise. This opened the lock but made it impossible to withdraw the key, because the first bit was no longer aligned with the vertical keyhole. At the same time, however, it brought the second bit into alignment with the horizontal keyhole. The only possible way to keep the door unlocked, therefore, was to insert the second bit into the horizontal keyhole. The resident then entered the apartment to find the first bit sticking out of the lock on the other side of the door. The only way to retrieve it was to turn the key again to a vertical position, which both locked the door and enabled one to remove the key. By the end of the procedure, the key had 'run through' the entire lock from one side of the door to the other.

Double-sided key 'no. 65' with carry cap, Kerfin company, Berlin, *c.* 1930.

The close-constraint key therefore forced inhabitants of tenements to lock their apartments and to carry their key with them, instead of leaving it lying around or in the door. It also meant that access to an apartment was limited to the one person carrying the key. The System Schweiger allowed greater surveillance, prevented people from being forgetful and proved an enormous success; at the peak of its popularity in the two decades following the Second World War, the run-through key was employed in more than 20,000 addresses. But by the 1960s, because of the Berlin Wall, the eastern part of the city could no longer be served. Wedding, the operating base of Kerfin & Co., was in West Berlin, and many of its former and future clients in Prenzlauer Berg and Friedrichshain had become unreachable. In the mid-noughties, some 8,000 locks were still in use, but this number has declined rapidly ever since.

To understand why such a specific type of key to promote security and guard against carelessness was widespread in Berlin, but never took off outside the German capital, is also to expose the particular nature of the speculative housing market in that city in the late nineteenth century. After German unification in 1871, the city attracted labour forces who left the countryside, particularly after the agricultural depressions of the 1880s, to work in the factories of the young capital. Tenement houses or *Mietskasernen* ('rental barracks') for these new urban employees shot up like mushrooms in the city centre over the following five decades. In the absence of a central planning authority, property developers were only too happy to exploit the situation. The newly built rental flats in Wedding, Moabit, Neukölln and Kreuzberg were designed to generate maximum profit from what were, in most cases, very modest and even cramped quarters.[1] The courtyard of a rectangular *Mietskaserne* led in many cases to a shared staircase that served a variety of residences, from a few multi-room apartments overlooking the street to a majority of single-room units in the rear ranges. Large parts of the courtyard were crammed with workshops and enterprises including woodturners, brush-makers, print shops and even gas factories.

These complexes were consistently overcrowded, with several hundred people living and working in a limited space. The German sociologist Werner Sombart found that in 1906 some 43 per cent of the Berlin population were living in flats with only one room. Often, apartments had six or more people occupying one room, or even eleven or more occupying two rooms that could barely be aired. One in three Berlin citizens could not afford the inflated rents in the *Mietskasernen*, which made people look for ways to generate money from their apartments, mostly by subletting them to prostitutes, seasonal workers and lodgers of all kinds during the day.[2] Sombart and others concluded that such circumstances did no good to a family's *Häuslichkeit*, or domesticity. As the psychiatrist Hans Kurella stated in 1900, to safeguard a family's unity, an apartment must provide a minimum amount of space and 'establish a home that is closed off against the intrusion of elements alien to it and against observation from outside'.[3] Another critic of the cohabitation of families and ad hoc lodgers warned of the 'dangerous opportunities' and the 'erotic situations that present themselves to any subtenant who lives in such close proximity to the householder's family'.[4]

The close-constraint key came on to the market at exactly the moment that such concerns about the defence of domesticity surfaced. In that sense, it was not only a locking mechanism for those with poor memories. Its association with safety, control and security is comparable to a *stille Portier* ('silent housekeeper'), the name given in Berlin to a framed list hung on the wall of the communal hallway, detailing who lived on which floor and in which apartment of a rental block. Unlike in Paris or Vienna, actual house-keepers were an exception in nineteenth-century Berlin, but for a long time the close-constraint key played the role of the watchful eye and the firm hand of a housekeeper – who enters which apartment, with whom? Who wanders around fruitlessly in the corridors? – with verve.

# Concorde
## Thomas McQuillan

In extinction, it's not the objects that fail. It's the world that supported them that has gone. Consider Concorde. In the years since the last flight of that supersonic marvel, in October 2003, it has become commonplace to bemoan its loss – it was beautiful and odd, faster and more sophisticated than any other form of commercial transportation, a realized dream of the modern world. It is sobering that what was possible in the 1960s – travelling from London to New York in three hours – is no longer attainable.

The seed of the idea came from the Cold War. The British Air Warfare Committee concluded in the early 1950s that a new enemy jet bomber could easily reach London with its payload before anyone on the ground could shoot it down. One solution was a supersonic interceptor, travelling more than three times as fast, at Mach 3. But the civil market revealed another, greater potential. The British had fallen behind the Americans on the commercial jet, but the supersonic range was still up for grabs. One hundred passengers travelling beyond the speed of sound was the bold dream of a technologically advanced society: progressive, democratic and *fast*.

British aeronautics of the period were excellent, bolstered by the brilliant German aerodynamicist Dietrich Küchemann, who came to Britain after the war. The science was there. But it would be expensive, and the government wanted international partners to share the cost. The Americans had plans of their own and declined. Happily, the French were interested – and had the resources. A deal was struck, a name suggested (by the eighteen-year-old son of one of the engineers) that worked in both tongues, and the phenomenally complex and politically sensitive as well as very expensive project was begun in 1962.

Concorde may have been exemplary in its application of science to master its mission, but it faced stiff opposition from society when its details became known. The cost to taxpayers in both countries was exorbitant: $6 billion in development costs, or $2,400 for each supersonic traveller carried over the fleet's lifetime. (For comparison, the Boeing 747 was estimated at $1 billion in development – a mere 28 cents per passenger.[1]) In travelling at supersonic speed, the aircraft created a sound wake, a sonic boom whose crack was terrifying on the ground. It guzzled fuel. And the ticket price? You couldn't afford it.

Concorde, maiden UK flight, Fairford, 10 April 1969.

Everything about Concorde was deluxe. The royal couturier Sir Hardy Amies designed the stewardesses' uniforms, classic and well cut; the industrial designer Raymond Loewy contributed the cutlery, geometric and instantly iconic, which Andy Warhol encouraged everyone to steal ('It's a collectors' item'); and the chef Paul Bocuse designed the menu, originating in the process *la nouvelle cuisine* – a seven-course meal including lobster with truffled eggs, duck foie gras and *une sélection de fromages des regions françaises*. From its first flight in 1976, it became the province of the elite: rock stars, heads of state, financiers. And there it was, a supercar of the skies, a blend of refined technology and luxury. But it was felled in the most ignoble way.

On 25 July 2000 a McDonnell Douglas DC-10 dropped a small strip of metal cowling on the runway as it took off from Paris Charles de Gaulle airport. Five minutes later, tyre no. 2 of Air France Concorde F-BTSC ran over the strip and ruptured, striking the underside of the fuelled-to-capacity delta wing and rupturing the tank. Jet fuel gushed and ignited as the craft lifted from the runway, streaming fire. The pilots lost control. Concorde rolled to the right and crashed into the Hôtelissimo Les Relais Bleus, exploding into a fireball and killing 109 passengers and crew.[2]

This, Concorde's only crash in 24 years, stunned the aeronautic community. The fleet of fourteen were grounded, and over the next year intense investigation and engineering were invested in retrofitting the entire fleet. Some would say this was the death knell for the project. But it wasn't dead yet: British Airways and Air France immediately set about upgrading all the planes with Kevlar-lined fuel tanks and new burst-resistant tyres. The new spec was intended to keep Concorde in service until about 2015. After a year of modifications the fleet was ready for service. British Airways had flown a few test runs over the Atlantic, circling back at Iceland. And in the days leading to its reintroduction, the airline arranged a celebratory flight for those involved with its comeback. The mood was joyous on board. But when the flight returned to Heathrow airport, the passengers were met with terrible news. The date was 11 September 2001.

Concorde struggled on, but the battle was lost. Air travel was in turmoil, and, tragically, the World Trade Center had housed some of Concorde's best customers – some forty people who travelled dozens of times a year. Within two years both British Airways and Air France had discontinued the service. The French maintenance partner, formerly Sud Aviation, now Airbus, was building its own Concorde-killer, the A300, and declined to provide further service or parts for Concorde. In this symbolic stroke, the new norm – wide-bodied jets – forced the slim supersonic out of the market. Modern egalitarian air travel, with its security protocols and decided lack of luxury, had triumphed.

But all these factors – the crash, 9/11, budget travel – were simply a confirmation of what had become a fact long ago, that Concorde was dead on arrival. Only twenty were ever built – fourteen commercial and six technical prototypes – and the early excitement among potential airline customers evaporated rapidly. The studios and labs were dissolved, and the engineers moved on to other things. No more were built, and there were no plans to do so.

The world had already changed, already closed the niche that Concorde sought to fill. The idea of the future that once sustained it melted into air, and the great national enterprises of technical progress seemed quaint. The innocence of burning tons of jet fuel scot-free has evaporated as climate change has become the issue of our time. The rich have their own planes, and for the rest of us, flying is a drag.

# ConvAirCar
Emily M. Orr

On 12 July 1946 the pilot Russell Rogers took flight in the prototype
ConvAirCar Model 116 over Lindbergh Field outside San Diego,
California. The two-door Crosley coupé glided through the air on its wing-
span of 12 m (nearly 40 ft) for an hour and performed well, consuming
5 litres (1 ⅓ gallons) of fuel over 72 km (45 mi.). Lindbergh Field, named
after the famous Charles who had flown nearby, had witnessed many flying
experiments since its founding in 1928. It was the first certified airfield to
serve all aircraft types, including seaplanes, establishing its association
with experimentation in flight. The ConvAirCar carried on this legacy.

It was in 1940 that the designer Theodore P. Hall had attached a
single-engine carrier aircraft to the roof of the most lightweight car possible.
This was the first prototype of the Model 116, which he then sold to the
Consolidated Vultee Aircraft Company (later known as Convair). Hall followed in the footsteps of like-minded inventors who dreamed of facilitating
a traffic-free commute in the sky that would bring efficiency, innovation and
speed to modern daily life. Work on the Model 116 flying car was halted by
the start of the Second World War, but resumed afterwards, when the design
process benefited from the technological transfer of materials and techniques
from wartime production. With the increased capability of the assembly line
and access to industrial materials (such as aluminium, plastics and wood
laminates), many manufacturers, including Convair, propelled themselves
into new product categories to meet the needs of the affluent and aspirational
post-war public. Within this new paradigm, industrial design firms served
as the mediator between manufacturer and consumer in order to gauge
and cater to the desires and needs of the marketplace.

In 1944 the leading American industrial design firm Henry Dreyfuss
Associates was brought in to help develop Convair's next-generation roadable
plane, the Model 118, into a new product for a new consumer. A more
style-conscious and mobile American workforce, epitomized by the figure
of the travelling salesman, was the invention's target audience. Promotional
photographs in the Henry Dreyfuss archive show the car body (with the
flying gear removed) parked casually in front of a suburban home, conveying
how easily the ConvAirCar could be integrated into everyday life.

The car body's dramatically aerodynamic shape embodied the spirit
of post-war futurism. At Dreyfuss's office in Pasadena, California, Strother
MacMinn, an experienced car designer, and Charles Gerry, a draughtsman

ConvAirCar four-seater
flying automobile,
designed by Henry
Dreyfuss, manufactured
by Consolidated Vultee
Aircraft Company,
USA, 1947.

and models specialist, built car models in clay, from quarter-scale to full size. The combination of this handwork in the model-making and the development of engineering drawings, with a sophisticated application of industrial materials, created a cunning, streamlined car body with wings. The brochure for the Model 118 promoted the affordability and structural efficiency of the sleek flight section: 'one large seamless aluminum alloy tube is used for the wing span, one for the tail boom and smaller tubes for the empennage spars.' The vehicle body was fibreglass, which was advertised as 'tougher than steel yet 20% lighter than aluminum'; fibreglass also offered 'inherent sound deadening properties' and allowed 'quiet operation'.[1] Such technological facts were leveraged as selling points in the promotional literature and accompanied by detailed design drawings, encouraging consumers to take an interest in the design story of the flying car.

The comfort of the ConvAirCar passenger relied on the expert design of the pilot and driver's seat, which was in line with the strong reputation that Dreyfuss's firm had built in developing user-centred products that balanced safety and ease of use. Dreyfuss's important industrial design projects that foregrounded human factors included the Model G telephone handset (patented in 1948); an adjustable John Deere tractor seat, developed in consultation with Dr Janet Travell, a specialist on muscular-skeletal disorders (1950s); the anthropomorphic charts of the average Americans Joe and Josephine (1950s); and the publications *Designing for People* (1955) and *The Measure of Man* (1960). The firm's full consideration of the human form in relation to design set new industry standards for ergonomics.

From the first prototype model, the design of all flight gear and controls as a single removable unit made it possible for the ConvAirCar's aeronautical apparatus to be serviced while the car remained in use. The conversion from plane to automobile was executed with the disconnection of simple pins, leaving neither systems nor circuits to be untangled. During the war the Dreyfuss firm had excelled at the design of cockpits, improving the legibility and tactility of the dashboard and controls while optimizing the space for the comfort and ability of the pilot. The ConvAirCar had separate automobile and flight controls; the set for flight was on a control stick that swung down from above.

Following a failed test flight in 1947, when the plane unexpectedly ran out of fuel, the ConvAirCar experienced a sharp downfall in its fortunes. Investors who had previously agreed to finance the project withdrew. In January 1948 the car was restored and test flights resumed with two prototypes, but the design never went into production. Convair and Dreyfuss achieved success with ConvAirCar's on-ground performance; the vehicle boasted sophisticated style and took easily to the well-established systems of vehicular traffic, offering a 'boulevard ride' with operation and economy superior even to that of Cadillac and Ford models.[2] However, it was the rather untried territory of the skies that proved a more challenging environment for daily, let alone reliable, use of this new technology of personal mobility in flight.

The ConvAirCar came of age in the years leading up to the launch of Sputnik; a promotional brochure shows a couple on the ground with the car in road mode, waving to the car with wings. In the distance appears a rocket pointing straight to the sky, implying the technological link and shared cultural context between these two airborne advances. The illustration projects the ConvAirCar's ability to make fantasy into reality for consumers.

Even though the ConvAirCar and other artefacts of the history of flying may no longer survive, the idea of – and quest for – readily available and personally customizable flight certainly persists in the dreams of designers and engineers who look to the future. At NASA this project is known as the development of 'on-demand mobility', catering to the present preoccupations with the amenities of personalization and immediacy, and as yet mostly implemented in the form of unmanned, remotely controlled devices such as drones. At the same time, most space travel no longer carries human passengers. As Henry Dreyfuss helped Convair to realize decades ago, it is the human factors that are the most difficult to accommodate in aeronautical design. Nevertheless, the imaginative lure of flight remains strong, and the CEO of Virginia-based Aurora Flight Sciences – 'a world leader in the push to make on-demand flight accessible to everyone' – recently announced to a NASA audience: 'We have pretty good evidence that once you cross some threshold of ease of use, people really want to fly.'[3]

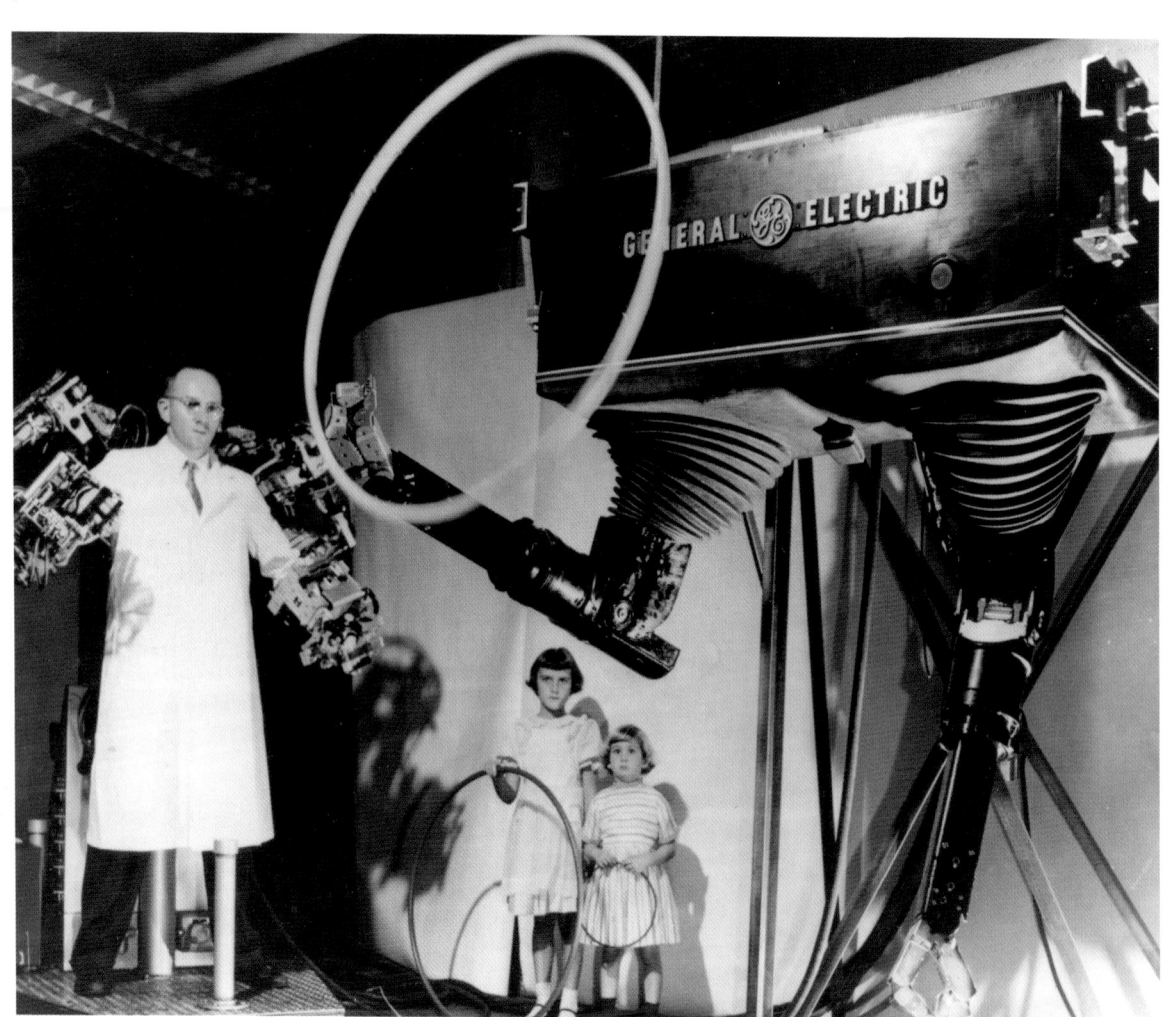

# Cybernetic Anthropomorphic Machines
Lydia Kallipoliti

If 'we judge every object by analogy with our own bodies,' as the art historian Heinrich Wölfflin once argued, the story of Cybernetic Anthropomorphic Machines (CAMs) is an unsettling case. It operates in many ways as a type of self-mirroring of how we, as a species, envision physical and mental sovereignty over bodies and territories.[1] Supported jointly by the engineering psychology programmes at the United States Office for Naval Research and the Army Mobility Equipment Research – a Development Center in Washington, DC – CAMs were developed at the General Electric (GE) headquarters in Schenectady, New York, as human–machine integration prototypes to augment strength and endurance in extreme conditions.[2] In one type of CAM advanced at GE between the 1950s and 1970s, the body was encased in an exoskeletal armature, also known as an 'exoframe', 'exosuit' or 'powered armour'. Motorized by a system of hydraulic engines or pneumatics, the exoframe enhanced the wearer's physical capabilities by increasing energy for limb movement. These were known as 'master–slave' machines. Rather than operating autonomously, they were mechanical replicas of the 'master' human operator, echoing their movements in an act of orchestrated puppeteering.

This species of robotic machines projected not only the ideal of technological supremacy – that is, the amplification of the body's physical strength using extra robotic arms and limbs – but the outsourcing of labour and the instrumentalization of servitude. In many ways, CAMs reflect a modern subject, which is simultaneously enhanced by telecommunications technology, virtually connected to the necessity of labour, and yet detached from the coarse, dirty and rough actions that corporeal labour sometimes necessitates. The fascination of these master–slave CAMs lies not in their ability to accomplish demanding tasks precisely, but in their uncanny resemblance to the features and movements of living organisms, such as the chicken's reflex to keep its head straight while the rest of its body is moving. In fact, these machines are disquieting life forms, because they mimic dexterity, distress and resistance in a way that goes beyond the execution of the task.

The father of GE's CAMs programme was the American engineer Ralph Mosher, who pioneered a series of human–machine integration experiments. With the joint support of the U.S. Navy and Army, Mosher developed the 'Man Augmentation' programme at GE with bionic electromechanical body

American engineer Ralph Mosher, the inventor of Cybernetic Anthropomorphic Machines (CAMs), uses the Handyman manipulator to control remotely the swirling of a hula hoop and twirling of a hammer, at the Debut Conference of CAMs at the General Electric Headquarters in Schenectady, New York, 1958.

parts. The aim was to design exoskeletal armours for war and for use in space exploration. Mosher's CAMs included 'Yes Man' in 1956; 'Handyman', a set of robotic arms controlled at a distance by an operator in 1958 (pictured here); 'The Pediculator', a walking machine, in 1962; the four-legged 'Walking Truck' in 1969; and finally 'HardiMan', which evolved between 1965 and 1971. 'HardiMan' was an acronym for Human Augmentation Research and Development Investigation, plus Man from MANipulator. It evolved as a series of experiments with powered exoskeleton frameworks enveloping humans; whereas Handyman was remotely controlled, the user would sit inside the skeleton of HardiMan, which directly amplified its wearer's lifting ability and strength.

Although the original intention was that these robotic prototypes might perform mechanical tasks in hazardous radioactive areas, Mosher introduced additional features to make them more lifelike and to give them a capacity for error, typical of human actions. To this end, he worked on machines that were tied to the human nervous system, to replicate the logic of hesitation. Mosher envisioned the human–machine union – our neurons translating desire into kinesis – as a wedding of sorts: human and machine combined into an intimate, symbiotic unit that performed as one integrated system. To further consummate this union, he applied the feature of 'force feedback' to all his CAMs, consequently relabelled as 'sensory feedback'. In Mosher's machines, a mediated level of force was fed back to human operators to give them a sense of the environmental interaction of the limbs under their control.

A second feature that GE broadcast widely was control and dexterity. In Mosher's words, CAMs 'will respond to irregular force and position patterns with the alacrity of man's information and control system coupled with the machine's power and ruggedness'.[3] Despite their strength, CAMs were also deft and polite. The 28 May 1956 issue of *Life* magazine featured a photo of Yes Man helping pretty Ruth Feldheim gently into her coat. In various papers at GE, Mosher created a series of diagrams dramatizing the detrimental effects of strong robots lacking human sensing and thus breaking doors, demolishing walls and engaging in destruction. In his sketches, human motor systems would help robots to respond to feedback forces that would then be interpreted to control the exertion of force. Behind the scenes, however, he was distressed by the laborious and demanding efforts needed to control his mechanical children. Archival photographs of him perspiring heavily as he endeavoured to manage his slaves beg the question: who is the slave in the human–machine relation?

Although CAMs preceded Boston Dynamics' robot series (acquired by Google in 2013), the literal kind of enhancement envisioned by Mosher – the creation of an enslaved replicate body – is no longer prevalent. Significantly, although wearable devices that enhance the human and prosthetics are in wide use, they do not depend on force feedback, which has largely disappeared from contemporary design. The elimination of force feedback from robots that perform laborious tasks not only blurs the sense of control and precision from the human operator, but obliterates empathy: our direct connection with the machine. Instead, we have virtual reality and augmented reality, which project a realm based on simulation and mediation and which are not tied to human nervous systems and corporeal interactions. The ubiquity of mediation also projects a modern subject that is indifferent and untouched, against which the fragility of human life is increasingly miscalculated.

The way in which the design of CAMs was tied to and replicated human performance and behaviour suggests a future forecast in vain. CAMs also divulge the deeply rooted sentimentality that drove the early experiments of anthropomorphic cybernetic machines designed as replicas of humans with augmented capabilities. As the philosopher José Ortega y Gasset saw it in the 1930s, the civilized world had given birth to a new kind of subject: *Naturmensch*, the naturalized man: 'This new man wants a motor-car, and enjoys it, but he believes that it is the spontaneous fruit of an Edenic tree. In the depths of his soul he is unaware of the artificial, almost incredible character of civilization, and does not extend his enthusiasm for the instruments to the principles that make them possible.'[4] In the case of CAMs, the desire to re-create 'natural' human movements mechanically, using neural connectivity, reflects a similar 'Edenic' impulse. Nevertheless, the translation process and tools enabling the transformation of the physical action to its mechanized counterpart did not endure. As empathy yielded to optimization, the resistance of the machine, felt on the skin of the perspiring master, became extinct.

# Cybersyn
## Hugo Palmarola and Pedro Ignacio Alonso

A rare picture of the Chilean 'Cybersyn' (Cybernetics Synergy) project from the early 1970s shows the Operations Room. Created during Salvador Allende's presidency (1970–73), the Cybersyn system attempted, by means of the Cyberstride software, to connect all the state companies and industries in real time, providing access to various levels of information, to allow an unprecedented degree of centralized management control. The room was designed on a hexagonal plan, and contained seven rotating armchairs with remote-control commands in their armrests so that ministers, experts and factory workers could interact with the screens arranged on the walls, via a set of specially designed visual codes. Looking like a set from a Stanley Kubrick film, this was on the contrary a fully functioning real space, the centre of operations for the entire national economy of a new state transitioning to Socialism.

The president was updated regularly on the progress of the project, and supplied with data processed by the system. A fully functional prototype of the Operations Room was initially and provisionally built in a warehouse in Santiago. From there, Allende recorded a speech setting out the ambitions of the project:

> I have decided to take charge personally of this presentation because I have devoted continuous and profound attention to the development of this Room . . . What you see around you is the result of eighteen months of intense work by a group of Chilean engineers specializing in management problems. They were able to create completely new tools to help us in the task of controlling the economy. The Government, for the first time, has the opportunity to handle very complex problems with the help of modern science and, in particular, of the electronic computer . . . What you are going to see today is revolutionary not only because it is the first time this has been done in the world, but mainly because we are making a conscious effort to give the people the power of science.[1]

Project Cybersyn, Operations Room, fully functional prototype, scale 1:1, Santiago, Chile, 1973.

A massive programme of nationalization under Allende's government created an unprecedented need for new techniques of organization. As part of the state's drive towards the construction of a socialist economy, the government promoted an experimental cybernetic model of industrial management to improve its decision-making process.

Cybersyn was based on the studies of the English scientist and professor Stafford Beer. A psychologist and philosopher, Beer is considered one of the pioneers of modern cybernetics. He arrived in Chile at the invitation of Fernando Flores, then Technical Director of the Production Development Corporation (CORFO). Beer had a meeting with President Allende in the first week of November 1971, and travelled periodically to Chile thereafter to work on the project. For years, his studies had focused on the creation of a cybernetic organization model called the Viable System Model, which applied the principles of the human body's nervous system to the control of information and decision-making processes in the management of organizations. One of the most important features of the system proposed by Beer and Flores was the ability to use current data to coordinate high-level management decision-making across different state industries and factories.

The implementation of Cybersyn was entrusted to a multidisciplinary team of the Institute of Technological Research of Chile (INTEC), and to the National Enterprise of Computers and Computing (ECOM). For the project, the Industrial Design Area of INTEC, led by the German designer Gui Bonsiepe of the Hochschule für Gestaltung in Ulm (HfG Ulm), which included foreign graduates from HfG Ulm and groups of Chilean students of Industrial Design from the University of Chile and of Graphic Design from the Pontifical Catholic University of Chile, developed an Operations Room where users could interact with the complex industrial information system; the room was to be installed in the Presidential Palace, and duplicated in various different ministries. The strategic significance of the Operations Room was that it was the point of interface between all the various elements of the planned economy. Beer explained:

> If this name reminds some of a war headquarters, it is because the reference is completely intentional. This is so because in the Ops Room the information is displayed in a very graphic way and in real time for immediate decision-making and because in this room the synoptic view of the whole battle is clearly presented.[2]

In March 1972 the computer processed the first batch of data, and at least two-thirds of the country's state companies were integrated into the system. Cybersyn and its Operations Room were finally inaugurated in December that year. However, growing political and economic instability would prevent it from becoming a tool for anything other than managing urgent matters, and it never became effective as an instrument of economic planning.

Although other countries developed technology that bore some similarity to Cybersyn, the concept and objectives of the Chilean project were more or less unique. In the early 1960s, a similar project was attempted in the USSR under the name OGAS (the Russian acronym for 'State Automated System for Gathering and Processing of Information for the Accounting, Planning, and Governance of the National Economy'), which promised to improve state operations by eliminating corruption, identifying inefficiency, increasing the volume of information collected and analysed, and providing better access to that information. In addition, it offered new forms of surveillance and of state control.[3]

Thousands of computing centres were opened in the USSR to analyse information, but they were never interconnected. Unlike the hierarchical Soviet model, the Chilean project preserved the country's democratic institutions and its civil liberties. Another system, ARPANET (Advanced Research Projects Agency Network), in the United States, was conceived in 1963 as a nationwide computing network for sending messages securely and efficiently in a context of war. Cybersyn combined elements of both systems, by being both a network and an instrument for collecting economic data and for economic planning, and in that respect it was unparalleled.

On Monday 10 September 1973 Bonsiepe and part of the Cybersyn team went for a meeting with Allende at the La Moneda Presidential Palace to prepare for the installation of the Operations Room there. The president did not attend the meeting; that very day he had called an extraordinary meeting of ministers before announcing his intention to hold a plebiscite. The following day the military staged the *coup d'état* that brought Allende's government to an end. The Operations Room in La Moneda never opened, and the dictatorship of Augusto Pinochet dismantled and threw out the entire Cybersyn project, suspicious of the political motives for its development. This and a few other photographs are the only surviving relics of this promise for armchair, and fingertip, management of an entire national economy – a vision of centralized data collection and of economic planning as yet unrealized, and indeed never now likely to be realized in any part of the world.

# Cyclegraph
## Barbara Penner

Few images encapsulate more perfectly the ambivalent and occasionally absurd spirit of modernity than cyclegraphs. Commonly deployed from the 1910s to the 1950s, they were a favoured tool of the American scientific management movement and a staple of efficiency studies at work and at home.

To produce a cyclegraph, a single skilled worker was set up in a darkened lab and given an electrical bulb to wear, usually on his or her finger. A photo with a very long exposure was taken of this worker doing a particular task. The arcs of light in the resulting photo captured the range and distance of movements involved in performing a task from start to finish. For the purposes of measurement, the worker's actions were set against a cross-sectioned screen. (The screen was often created through double exposure: the grid was photographed first and the task was photographed afterwards, using the same negative.) These images aimed to make visible the entire orbit of motion used by a skilled worker to complete a job, whether laying bricks or washing dishes, so it could be studied, replicated and perfected.

Such techniques were the hallmark of the new 'science' of industrial management, first developed in the United States. The originator of this field, Frederick W. Taylor, had called for systematic studies of work in the late 1890s. Taylor's own techniques, however, focused on time. In his quest to stop the practice of 'soldiering' (where workmen deliberately slowed the pace of production), he surveyed labourers in factories, breaking down their tasks into separate elements and timing them: shovelling dirt alone, for example, consisted of between fifty and sixty discrete timed actions by his reckoning. Added together, these elements suggested the 'proper' time each job should take – information that, Taylor claimed, allowed managers to offer more generous pay to workers who achieved a set output. But skilled labourers, subject to what they saw as arbitrary targets set by unskilled observers, did not agree. Instead, with remarkable rapidity, Taylor and his studies became a locus for protest against capitalist exploitation.

Tellingly, cyclegraphs were not used by Taylor himself, but rather by his protégé and later competitor, Frank Gilbreth, and his wife, the psychologist Lillian Gilbreth. Keen to humanize the despised Taylorist regime, the Gilbreths believed that working *with* labour to improve the performance of tasks was a more rewarding way of achieving efficiency. Instead of trying to speed up the pace of work, they focused on optimizing motion, identifying wasted movements and teaching labourers easier and less fatiguing ways of

Efficient ironing strokes, cyclegraph set up by Motion and Time Study Laboratory of Purdue University, photographed by Gjon Mili, from *Life*, 9 September 1946.

doing the job. As the design historian Sigfried Giedion noted, the Gilbreths'
pedagogical aims required that they devise new film and photographic methods, in order to analyse and intervene in the work process.[1] The cyclegraph
exemplified their approach; once photographically captured, the skilled
worker's path of motion was shown to less skilled workers, who learned to
reproduce it by sight – described by Frank Gilbreth as 'learning through
[the] eye' – and by feel through three-dimensional wire motion models.[2]

Cyclegraphs were not only used in the factory, but also quickly mi-
grated into the domestic realm. In the 1910s household engineers such as
Christine Frederick transposed scientific management techniques into the
American home, and they were very soon afterwards adopted in Europe.
These techniques attempted to recast the housewife as factory worker, with
the important but rarely mentioned difference that she was unpaid. Once
Lillian Gilbreth herself entered the field of home engineering, in the 1920s,
motion-minded studies began to be deployed systematically, most often in
kitchens. Rationally planned domestic workspaces, equipped with the correct
tools, aimed to create a proper workflow for the 'operator' – the housewife
– who could perform everyday tasks with a minimum of wasted effort. The
payoff was that the housewife would recover time and energy for her own
pursuits, referred to unironically as 'happiness minutes'.

The promises of scientific housekeeping were attractive and much hyped.
The cyclegraph reproduced here was commissioned by *Life* magazine, which
then had a circulation of 13 million. Such studies derived their authority from
their apparent scientificity, evoking a long line of human motion studies
from Étienne-Jules Marey to Eadweard Muybridge. Yet, even at the height
of their popularity, it was observed that their usefulness was limited. It turned
out that domestic work was really not at all like that done in a factory; rather
than a single task performed repeatedly, most domestic tasks, such as preparing a meal, involved many different actions, often carried out simultaneously
and with children underfoot. And even when a single household task could
be isolated for study, cyclegraphs did not impart skill-specific knowledge, for
instance how to iron an acetate rayon top with sequins (inside out, at very
low heat) or ruffled curtains (ruffle-first, inch by inch).[3]

But transmitting specific know-how was not where the cyclegraph's effectiveness was thought to lie. The Gilbreths believed their studies had identified
a universally efficient motion that largely transcended any specific task. And,
more than any other form of image, cyclegraphs were able to transmit the
essence of this ideal motion to workers of all kinds. As the article accompanying the *Life* study stressed, in all household tasks, good housewives were
to use 'sweeping strokes' – never 'jerky' or 'scrubbing' ones – highlighting
the energy-conserving power of smooth motion.[4] This overriding emphasis
on smooth motion highlights that, in the same way that domestic products,

from refrigerators to vacuum cleaners, were being streamlined, so too was domestic labour. It was also being aestheticized. Cyclegraphs made efficient domestic labour beautiful, as serene housewives performed tasks in elegant, flowing arcs like the dancers of Rudolf Laban. Here, however, housewives' bodies were being artfully integrated into a larger system of (re)production.

It was ultimately the way cyclegraphs enshrined a model of smooth and emotionally satisfying labour that led to their fall. In the 1960s and 1970s feminists led the charge. Even though they dismissed home-engineering research as pseudo-science, they believed that its seductive vision of happy homemaking had nonetheless contributed to women's self-alienation and distracted them from the inequities of household labour. Feminist scholars fiercely attacked the home economists' techniques, while feminist artists actively disrupted their streamlined visual regime. In *Semiotics of the Kitchen* (1975), the artist Martha Rosler went through a lexicon of everyday kitchen implements, from A to Z, her movements becoming progressively jarring and irrational – who can forget her malevolently snapping a hamburger press? – a riposte to the cyclegraph's abstract image of labour as logical flow. Under the pressure of constant protest and critique, the cyclegraph's model of rationalized domestic work and home engineering as a discipline was discredited and mostly collapsed in the 1970s. It was replaced by the post-Taylorist model of supposedly flexible domestic labour with which we live today, complete with ideals, aesthetics and inequities of its own.

# *Cyclops 1*
## Thandi Loewenson

On 24 October 1964 thousands of people gathered in Lusaka to celebrate Zambia's Independence. Had UNESCO responded to funding requests from one of the country's most ingenious freedom fighters, Edward Festus Mukuka Nkoloso, the day might also have been marked by the launch of a spacecraft, *Cyclops 1*. In Nkoloso's own words, '*Cyclops 1* will soar into deep abyssimal [sic] space beyond the epicycles of the seventh heaven. Our posterity, the Black Scientists, will continue to explore the celestial infinity until we control the whole of outer space.'[1]

To the uninitiated, *Cyclops 1* would appear to have been a rudimentary spacecraft, a painted metal barrel with a single porthole. Standing alongside the craft on 14 November 1964, a reporter for the British television outlet ITN described it with derision as the work of 'a bunch of lunatics'. The crew seemed equally unlikely: the 'especially trained space-girl' Matha Mwamba, Zambia's 'No 1 astronaut candidate' Godfrey Mwango and ten cats 'also specially trained'.[2] Undeterred by limited finances, insufficient propulsion systems and absconding cadets, preparations continued after Independence at the Zambian Academy of Science, Space Research and Philosophy. Self-titled Director General Nkoloso devised a rigorous training programme under which the astronauts were placed in the spacecraft every day: 'a 40 gallon oil drum in which they sit and I then roll . . . down the side of a hill. This gives them the feeling of rushing through space. I also make them swing from the end of a long rope. When they reach the highest point I cut the rope – this produces the feeling of free-fall.'[3]

Black-and-white footage thrusts us back to Lusaka's Chunga Valley. We watch from a well-kept lawn as a boy – aged about twelve or thirteen – sprints towards *Cyclops 1*. Bending his knees as he approaches, he is plunged feet first into the spacecraft. Nkoloso places his hands protectively on either side of the boy's head before the group lifts the spacecraft into the air. The ground team then gently shimmies the rocket from side to side. Later, we are part of a group pushing Mwamba and Nkoloso on a rope swing higher and higher, to Mwamba's visible glee. Perhaps these are tests for the reported 'Mulolo system' for launching *Cyclops 1*: 'we have tied ropes to tall trees and then swung our astronauts slowly out into space. Thus far we have achieved a distance of ten yards. But of course, by lengthening the rope we could go farther.'[4] They were indeed to shoot for the Moon, and then on to Mars, but this rocket was built in the context of significantly loftier ambitions.

Zambian astronauts train for Moon trip, still from news film, ITN, 14 November 1964.

Zambian astronauts
train for Moon trip,
stills from news film,
ITN, 14 November 1964.

By 1965, as well as acting as Director of the Academy, Nkoloso was
also appointed as President Kenneth Kaunda's personal representative to
the newly formed African Liberation Centre, a headquarters for exiled
freedom fighters from the region. A veteran of both the Zambian struggle
for liberation and the Second World War, Nkoloso was responsible for
supporting combatants in coordinating activities and distributing aid to their
movements, and for monitoring their activities in Zambia. While the sites
of the Academy and Centre were separate, Nkoloso's concurrent leadership
of both suggests that the two were ideologically connected. The memories
of Cuthbert Kolala, a Zambian freedom fighter, further indicate that *Cyclops
1* and the manoeuvres of the Academy played a physical role in liberation
training, too: 'I saw his wisdom when he made a drum . . . put it on the ant
hill and rolled . . . It was so exciting.'[5]

The writer Namwali Serpell suggests that the space programme was 'both
a real science project and a cover', and Nkoloso's son describes the cadets as

gaining skills in 'readiness for independence': 'He was teaching for the program, but hidden from the British government. Teaching the youth so they could be active.'[6] Alongside anti-gravity training, Nkoloso's son describes a parallel programme of political action in which cadets went to Tanzania to broadcast political propaganda, build explosives and burn bridges.

There is overt racism in articles about the Zambian space programme, at times accompanied by cartoons that depict Nkoloso and his astronauts in states of undress and with animal-like features. *Cyclops 1* is shown in pieces and, incorrectly, as a wooden barrel. These depictions are explicitly revealing of the dominant attitudes towards African intelligence and bodies at the time – seen as primitive and farcical – which the liberation movements sought to combat.

While Mwamba may not have made it to Mars, the space programme was perhaps not so single-minded. Instead, bound together with the idea that a Zambian could break from an Earthly reality constructed with their subjugation in mind was the potential that a flight into a space of independence might occur. The fate of *Cyclops 1* is unclear; communications from the Zambian Ministry of Power, Transport and Communications, dated July 1988, tell us that the programme received no official backing and 'died a natural death'.[7] However, at a time when radical changes were taking place in the country, perhaps the extinct programme served its purpose. The very presence of a space programme, with a name, a crew, a spacecraft and a serious training regime, required that one reconceive of the possibilities available to Africans. *Cyclops 1* was not only a barrel and a rocket, but a vehicle of liberation through which freedom of the mind was to be won. This was independence gained not only through the control of ground and territory, but through pioneering space in which to imagine possibilities of presence and futures beyond those previously within reach.

# Dougong
Guang Yu Ren and Edward Denison

There are some things for which extinction is a mere blip in a broader existential experience that long outlives the subject's original function. The humble serif, for example, born of the painter's brush or stonemason's chisel, lives on in print and digital typefaces that possess none of the original functions that inspired its creation more than 2,000 years ago. The same is true of the dougong, a vital load-bearing element in traditional Chinese building that united the roof and columns and was responsible for creating the distinctive upturned eave. The dougong was chiefly a structural component that, vitally, reflected social status through the creation of the characteristic Chinese roof – an orientalist motif that has become as globally ubiquitous in conveying 'Chineseness' as the calligraphic brushstroke script. So essential was the dougong's combined socio-structural role that, long after its functional demise, it has continued to dominate the search for some kind of essence in modern Chinese architecture.

To understand the dougong, its extinction and enduring legacy, it is necessary to recognize the essential and exceptional relationship between China's incomparably long building traditions and its strict social hierarchy founded on Confucianism. This relationship encapsulated a system of societal organization that was unprecedented in its bearing on the built environment, where the imposition of strictly regulated measurements based on social status determined the physical form of everything from city planning to the arrangement of domestic space.

Emerging in the Western Zhou Dynasty in the second millennium BC, the dougong reached its apogee in the Tang Dynasty (AD 618–907), whose massive wooden structures represent the zenith of the craft of timber building in China. Physically, the dougong comprised a number of interconnecting units, the two principal ones being the 'dou' and the 'gong'. The 'dou' was a square base element on which the lintel-like 'gong' formed a cantilever that supported either a further interlocking gong or a roof beam. The structural integrity of a dougong relied not on fixings or fastenings but on precise carpentry and gravity, with each piece cut to fit perfectly with its neighbour and the whole being kept in place by the weight of the roof. The inherent flexibility of this method of construction was also an effective response to seismic activity, providing stability to the entire timber frame.

The dougong was always the embodiment of the essential synergy between building and social status in China, but this became explicitly

Dougong in the Great Hall of the Forbidden City of Shenyang (now a UNESCO World Heritage Site), 1625–36.

codified in the publication of the building manual *Ying Zao Fa Shi* at the start of the twelfth century. Designed to improve the efficiency of the building trades, this laid out in extraordinary detail the relative size of building components based on the user's social standing. Given the systemic nature of Chinese construction, the size of dougong, for example, determined the size of the roof, which in turn revealed the status of the owner. Only the emperor could use the largest dougong, which could be stacked in five tiers to create the deepest eaves and the largest roof. Conversely, ritualistic buildings such as temples could use only smaller dougong in up to three tiers, creating shallower eaves and lesser roofs.

From the zenith of the Tang Dynasty, when China's building traditions were exported to Japan and flourished, the dougong diminished in size, becoming increasingly delicate and ornamental during the Song Dynasty (960–1279). As China's feudal system waned, so too did the dougong's rationale in physically representing authority. Britain's international opium dealing and subsequent invasion of Shanghai in 1842 precipitated China's torturous encounter with modernity that ultimately brought about the extinction of many seemingly essential cultural attributes: classical script, dynastic rule, imperial examinations, seal carving, feet binding, the male pigtail, timber construction and the dougong.

As steel and concrete replaced China's enduring timber frame at the start of the twentieth century, the architect replaced the artisan builder. The demise of the traditional Chinese roof on its sophisticated raft of dougong was mourned by many, but not all, of China's architectural profession, whose ranks were unmatched in their cultural diversity and included almost every European nation, as well as many Americans, Asians and the first generations of Chinese architects. In seeking to reconcile the irreconcilable problem of what form modern Chinese architecture should take, all except the most diehard modernists drew inspiration from the Chinese roof, while some even sought to revive it, albeit in concrete.

Such an approach was common throughout Asia. In China, many young Chinese architects adorned their designs with non-structural representations of the dougong. Some of China's foreign architects even sought to revive the Chinese roof, including the American architect Henry Murphy, whose 'adaptive Chinese renaissance' style became his chief legacy. In Japan, the roof assumed a political significance in its adoption by architects such as Shimoda Kikutaro, who promoted the 'Imperial Crown' style that was effectively deployed throughout Japan's growing East Asian empire. China adopted a similar approach to the design and construction of public buildings in the areas it administered. The design competition for the modern capital of Nanjing after 1927 stipulated that certain buildings had to be designed in a 'Chinese style', much to the chagrin of some local architects, such as Chen

Revival of dougong on the Bank of China's former headquarters, Shanghai, 1939.

Zhanxiang, who in 1947 described such designs as looking 'alarmingly like peacetime Piccadilly Circus dressed up in Chinese clothes as tawdry as a music hall mandarin's'.[1] Such views were also aired by the Chinese architect and University of Pennsylvania graduate Tong Jun, who ridiculed the retention of the big Chinese roof on modern buildings by likening it to the preservation of the Manchu pigtail. 'The Chinese roof,' he scoffed in 1937, 'when made to crown an up-to-date structure, looks not unlike the burdensome and superfluous pigtail, and it is strange that while the latter is now a sign of ridicule, the Chinese roof should still be admired.'[2]

The debate over a modern 'Chinese style' continued into the Communist era from 1949, and has waxed and waned ever since. Today, the legacies of this debate appear fossilized in the facades of buildings all over China in the form of stone or cast-concrete dougong relieved of their original function while designed to convey a sense of Chineseness. Even among those architects for whom the superficial reproduction of the dougong is an abhorrence, the legacy of this now redundant element remains in the form of the big roof. Whether in the prosaic municipal buildings of China's booming cities, such as Shenzhen, or in showcase structures such as the celebrated China Pavilion at the 2010 Shanghai Expo, or in the innovative designs of China's only Pritzker Prize-winning architect, Wang Shu, the spirit of the dougong lives on.

# Dymaxion House
## Barry Bergdoll

Rapid extinction seems somehow to be encoded in the DNA of prefabricated housing, despite the animating force of definitively solving the shortage of housing that has repeatedly driven modernist architects to propose ideal prototypes. Indeed, the history of designs for housing to be produced in an off-site factory and shipped to a building site – often being prepared even as the house is being fabricated – is a history of false starts, bankruptcies and a litany of failed experiments. The decoupling of such progressive experiments from history as a linked chain of successes is all the more poignant in the case of some of the twentieth century's prefabricated projects, which were launched on to the architectural stage with great fanfare, publicity and optimism. So it was for the 32-year-old Richard Buckminster Fuller, an indefatigable American architect and inventor, when he posed in 1927 with a model of his prototype for a prefabricated factory-producible house of unprecedented design and unsurpassed lightness, baptized 'Dymaxion'. Dynamic, maximum and tension were merged in both the name and the concept as Fuller proclaimed an utterly unprecedented invention, one that could rival the car in dynamic design (he also invented a Dymaxion car), provide a retort to the European idea of efficiency through the minimal dwelling, and present a structural system taking advantage of tension cables used primarily in the design of bridges. 'I just invent, then wait until man comes around to needing what I've invented,' Fuller later exclaimed. All was to be suspended from a central mast, leaving a floor plate free of structural supports to a greater extent than any of the designs put forth in these years by Le Corbusier or after the war by Ludwig Mies van der Rohe.

Despite the fact that Fuller returned to the drawing board-cum-laboratory after the initial failure to find public support for his idea, and developed a version 2.0 after the Second World War, neither of his prototypes went into production. They remain famous as eccentric inventions, their untapped potential largely unrecognized even amid the recent renewal of interest in paradigms for industrially produced buildings, spearheaded by the popular success of *Dwell* magazine or kit houses proposed by Ikea in Sweden or Muji in Japan, as twenty-first-century reprises of the early success in marketing houses of the American mail-order pioneer Sears Roebuck. 'There is nothing in a caterpillar that tells you it's going to be a butterfly,' Fuller quipped. But, although he thought his advanced version of the Dymaxion, known as the Wichita House and produced in 1944, was more beautiful than

Exterior, front view of the Wichita House, constructed near Wichita in Rose Hill, Kansas, c. 1949.

the first, it failed to take off after Fuller Houses, Inc., was set up after the war, accompanied by much hype. Here was a house, asserted the publicity in magazines and promotional films, that might be priced by weight rather than quality of materials, one whose cost could be paid off in five years on the model of car payments. But how many consumers were willing to purchase a property made of aluminium, or to park in front of a house made from the same metals as their car? It was acceptable, perhaps, for an Airstream caravan for holiday use, but did not have the right 'feel' for a permanent home.

The prefabricated house was the holy grail of modernist architects on both sides of the Atlantic, a leitmotif of Le Corbusier's applications for patents and the manifesto polemic of his *Vers une architecture* of 1923, in which he compared the production of cars to the need for an equally efficient means of producing houses. This was published in the same year that Walter Gropius and his Bauhaus colleagues exhibited projects for prefabricated designs in the first Bauhaus exhibition in Weimar. By then, American advances in factory-made houses were celebrated in European publications, notably the architect Grosvenor Atterbury's experiments with a panellized system in concrete at Forest Hill Gardens in Queens, New York, in 1909. But what distinguished Fuller's efforts from all previous experiments in either the American building industry or the European avant-garde was his openness to imagining that a factory-made house might break free of the traditional imagery of home. Even more consequential was his attempt to incorporate all the subsystems necessary to turn a building into a dwelling, from electricity to heating and ventilation; indeed, he imagined that these might even be optimized by the design itself rather than fitted after delivery and assembly. In addition, built-in mechanical storage systems for clothing, advertised in films showing enchanted potential buyers, were but one of the unprecedented modern conveniences that the new house type made available.

In its original version of 1927, the whole Dymaxion House was suspended above the ground – liberated from expensive and time-consuming foundations, as well as from ground humidity – as the lightweight aluminium dwelling was shaped to maximize air circulation and even to be naturally air-conditioned by shape alone: a huge selling point, Fuller imagined, given the sometimes sweltering summers of the United States. A heat-driven vortex would suck cool air into the house, the whole to be ventilated by a single overhead vent and secondary vents at the house's periphery. With today's environmental problems, air-conditioning involving the minimal use of electricity seems all the more compelling. Efficiency indeed drove every aspect of Fuller's invention: the house could be erected in a matter of hours, since the structure was winched into place up the central mast; shipping costs were minimal for a building of such light weight; and the efficient design of component parts meant that the whole kit could fit into a truck. It was even light enough to be delivered by aeroplane or Zeppelin. The house came with much of its storage built into the structure, as well as with a prefabricated bathroom unit complete with 'packaging toilet' and 'fogger' showerhead. Fuller imagined a grey-water system that conserved a resource whose preciousness we realize urgently today. He claimed that the occupants could bathe using less than a cup of water a day.

At the end of the war, as the United States government began to subsidize companies that could convert the wartime production economy from armaments to fulfilling urgent housing needs, Fuller turned again to his research. His Wichita House model was radically different from the Dymaxion, revealing the extent to which he was eager to allow the parameters of the problem of modern dwelling, rather than nostalgia, to generate the form of a house that could be delivered by truck and assembled using easy-to-follow instructions by the new owner. But, whereas some thought his inspiration derived from Central Asian yurts, in fact the new curved form with its central vent rising like the stem of a giant pumpkin was calculated for its aerodynamic form, only to improve ventilation and natural cooling rather than to lift bomber jets into the air. Indeed, the whole was produced in the repurposed Beech Aircraft factory in Wichita, Kansas, and advertised as available for the price of 50 cents the pound – approximately $6,500, if Fuller's production goals could be achieved. He hoped to produce some 50,000 to 60,000 a year, although in the end only two prototypes were ever built.

The initial response was favourable, suggesting that the publicity film produced in 1946 had won over a great number of Americans to the idea of a house that could be paid for in instalments and which would never need repainting. Even if Fuller Houses, Inc. was successful in attracting stockholders – and, within six months, more than 35,000 unsolicited orders – Fuller was reluctant to let production begin until every detail was perfect. As the counter-model to the heavy all-steel Lustron House, an industrialized version of the type of Cape Cod Colonial, Fuller's Dymaxion House might have provided an image of the future rather than an industrialized replica of the past. Whether due to the perfectionism of Fuller, which led to his buying back many of the stocks from frustrated investors, or the U.S. government's rapid phasing out of post-war subsidies for industry to supply housing, which also pushed the Lustron company into insolvency, only two Dymaxion houses were ever produced, one of which lives today – ironically enough – at the Henry Ford Museum near Detroit, Michigan. 'You never change things by fighting the existing reality. To change something,' Fuller quipped, 'build a new model that makes the existing model obsolete.'

EDISON'S ANTI-GRAVITATION UNDER-CLOTHING.

ENABLES THE WEARERS THEREOF TO SUSPEND AT WILL THE FORCE OF GRAVITY, SO THAT THEY CAN FAN THEMSELVES GRACEFULLY ABOUT THE ROOM.

# Edison's Anti-gravitation Under-clothing
## Bob Nicholson

The opening of the Royal Academy's annual Exhibition was always a landmark event in Victorian London's summer season. Thousands of visitors paraded through its galleries and ambitious artists competed for the most coveted spaces on its overcrowded walls. Pictures that secured prime position 'on the line' could be viewed comfortably by the crowds at eye level, but less fortunate paintings were hung so high that spectators were forced to crane their necks and squint towards the ceiling. The ignominy of being 'skied' by the Academy's Hanging Committee haunted British artists for more than a century until, in December 1878, *Punch* forecast that the coming year would bring a surprising solution: 'Edison's Anti-gravitation Under-clothing' would soon 'enable the Wearer thereof to suspend at will the Force of Gravity.'[1] Anybody who donned a set of this high-tech American underwear would supposedly be granted the power to fly effortlessly around a room, propelled only by the gentle wafting of a handheld fan. An accompanying cartoon by George Du Maurier depicted fashionable Victorians floating gracefully towards the ceiling of a gallery, examining even the loftiest paintings.

This was, of course, a joke. Alas, anti-grav undies existed only in the imagination. It was one of many fanciful contraptions dreamed up by Victorian humorists in the late 1870s – a period in which the world seemed to be caught in the grip of 'discovery mania'.[2] In 1877 demonstrations of Alexander Graham Bell's telephone sparked a frenzy of press coverage in which journalists, scientists and readers speculated about how this astonishing new invention would transform human communication. Stories about remarkable new technology – and reverential profiles of the 'ingenious inventors' behind them – became a fixture of Victorian newspapers and magazines. Even advertisers of everyday consumer goods such as coffee makers and clocks caught the public mood and proclaimed their products to be revolutionary 'NEW INVENTIONS!!!'

At the heart of this cultural moment was the American inventor Thomas Edison. He rose to prominence following the invention in 1877 of the phonograph, the first device capable of capturing and replaying sound. Within months, news of this breakthrough had been reported throughout the world and made its inventor a household name. Newspapers on both sides of the Atlantic giddily described Edison's discovery as a feat more akin to magic than to science. They dubbed him a modern-day 'wizard', and

Edison's Anti-gravitation Under-clothing, *Punch's Almanack for 1879*, 9 December 1878.

reporters flocked to his research laboratory at Menlo Park in the hope of catching a glimpse of the Yankee Merlin at work. His every pronouncement was dissected forensically and widely recirculated by newspapers that waited 'in daily expectation for the announcement of new marvels to be performed by him'.[3] Excitement reached fever pitch in the autumn of 1878, when Edison unveiled plans to develop a long-lasting electric light bulb that promised to replace indoor gas lighting at a fraction of the cost. By the time *Punch* magazine published its cartoon that December, Edison was firmly established as an international celebrity, a Victorian version of a Silicon Valley tech messiah who seemed destined to change the world with revolutionary new technology. 'With such a genius,' proclaimed one Scottish newspaper, 'it is impossible to say what may *not* be produced.'

This 'discovery mania' provided fertile ground for rumour-mongers, pranksters and satirists. The *New York Daily Graphic*, for instance, reported mischievously that Edison had invented a 'Food Creator' capable of 'manufacturing biscuit, meat, vegetables and wine out of air, water and common earth'. It was hardly the most subtle of hoaxes; the story was published on 1 April and ended with the revelation that the author had dreamed the whole thing after falling asleep on the journey to visit Edison's lab. Nevertheless, several credulous newspapers reported it as fact, and Edison was amused to receive letters asking when his miraculous new invention would be available to purchase. In their defence, genuine interviews with Edison often featured outlandish promises and boastful predictions about the transformative power of his inventions, long before they were ready for commercial use. He once suggested to an American journalist that his 'aerophone' (a giant megaphone) could be installed inside the mouth of the Statue of Liberty and used to broadcast the Declaration of Independence 'so loud that she could be heard by every soul on Manhattan Island'. This provocative, but highly improbable, proposal was reported widely in Britain and the United States.

Edison was a slick self-publicist with an eye for the angles that excited reporters – an enterprising quality that British observers regarded as quintessentially American, but which also conjured images of the snake-oil salesman and the showbiz 'humbug' of P. T. Barnum. Indeed, Edison's nationality was a vital component of his celebrity. He rose to fame at a time when the United States was increasingly imagined by European observers as a crucible of the new, a young country bristling with fresh ideas, practices and technology, whose citizens were imbued with a restless 'go ahead' energy that seemed to propel them headlong into the future. While some Victorians enthusiastically embraced many of the things modern American culture had to offer, others bristled at its disregard for tradition and were sceptical about the virtues of the new. *Punch*'s humorists typically fell into the latter group and were quick to pounce on Edison when some of his more outlandish promises failed to materialize. By the end of 1878 his much-touted electric light had yet to be unveiled, and it was the Yankee inventor's inclination for 'tall talk' that inspired *Punch*'s satirical predictions of his implausible innovations. These cartoons were not another paean to Edison's genius, but a satirical jibe aimed at everybody who had succumbed to the hysteria of 'discovery mania'.

Imagined objects, such as 'Edison's Anti-gravitation Under-clothing', are by their nature ephemeral. Most of them existed only long enough to service a punchline or sustain a rumour, before fading into extinction. Some, however, have since returned to life. Many of the satirical inventions that were deliberately absurd in the 1870s bear a striking resemblance to technology that is now part of everyday life. A *Punch* cartoon from the same almanac featured an 'Edison Telephonoscope' that would 'transmit light as well as sound' and allow people to see and hear each other at great distance. It depicted a pair of elderly British parents conversing with their children in Ceylon (now Sri Lanka), using a device that is remarkably like a Victorian version of Zoom. Other nineteenth-century humorists imagined Edison's phonograph inside a range of talking objects, such as hatstands (to scold errant husbands) and lamp posts (to move on vagrants by using the voice of a policeman), that would not be out of place in modern surveillance systems or the Internet of Things. A contributor to *Fun* magazine, one of *Punch*'s rivals, imagined visiting Edison's laboratory and being confronted by a robotic 'house-dog (powered by electricity)' that now exists in the form of Boston Dynamics' Spot. One British satirist even claimed that Edison had invented a joke-writing machine named the Funograph – a fiendish challenge that is nearly within the reach of modern computer science. Sometimes, it seems, even the most comic visions of the future can eventually come to pass. Perhaps we should stop waiting for jetpacks and hoverboards and pin our hopes to anti-gravity underwear instead.

# Electrotype Pattern

## Angus Patterson

On 17 May 1854 Queen Victoria, Prince Albert, the Princess Royal (Victoria) and Princess Alice visited Gore House, Kensington, now the site of the Royal Albert Hall, for a glimpse into the future. There, in recently completed workrooms, they spent two hours watching demonstrations of two revolutionary new combinations of science and art: photography and electrotyping.[1]

Their host was Henry Cole, General Superintendent of the Government's Department of Science and Art and, three years earlier, commissioner of the 'Great Exhibition of the Works of Industry of All Nations' held in nearby Hyde Park. Cole had just appointed Charles Thurston Thompson, the world's first professional museum photographer, to oversee the photographic studio. He had outsourced the electrotyping workshop to the great art-metalwork manufacturer Elkington & Co., a pioneer in the use of electricity to create artworks. For the next three years, both workrooms produced perfect likenesses on paper or in metal of works of art either acquired for the department's collection or borrowed for temporary exhibition. The copies made at Gore House put hitherto little-known works of art into the public domain. Multiple copies enabled artists and visitors in art colleges and museums around the world to study the same artworks simultaneously, long after the originals had returned to private ownership.

This type pattern, or copper model, for a plate was made around the time of the royal visit. It was the halfway stage in the reproduction of an ornamental dish. The original, made in Hamburg of silver-gilt in 1698, was an early purchase by the department in 1853 (Victoria and Albert Museum no. 1153-1853). The type pattern was formed in a mould cast from the original. Copper was deposited from solution into the mould using an electric current. Future moulds could be made from this solid copper deposit, to save going back to the original. These secondary copies were then trimmed and electroplated or electrogilded using the same technology, to create electrotypes. The type pattern is the original copy from which other electrotypes could be made.

Likened to alchemy when first patented in 1840, electrometallurgy offered Cole an industrial solution to the challenge of finding robust, accurate copies of artworks for artists and students in the new Government Schools of Design. Like photography, electrotyping was an imaging technology that perfectly reproduced the outer appearance of an original. Elkington's electrodeposits were the first commercial applications of

Type pattern for a reproduction plate, electrotype (electroformed copper), Elkington & Co., UK, 1854.

the electrical revolution, and its Birmingham factory became a magnet for industrial tourism.

A few hundred metres from Gore House, the South Kensington Museum – showroom of the Department of Science and Art and forerunner of the Victoria and Albert Museum – was emerging. Original works were displayed alongside plaster casts, electrotypes and photographs to create an encyclopaedia of international art. The collection aimed to be a three-dimensional sourcebook for inspiring artists and designers, and to educate public taste to help develop domestic and international markets. Its most vivid expression today is the Cast Courts, opened in 1873. Less visible is the 1,000-strong collection of type patterns, which survive as a relic of an educational programme that lasted until the First World War, when it was brought to a sudden halt by the forces of a new modernity.

In 1910 the Vienna-based architect Adolf Loos presented a lecture titled *Ornament and Crime* – an early manifesto for modernism. Published in several languages, it reflected growing international consensus about the desirability of minimalism, following decades of debate over theories of ornament. After the cataclysm of the First World War, Loos's influential lecture cemented a widespread rejection of traditional art teaching focused on the study of the past. The South Kensington System, adopted by art colleges and public museums around the world, had harnessed the reforming zeal of industrialization but remained wedded to the emulation of historical styles. For Cole, the twisting foliage on this copper type pattern was 'suggestive' for art students seeking inspiration, but for Loos it was superfluous, immoral and degenerate. His approach to design rendered the large-scale collection of historic examples of ornament redundant.

In 1899 the South Kensington Museum became the Victoria and Albert Museum, and its identity shifted to become more a treasure house of original works than a teaching institution equipped with exemplars. The legitimacy of the copy was challenged. Much as digital scans are today, the type patterns were the subject of evolving discussions around copyright, and in 1913 a bitter row between the museum and Elkington over the ownership of the patterns brought the electrotyping programme to a halt. Complaints from owners of originals about illicit copies prompted the museum to recall all the copper type patterns from Elkington's factory. The V&A's huge electrotype collection and their type patterns were relegated to the basement. After the Second World War, in a bomb-damaged museum short of space, 879 electrotypes described as 'quite useless' and 'in a disgusting state' were sold to MGM Studios, where their accuracy as copies provided authenticity for a range of feature films, including *Ben-Hur* (1959).[2]

The technological novelty of electrotyping was also short-lived. As early as 1848, the scientist Robert Hunt had proclaimed, 'The practice of

Elkington's plating workshop, showing the magneto-electrical generator and plating tanks, engraving from the *Illustrated Exhibitor and Magazine of Art*, vol. I (1852).

electrotyping has passed from the hands of the man of science into the hands of the manufacturer.'[3] Birmingham MP Harry Howells Horton wrote five years later, 'scarcely is the period passed when specimens of electro-depositing, such as medals, &c, were shown as curiosities.'[4] As patents for electrotyping and electroplating were licensed worldwide by Elkington and the resulting luxury goods became more commonplace, industrialists sought other applications for the technology. In 1865 James Balleney Elkington patented methods for electrolytic copper refining that enabled the creation of copper of almost 100 per cent purity. In 1869 the world's first electrolytic copper refinery was built at Elkington's copper-smelting works in Pembrey, near Swansea, and their method of electrolytically refining metals is effectively the one still used today. If electricity was the lifeblood of the late nineteenth-century electrical revolution, pure copper wire was its arterial system.

The copper type pattern epitomizes industrial modernity, the product of a scientific, industrial and electrical revolution that transformed Victorian Britain culturally, economically and socially. It was one of the first commercial applications of a technology that now appears more futuristic and culturally relevant than ever. Deconstruct an electric car and much of the technology that produces the circuitry, the microprocessors and even the moulded body originates directly in the discoveries and developments of Elkington's factory, which found a ready market and public advocate in the fledgling South Kensington Museum.

# Fisher-Price Peg Figures
## Mark Morris

If you grew up between the late 1960s and the early 1980s, you probably played with Fisher-Price toys: specifically, the Play Family toy sets. And if you were asked how the figures that populated these toys *smelled or tasted*, odd though that might seem, somewhere in the depths of your childhood memories you would catch the whiff and tang of the 'classic' small peg figures with simplified bodies made out of turned ponderosa pine capped with spherical heads, often of plastic. These were kinaesthetic learning objects par excellence – about the size and shape of a champagne cork – abstract and tactile and mildly tasty. The figures had no operable or even illustrated limbs. The turned profiles of the 'mother' and 'daughter' figures included a flare suggesting dresses, while the 'father', 'son' and 'dog' had basic cylindrical bodies in the manner of clothes-peg dolls. It would be hard even to refer to the Fisher-Price family as dolls, and this was part of their appeal.

Founded in 1930 in East Aurora, New York, by Herman Fisher, Helen Schelle, Irving Price and Irving's wife, the children's book illustrator Margaret Evans Price, the Fisher-Price Toy Company owed its success largely to its uncomplicated but well-crafted wooden and plastic toys paired with the bright and evocative lithographic decals designed by Margaret. Young architects-to-be might have been particularly fond of Fisher-Price given the Play Family line's focus on buildings. The figures, later branded 'Little People', featured in the first 'play-and-carry' set, the Play Family Farm of 1968. The Play Family House followed a year later. Its success would lead to the Play Family School, Hospital, Airport, Castle and Village, and then, inevitably, to commercial partner tie-ins: the *Sesame Street* set and the Play Family McDonald's restaurant.

The Fisher-Price Play Family House was conceived – as were most Fisher-Price products – as gender neutral, and its original packaging featured a young girl and boy playing with the set. The plastic pieces of furniture included with the house were minimal, in contrast to the faithful miniature copies valued by doll's-house collectors. The squat seats of the dining and easy chairs were scooped out as round buckets to hold the base of each figure. The furniture was not decaled or detailed beyond moulding, but was designed in such a way as to complement and visually connect to the decals of the rooms' walls and floors. The lithographs that covered the interior and exterior of the house identified and extended the character of the house and its rooms. The architectural and interior-design elements featured are straight

Fisher-Price 'Little People' figures, *c*. 1970.

representations of popular domestic spaces of the 1960s. Blue carpet and a large fireplace signified the living room, while the kitchen had a floor resembling faux-brick linoleum along with knotty pine cabinetry and double ovens set in a stone wall. Wood panelling lined the parents' bedroom, and a chest of drawers and walk-in wardrobe were included on flanking walls. The kids' room was bright yellow with a braided rug. Windows were highlighted with either shutters or frilly curtains tied with red bows. The world represented by the lithographic decals was more detailed and evocative than the one suggested by the simplified figures and plastic furniture. It was only with later designs, such as the Play Family A-frame House (1974–6), that the incongruence of the toy and its illustrations was worked out. The A-frame kitchen was notably realized as a three-dimensional built-in complete with saucepans and spice jars forever fused to the splashback; the whole kitchen was one piece of moulded golden-coloured plastic. Likewise, the hearth was partially modelled with decals indicating a firebox and chunky stone chimneypiece.

The Play Family toy lines were exceptionally long-lived, and occasionally updated in minor ways; for example, the decals for the Play Family House shifted noticeably in 1980 to represent a mock-Tudor exterior rather than the original Dutch Colonial. The material of the peg figures changed, the bodies becoming plastic like the heads, and printed, but the shape remained as before. Ponderosa and plastic figures could be used interchangeably with the toy sets until the mid-1980s, when the Boston trial attorney and consumer protection advocate Edward M. Swartz published *Toys that Kill* (1986), with the tagline 'Make Sure Your Child Is Safe, Avoid Thousands of Life-threatening Toys'. *Toys that Kill* included thousands of toys, but Swartz took special aim at Fisher-Price. Not only were several of its products included in the book, but Fisher-Price's Little People were the *only* toy featured on the cover. The photo of the primly smiling Fisher-Price Little People under Swartz's dramatic title was clearly intended to make a major impact in the vein of Ralph Nader's crusading *Unsafe at Any Speed* (1965). The choking hazard presented by the Fisher-Price figures was highlighted in the book with a diagram for a modified Heimlich manoeuvre developed specifically for ejecting the unscrewed peg figures' heads out of a child's throat.

For all its sensationalism, Swartz's book did highlight a real safety concern. In one recent post on Amazon, Steven G. of New York recalls his encounter with the Little People as a seven-year-old in 1988:

> I ate the dog, the grandpa, the two little kids, the mom, and had the bus driver halfway down my throat when I suddenly couldn't breathe! I was already starting to black out when my mother discovered me. Fortunately mom had Swartz's 'Toys that Kill' on the shelf in easy reach – she quickly turned to the section on Little

Kara K. Bigda, *Fisher-Price Family*, 2018, watercolour on cradled aquabord.

People to read the details on a version of the Heimlich maneuver specific to Little People choking accidents.[1]

Although apparently a saviour for choking children, the book was a public-relations nightmare for the toy company, by then owned by Quaker Oats. A comprehensive redesign changed the peg figure radically. The Chunky Little People were made doubly wide – too wide for a child to put in their mouth – and with that all the accessories and buildings became squatter. Gone was the architectural realism seen with the A-frame House; gone was any ambiguity about the 'appropriate' ages for the toy line. In their new, safe form, the toys took on a cartoonish appearance that put them squarely in the toddler aisle.

The revised figure type would be short-lived. Quaker Oats sold the company to Mattel, which speedily let go of Chunky peg figures in favour of soft plastic and rubbery monolithic dolls that maintained exaggerated squatness but recalled nothing of the original Fisher-Price figure. Fisher-Price toy sets were now entirely plastic, and the costlier pine that had linked the mass-produced toys to older wooden toy traditions and crafts was gone. Decals play only a minor role in the contemporary brand, with none of the artistry of Margaret Evans Price. The real tragedy of the demise of the peg figure was the associated end of Fisher-Price's architecturally inspired toys, which crept into the imagination through eyes, fingers and, unfortunately, little mouths.

# Flashcube
Harriet Harriss

The Kodak Flashcube – a rotating cube with a miniature flashbulb incarcerated within each of its four mirrored compartments – made amateur photography of the domestic interior possible from the mid-1960s onwards. It also reduced the risk of injury presented by its forebears. Its mother, the single-use luminescent flashbulb, resembled a domestic light bulb and would project shattered glass as well as light. Its fragility disguised its ferocity. Containing magnesium filaments, the oxygen gas, once electrically ignited by the click of the camera shutter, would generate substantial residual heat and often cause painful burns. Memories should burn brightly, not painfully. The Flashcube's grandmother, the flash lamp, carried an even greater risk of violence. Upon triggering the shutter, both photographer and subject risked being cut or even blinded should some stray glass enter the eye. Worse, photographers sometimes died when preparing the flash powder, a composition of metallic fuel and an oxidizer such as chlorate. Flash lamps maintained their market dominance for some sixty years before flashbulbs replaced them in 1929. In 1965, however, the need for greater safety and simplicity urged the Flashcube into existence.

Eastman Kodak's invention of the Flashcube also emerged from the company's specific wish to offer a flash that would work with its newly ubiquitous amateur photographers' cameras, made popular by the white middle-class families targeted in 1960s advertising campaigns. Partnered with Kodak's Instamatic camera, the Flashcube's adaptability, portability and ease of use made interior photography possible for the masses, without prerequisite skill or expertise. The impact on interior behaviour as well as interior spaces was substantial. Interiors and normative domestic relations could be captured and shared, exposing the aspirational aesthetics and social arrangements of this once most private of architectures. In magazine advertisements, a woman's hand was shown affectionately caressing a Flashcube, positioning it as a feminine technology to capture home life. In the Flashcubes' dazzling light, families staged domestic tableaux in an effort to display their nuclear family credentials.

In observing the phenomenon of staging, Susan Sontag observed presciently that 'needing to have reality confirmed and experience enhanced by photographs is an aesthetic consumerism to which everyone is addicted.'[1] And, since addiction is characterized by repetition, Flashcube 'users' could click/shoot, click/shoot, click/shoot, click/shoot, discard, reload, click/

Flashcube for Instamatic camera, Eastman Kodak, c. 1977.

shoot, click/shoot, click/shoot, click/shoot, discard, reload and so on, until
the three-pack was spent. Without agency over light levels, contrast or glare,
red-eye was inevitable, casting an unintentionally malevolent air over many
tableaux. And with each photograph, the Flashcube would make a small
snapping sound, as sharp and discreet as a breaking wishbone. The four-
compartment casing would feel warm for an instant, then cold, and then
forever silent. With each aluminium ignition the Flashcube's explosion
remained contained within its interior, the plastic housing intact. The subjects would freeze in anticipation, to avoid the image being blurred, only
to disperse their familial tableau seconds later, momentarily blinded.

   If they ever looked at the used Flashcube before discarding it, subjects
would have noticed the scorch marks inside, resembling the remnants of a
chip-pan fire in a doll's house. Aluminium, the element in the Flashcube
that helps the magnesium combust, must be one of the key material symbols
of modernity, having shaped the twentieth century through domestic and
industrial advances, air power and Moon landings. It is lightweight, strong,
non-magnetic and resistant to corrosion. Yet aluminium's shiny utopia
has a dark side.[2] Aluminium is a neurotoxin that, when ingested, causes
Alzheimer's. That an object designed to capture memories is contrived
from a substance that corrodes memories speaks of both alchemy and
irony. Aluminium in its raw form is contrived from bauxite, an amorphous
clayey rock whose strip-mining extraction requires the removal of all native

Kodak Instamatic
277 X 'point and shoot'
126-format camera with
Magicube flashbulb
cartridge attached, 1977.

vegetation in the surrounding area, the destruction of habitat and food for local wildlife, and with soil erosion and river pollution thrown in. Since they first opened, bauxite mines have fuelled resource disputes in Africa, India and the Caribbean, and have elicited both the greed and the wrath of multinational corporations. Aluminium embodies the carcinogenic contradiction of our time: that affordable, playful and convenient consumerism is non-degradable toxic waste in waiting. Perhaps this is the reality that the spent Flashcube illuminates best.

None of this was on the register of consumers at the time, of course. They simply wanted a flash that wouldn't die after being fired only once. The solution was in fact being developed as early as 1931, thirty years before the Flashcube, by Harold Edgerton, a professor of electrical engineering. His Electronic Flash, a battery-powered device able to carry its own energy supply and integrated into the camera body, began to dominate the consumer market in the late 1960s. It was perfectly suited to several uses without the detriment of detritus, and quickly deposed the Flashcube as the device of choice for capturing domestic interiors. Consequently, the Flashcube's principal manufacturer, the Kodak subsidiary Sylvania Electric Products, ceased producing it in the 1970s. As a leading contractor for warfare research and other forms of surveillance, the company returned to its main product lines, principally, avionics systems for observation helicopters, and personal distress radios for downed pilots.

Despite the closed production line, is it disingenuous to claim that Flashcubes are extinct? What about the familial memories they captured? As Sherry Turkle observes, 'we consider objects as useful or aesthetic, as necessities or vain indulgences . . . [but] are on less familiar ground when we consider objects as companions to our emotional lives or provocations to thought.'[3] A Flashcube's aluminium filament takes one hundred years to decompose, and its plastic casing up to a thousand. However, the photographic prints it produced will degrade within half a century, making the interior, familial memories that the Flashcube helped to capture far more susceptible to disposability than the object used to create them. Consequently the Flashcube's simple provocation reveals that vanity and violence are essential accompaniments to un-disposability. It might have fallen out of use, but the abandoned and slowly degrading Flashcube acts as a dimly pulsating warning light, reminding us that the objects we design to indulge our narcissistic fumbling towards immortality only serve to end our life, rather than extend it.

# Flying Boat
## David Edgerton

In the late 1930s and into the 1940s, the largest long-range aircraft were flying boats, which took off and landed on water. Flying boats had a fuselage designed not only to fly, but to float and to move on water at speeds much greater than any normal watercraft. Operated on their longest routes by Pan American Airways and Imperial Airways (known from 1940 as British Overseas Airways Corporation, BOAC), they carried tens of passengers in spacious cabins, often with sleeping compartments. They were viewed as the future of long-distance aviation. It is telling that during the Second World War the director general of BOAC suggested that a giant lagoon for landing flying boats should be built alongside the runways at the new London Airport at Heathrow.[1] Only when we consider the might-have-been of continued reliance on the flying boat for civil aviation do runways – and thus landplanes – become visible as extraordinary creations rather than obvious features of aviation. Investment in the infrastructure of concrete runways, propelled by the military, swung the balance in favour of land-based planes. But for the war, this might not have happened, or happened so rapidly.

Many histories of aviation have been misled by the streamlined look of small 1930s landplanes such as the Douglas DC-3. Looks are not everything; the large flying boats of the 1930s were two to three times heavier (in terms of maximum take-off weight) and pioneered long-range flight with heavy loads. They were so heavy on take-off that the equivalent landplane would have required long hard runways that neither the military nor even busy airports had in the 1930s. (For example, Croydon, the UK's main airport of the interwar years, had grass runways.) Sheltered water provided free runways usually approximately 1.6-kilometres (1 mi.) long. Many parts of the world had harbours and lakes large enough – such as Lake Victoria in Africa, or Lake Habbaniyah in Iraq – or rivers to act as flying-boat bases. New York City had a flying-boat base right in the city: the spectacular Marine Air Terminal at the new Municipal Airport (LaGuardia, 1939). There were obvious disadvantages, however: suitable, uncluttered water was often far from where people were, and seawater corroded aircraft. Another problem was that a hull that floated on water was not optimized for flight through the air. But, ultimately, these problems were counterbalanced by the flying boats' avoidance of exorbitantly expensive runways.

Flying boats not only pioneered new long-range overseas routes, but in many cases *replaced* smaller landplanes. This was true of the most enthusiastic

Pan American Airways Boeing NC314 BC18603, the 'Yankee Clipper', c. 1939.

users of large long-distance flying boats, the British. The four-engined Empire Flying Boats of the late 1930s were about 20 tons take-off weight. By 1939 three flew to South Africa each week, and another three to Australia. By the war, Imperial Airways had about forty flying boats, far more than any other country or airline. By contrast, it later ordered only twelve landplanes of the same size (although these were advertised as 'the largest landplanes in the world'), which operated from grass with huge wheels.

Even though the British became the most significant users of large flying boats, they were not responsible for their greatest innovations. Germany and France both built bigger ones, albeit in smaller numbers. Pan American, the U.S. imperial airline, had the aircraft that flew furthest with the highest loads. From the 1930s Pan Am used ten Sikorsky s-42s to fly down to Rio de Janeiro and beyond. From 1935 it flew four Martin m-130s from San Francisco via Pearl Harbor and a series of U.S. islands to Manila and China. They were followed in the late 1930s by the huge 38-ton Boeing 314 (of which twelve were built), larger than any U.S. Second World War bomber, except the Boeing b-29, or any land transport aeroplane, including the Douglas dc-4.

The Second World War gave a massive boost to the large four-engined flying boat for long-range naval reconnaissance over vast areas of sea. Japan built more than 150, the USA over 200 and the UK over 700. It also prompted the design of new, extremely large flying boats for military and civil service. Germany built four very big six-engined ones, one of which, at nearly 100 tons, was the largest aircraft built anywhere in the war. France built eleven 70-ton, six-engined machines (the Latécoère 631) that had a disastrous safety record and were all withdrawn by 1955. The British built two Short Shetlands at 56 tons; the U.S. Navy got seven 75-ton Martin Mars machines. The most famous flying boat of all was Howard Hughes's 180-ton Spruce Goose: made of wood, it was the largest ever produced, but never went into service. Neither did the largest metal flying boat ever built: the British Saunders-Roe Princess, which flew with its ten turboprop engines in 1952.

However, the increasing role of aerial bombardment in modern warfare meant that the days of the flying boats were numbered. Once it was under-stood that hard runways increased the maximum take-off weight and hence the load or range of bombers, more long runways were built to accommodate heavier and heavier bombers. Flying boats could not compete; not only were they less efficient in the air than landplanes, but also there would never have been enough protected water for the huge number of bombers deployed (often well over a thousand a day from the UK). The transformation was extraordinary. In 1939 there were only nine hard military runways in the UK; by 1945 there were hundreds. The biggest were 1,830 m (2,000 yd) long, until in 1945 four 2,700-metre (3,000 yd) runways, of 30-centimetre (12 in.) concrete, were built, including one at Heathrow.[2] These vast works found their echo across the world, including the New York Idlewild airfield (now John F. Kennedy International airport), which opened in 1948 with runways of similar length, costing the vast sum of $60 million, over half a billion dollars today.

The decline of the flying boat was not immediate. After the war a new fleet went into operation, once again led by the British. BOAC flew a total of about sixty (although many fewer at any one time), converted Sunderlands of 27 tons (called Sandringhams, and subdivided into various classes such as the Hythe and Plymouth) and 35-ton Short Solents. But, thanks to the greater availability of runways and the lower running costs of landplanes, which did not require specialized pilots, BOAC had become a mainly land-based airline by the late 1940s, and it dropped flying boats altogether in 1950. Pan Am had given up earlier – it flew DC-4s across the Atlantic and the Pacific from 1946, but it was only in the late 1940s with U.S. aircraft – the DC-6, the Stratocruiser and the Constellations – that civil landplanes became heavier than the Boeing 314.

The history of the large flying boat was not wholly over, and it continued to find specialist uses in places with less well-developed infrastructure, or for firefighting. Airlines in Australia, New Zealand, Tasmania, Argentina, Uruguay and Norway continued to use 1940s British flying boats into the 1960s and beyond, in very specific low-traffic locations. In 1950 a new airline took over the BOAC Southampton base, flying mainly to airport-less Madeira, until 1958. The development of flying boats was not completely finished. The U.S. Navy built eleven 75-ton flying boats in the 1950s (although these were deployed for only two years). The Russians and Japanese have built 40-tonners. Most recently, a Chinese turboprop flying boat, for firefighting and other purposes (it can operate at sea) weighing 54 tons has come into service. But the flying boat was for the most important uses functionally extinct from the 1940s. It had represented the peak of aviation in a world with few flights and few runways, but it could never have sustained the phenomenon of mass civil aviation which emerged in a newly concreted world.

121 GIZEH—SPHINX AND PYRAMID OF KHAFRE

G 449 OLD KINGDOM— (2980-2475 B.C.)

9KC/ ECTICUT COLLEGE—ART DEPARTMENT

BESELER LANTERN SLIDE CO. Inc.

# Glass Lantern Slide
## Daniel M. Abramson

It took many hands to produce a glass lantern slide lecture on the history of art. In front of a darkened auditorium, a speaker commanded 'next slide' and pointed to a screen. On cue, a projectionist at the rear glided a metal carriage into a heavy projector, quickly removed the previous slide into a long wooden box, and inserted the next slide into the open slot to await the professor's subsequent call. When operating two projectors at once, as was common in art-history classes, the difficulty of the task doubled. Hours or days before, the lecturer or teaching assistant arranged on a light-table dozens of slides gathered from the art-history department's slide library. There the faculty congregated, assisted by a professional slide librarian, their staff and student helpers, who catalogued, labelled, mounted, masked, repaired, filed and refiled tens of thousands of slides housed in the drawers of wooden cabinets lining the capacious library. In this collection, each glass lantern slide might have been sourced from the slide library's own full-time photographer, shooting images from books, prints and other photographs; obtained from faculty who used their own images sandwiched between plates of glass; or ordered from commercial manufacturers marketing images of the world's masterpieces, artfully photographed, for about two shillings, or a dollar, apiece.

Long before the lantern slide entered the practice of art history in the 1870s, magic lantern shows entertained Europeans and Americans. Invented in late seventeenth-century Germany, lantern slides and their presentation were the province of religious instruction, small-scale family entertainment and itinerant peddlers using painted glass slides illuminated by oil lamps projected through simple lenses. Then, in the 1880s and 1890s, advanced projection and photograph technology combined with urbanization and commercialized leisure turned magic lantern shows into mass entertainment. Complex projector arrays produced spectacular visual effects for audiences in the thousands eager for images of everyday life and current events, the Bible and faraway places.

Art historians adopted glass lantern slides to fashion their discipline's future in a similarly public way. The first was Bruno Meyer in the 1870s at the Polytechnic Institute in Karlsruhe, followed in the next decade by other pioneers at Princeton and Oxford University's extension school. The roll-out of glass lantern slide technology in art history occurred over the following decades. Professors at Columbia University did not adopt it until 1912. Still, the glass lantern slide contributed fundamentally to the modernization of

Glass lantern slide of The Great Spinx and Pyramid of Khafre, formerly in the collection of the Art Department, Connecticut College, USA, probably 1930s.

art history. Photography made the discipline scientific. It enabled practitioners to study and sequence art and architecture with seeming objectivity across time and space, analysed by categories of authorship, form, material, content and style, in ways that could not be achieved with limited collections of original works or inexact printed reproductions. It is no accident that the introduction of photographic images into art history coincided with the field's first graduate programmes, signalling the discipline's modern status. Differing from smaller reproductions such as photographs or prints, which could be displayed only to a limited number of viewers at a time, glass lantern slides projected on to large screens allowed this new scientific approach to be demonstrated to bigger audiences. Famously, the art historian Heinrich Wölfflin paired projected images to institute art history's basic comparative method.

The glass lantern slide gave art history an element of theatricality. Its use in sizeable lecture formats engendered the subjectivity of participants. At the opening of the film *Mona Lisa Smile* (2003) an aspiring mid-twentieth-century art historian gazes into a glass lantern slide, synecdoche of her desire for professional command. For that century's audiences, an affinity for art and its history often began in a darkened lecture hall as an authoritative voice narrated a sequence of screened pictures. Precisely because of the force of this seminal encounter with art as dematerialized, projected image, the art-history slide lecture arguably produced an even higher value for its material counterpart. We crave unmediated experiences with objects themselves – in solitude and silence, outside time and *in situ* – in varied environments from the art gallery before it opens to land art in the desert western USA.

By the 1920s the glass lantern slide was firmly established in academic art history – even as it had already been rendered extinct for mass entertainment by moving pictures. Scores of companies on both sides of the Atlantic provided canonical art-historical images, improved slide clarity, advanced projectors and other accoutrements to art-history departments and museums across Europe and America. Famous firms such as Alinari in Italy competed with German, French, U.S. and English businesses – including some thirty in London alone, and half that in New York – offering tens of thousands of glass lantern slides, and in different national formats, too: including 8.3 cm (3¼ in.) square for the UK market and 8.3 × 10.2 cm (3¼ × 4 in.) for the U.S. Britons also tended to use double-image slides on single projectors, whereas in the USA single-image slides projected on pairs of projectors was the norm. In turn, collections, such as the Ryerson Library of the Art Institute of Chicago, would lend out their slides to the general public, disseminating art and its modern historiography still further. By the 1930s additional innovation in emulsion, dye and layering technology by Kodak scientists had produced the smaller, lighter 35 mm transparent slide in a 5 × 5 cm

(2 × 2 in.) cardboard or metal mount. One-third the size and one-thirtieth the weight, this slide was cheaper to produce, as easily available in colour as in black and white, and less cumbersome to store and project. It made art and its history even more democratic, enabling less prosperous institutions and faculty to assemble viable teaching collections. From the late 1940s the glass lantern slide was steadily superseded by the smaller slide. Moreover, carousel machines for these smaller slides incorporated wire remotes, which also made projectionists redundant.

Still, the glass lantern slide hung on. Teachers at London's Courtauld Institute and elsewhere abjured smaller slides into the 1960s, because the colour did not match the original objects. For their superior sharpness and projection size, slide librarians at Harvard kept purchasing black-and-white lantern slides into the 1970s. And as late as the 1990s, an architectural historian at the same university kept using the occasional glass lantern slide to better represent monumental form, necessitating slide booths with arrays of projectors for both formats, and a live projectionist, too. It was more than half a century after the introduction of the 2x2 slide, into the first decade of the next century, before the glass lantern slide was terminated altogether, and suddenly, by digitization, which simultaneously outmoded the 2x2 slide too.

Even as the world known through screened images proliferates, the social and technological milieu that produced and was produced by the glass lantern slide, first in mass entertainment and then in academia, is gone or faded. Now, an instructor anywhere sitting alone with the Internet can in minutes download images from global servers, insert them into PowerPoint, walk into a classroom to deliver the lecture, and store the file on their computer for later reuse. The slow, communal and laborious process – mediated through the institution of commercial firms, the slide library and its staff – is now an instantaneous, privatized, democratic and immaterial phenomenon produced on a laptop.

Nevertheless, the glass lantern slide's artefacts and after-effects linger. Some major university collections are archived, not just digitized or discarded, allowing interested scholars to research the metadata, for example, of the label names of photographers and companies that produced the images that helped to found the discipline of art history. In the same self-reflexive vein, the prominent contemporary Black American artist Theaster Gates uses in his installations a collection of 60,000 glass lantern slides (donated to him by the University of Chicago in 2009) to reappraise the Western canon and its artefacts.[1] The lantern slide lecture itself remains a primal scene for apprehending art, and not only in classrooms. In museums, audio guides mimic the experience of the auditorium with the lecturer's voice still firmly planted in our ear – even as we eye the object itself and listen isolated from the crowd. The glass lantern slide may be extinct, but its show goes on.

# Globe of Mars
## Lucy Garrett

This globe may look familiar at first, but, instead of showing the Earth, it shows the surface of Mars as visualized in the early twentieth century. It was made by a woman about whom relatively little is known: Emmy Ingeborg Brun, a Danish amateur astronomer. Brun grew up in an intellectual bourgeois household; her father, a former soldier, wrote on horticulture and beekeeping and was involved in contemporary politics. She was well educated, but was not allowed to go to university and spent much of her life in bed or in hospital with chronic health problems, described at the time as 'weak nerves'.

Throughout her life Brun was engaged in political, theological and scientific discourse. She was particularly interested in the theories of the American astronomer Percival Lowell, who was inspired to study Mars by the Italian astronomer Giovanni Virginio Schiaparelli. Schiaparelli had observed a network of dark lines on the Martian surface, and published his findings, along with a map of the planet, in 1878. His map was just one of several produced in that period, but his depiction was revolutionary, using blue shading to show the planet divided into land masses surrounded by water. He named the dark lines *canali*, which can mean both natural and artificial canals.

Schiaparelli was initially reluctant to extrapolate from his observations, but his map and ambiguous nomenclature were eagerly taken up by some contemporaries. Lowell became prominent among those who believed that these dark lines constituted a Martian network of artificial canals. He went further than Schiaparelli and constructed a theory about the history and politics of the entire planet based on his observations. Lowell identified six stages of planetary evolution: the sun stage, the molten stage, the solidifying stage, the terraqueous stage, the terrestrial stage and the dead stage. He placed Earth in the terraqueous stage, which would lead to the terrestrial stage when all water eventually disappeared. Mars, he argued, was entering the terrestrial stage, and was gradually being reduced to desert. The canal network had been built by the planet's inhabitants to bring water down from the polar ice caps as the planet dried up. Lowell's theories, given the paucity of knowledge about the surface conditions of Mars, were contentious but not incredible, and he was a compelling writer. He published three books on the subject from 1895 to 1908, fuelling existing interest in the red planet; the concept of a Martian civilization appears in every corner of popular culture in the period, from scientific debates in scholarly journals such

Emmy Ingeborg Brun, Globe of Mars, hand-coloured papier mâché with varnished ink and plaster surface, USA, 1909.

as *Nature* to high fantasy including Edgar Rice Burroughs's early work *A Princess of Mars* (1912).

Interest in Mars was not purely sensational, but also imbricated in contemporary socio-political thought, which posited Mars as a potential site for the creation of a socialist or communist society. This often took the form of fiction, such as Alexander Bogdanov's novel *Red Star* (1908), but also featured in scientific theories. Schiaparelli himself speculated about how Martian society might look in an article in 1895, although he stressed that his ideas were pure fantasy. He suggested an 'institution of a collective socialism, such that each valley can become a Fourierist phalanstery, and Mars a paradise for the socialists!'[1] On Mars, surmised Schiaparelli, the inhabitants focused on fighting nature, rather than each other. Lowell, on the other hand, rejected the idea of Mars as an egalitarian utopia. To him, the canal network indicated that Martian society was highly developed both culturally and technologically, with both achievements underpinned by a commercial economy, even if the motivation to build it had been 'desperation, not enlightened thought'.[2] In a lecture in 1911 he told miners that Mars was ruled by a benevolent oligarchy of technical experts. Schiaparelli's and Lowell's contrasting attitudes bear witness to the fact that we tend to project our own identity on to the unknown. Lowell, heir to a cotton fortune and a lifelong opponent of socialism and unions, saw a society where survival had been enabled by development through capitalism. Schiaparelli, a trained civil engineer, envisaged a society with a collective structural capacity for building, a 'paradise for plumbers' as well as for socialists.

The debate concerning Martian climate and civilization was a key part of contemporary astronomical discourse, and it is unsurprising that it would be of interest to Brun. She studied Lowell's work, adapting the maps of Mars he had published and fashioning them into globes. There are twelve known surviving examples, thought to have been made between 1903 and 1915. She offered them to institutions and sent one to Lowell himself in 1915. It was initially arraigned at customs, because it looked like a bomb, but when it eventually reached him he responded warmly, stating that it was 'a capital piece of work'. Brun's interpretation of what the canal network might mean, however, was very different from Lowell's, as indicated by the inscriptions on the base of the globe. One reads 'Free Land. Free Trade. Free Men' and the other 'Thy will be done on earth as it is in heaven.' The first phrase is a socialist slogan inspired by *Progress and Poverty*, a work by the political economist Henry George, who argued against a system of profit from renting land or property without contribution, advocating a single tax on the value of land. Brun was interested in George's work and corresponded with the Danish branch of the George Society, giving one of her globes to its president, Sophus Berthelsen. The second, a line from the Christian Lord's

Prayer, is indicative of Brun's spiritual hopes for Mars. She wrote a book on the religious aspects of Charles Darwin's theory of evolution, and in it reproduced Lowell's maps, noting that the civilization implied by the canals might be an *ishvara*, a theosophical term taken from Hinduism, where individual humans can achieve divine collective unity.[3] For Brun, therefore, Mars was the possible site of a Christian socialist utopia.

Brun's globe must have been a welcome gift to Lowell, since by that time the scientific tide had already turned against his theories. The possibility of an extraterrestrial, cooperative, planet-wide civilization that humanity might be able to emulate was slowly extinguished when scientific techniques allowed a closer examination of Mars. Maps drawn from observation were no longer considered definitive, and photography supplanted cartography as the acceptable proof for astronomical observations. Photographs failed to show the detailed geometric network that Lowell had claimed to see through his telescope, and he last defended his theory publicly in 1910, when he wrote an obituary of Schiaparelli.

The knowledge presented by Brun's globes was therefore in flux even as she was making them. It was increasingly clear that there was no civilization on Mars, whether socialist or capitalist. The Georgist single land tax espoused by Brun has only ever been implemented once, in the German colony of Jiaozhou in China, which ended in 1914. But, although the globes themselves became obsolete, Mars was and still is treated as a blank canvas on to which possible human futures can be projected. In Ray Bradbury's story 'The Martian Chronicles' (1950), Mars provides a refuge from a disaster on Earth. This idea has now moved from the realms of science fiction to a genuine, albeit still largely theoretical, possible solution to population growth and environmental damage on Earth. Whatever the motive, the idea of another planet where humanity can start again remains powerful.

# High-pressure Water Mains
## Adrian Forty

Set into pavements and roads all over inner London are small cast-iron plates marked 'LHP'. These plates, many now much eroded, are the only visible surviving evidence of the network of high-pressure water mains that ran beneath London's streets, installed from 1883 by the London Hydraulic Power Company; the plates covered the stopcocks on the pipes. From around the same time other cities too were equipped with high-pressure water systems – Glasgow, Liverpool, Manchester, Birmingham, Antwerp, Sydney, Melbourne, Buenos Aires and possibly more – but London's was the largest, reaching, at its fullest extent in 1930, nearly 300 km (186 mi.) of streets. The networks delivered pressurized water, a power supply adaptable to a variety of purposes, but principally those requiring great force exerted intermittently and slowly. Hydraulic power was best suited to functions of lifting and pressing.

As a public utility, hydraulic power anticipated – but only just – the arrival of electricity, from whose shadow it never escaped and to whose competition it eventually succumbed: the London system was finally cut off in 1977, the others earlier. Nevertheless, precede electricity it did, and the high-pressure mains were the first urban infrastructure to deliver a publicly available source of motive power – for, although there had been gas supplies in cities since the early nineteenth century, gas was not easily converted to motive force. Pressurized water brought to customers for the first time a constantly available, reliable, local source of power, an amenity that is now widely taken for granted, even if access to it is not universal. While hydraulic power was the first to make real to people the opportunities and convenience of such a resource, public electricity supplies began almost simultaneously: Thomas Edison's in New York from 1882, and Sebastian Ziani de Ferranti's at Deptford, southeast London, from 1889.

London Hydraulic Power's six-year start on electricity supply in the capital gave it an advantage at first; in 1887 it already had 43.5 km (27 mi.) of mains and 650 customers, and was able to claim into the 1890s that hydraulic power, per unit of energy, was cheaper than electricity. Still in 1910 the company was confident that electricity was not a threat, but in that year the previously separate electricity undertakings started to amalgamate, and the price of electricity fell progressively. By the 1930s hydraulic power cost two and a half times as much as electricity, and by the 1970s, ten and a half times as much.

Valve cover, London Hydraulic Power Company, early 20th century.

The first applications of pressurized water were for dockside cranes and goods hoists, and various pieces of dock equipment, capstans, lock gates, swing and lifting bridges. The technology for these had been developed in Britain in the 1850s, and by the 1870s it was widespread in British ports. The advantage of it over localized steam power was that it was less of a fire hazard, simpler to operate and less prone to carelessness, a concern in the heavily casualized British docks labour market. Each individual dock had its own system, with its own steam pump and accumulator – a tower housing a giant compressor to maintain the pressure. In London alone, there were at least 29 independent systems operating downstream of Tower Bridge.[1] In the 1870s the engineer Edward B. Ellington, a manufacturer of hydraulic machinery, hit upon the idea of a single public supply, to which anyone could be connected, as an alternative to the many independently operated private supplies. He tried out his plan first of all in Hull docks, then transferred his attention to London. In 1881 he set up a company that in 1884 became London Hydraulic Power, with statutory rights to lay mains beneath the streets. A subsidiary company operated a similar system in Liverpool. Water at just over 48 bar (700 lb per sq. in.) was supplied through specially designed cast-iron mains 15 cm (6 in.) in diameter, and the pressure was maintained by pumping stations, initially one at Bankside, but ultimately five distributed across the capital, from Wapping in the east to Pimlico in the west. The water supply to individual customers was metered, and each was charged according to their consumption.

Initially, LHP's largest market was for lifts. By 1887 the company was providing power for more than four hundred goods and passenger lifts in offices, hotels and warehouses, and at this date the majority of lifts in London were hydraulically operated.[2] For shorter distances, the lifts were powered by a hydraulic ram beneath the car; for greater heights, or where it was impossible to excavate a borehole beneath the lift, the car was suspended from cables that were powered hydraulically. The ram-type lifts were capable of carrying great loads, up to 40 or 50 tons, and so were especially suited to carrying goods. Later, more specialized lifts were developed, for car showrooms, theatre stages, cinema organs and the retractable floor over the swimming pool at Earls Court Exhibition Hall. Other applications included revolving stages, scenery hoists, fire hydrants and sprinkler systems, presses for baling raw materials, textiles or scrap metal, or for forging metal and, most unlikely of all, building-installed vacuum-cleaner systems. Despite the competition from electricity, hydraulic power flourished in the first three decades of the twentieth century, partly by playing to its strengths – silent operation, simplicity, reliability, low maintenance – and, in London, because LHP progressively took over the supply of power to the London docks, which by the 1930s represented two-thirds of its business.

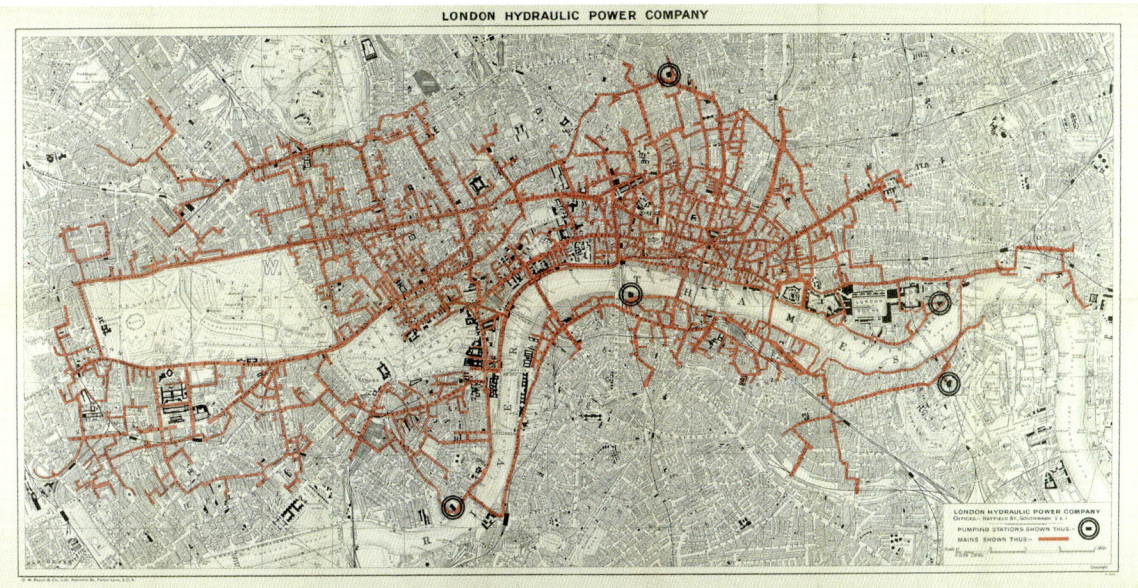

Map of the mains and pumping stations of the London Hydraulic Power Company, 1924.

The first serious blow to LHP was the Second World War, when bombing damaged the networks and equipment, and the combined London and Liverpool customer base fell from 7,887 connections in 1938 to 5,577 in 1945. Nevertheless, the company remained resilient, offering new services and in the 1950s converting its pumping stations from steam to electric power to reduce running costs. Even in the late 1950s extensions to the mains in London were still being made, to provide additional services in the railway goods yards at King's Cross and Broadgate, and in 1961, when the new Smithfield Poultry Market building was under construction, LHP was invited to tender for the lifts that were to connect the market floor with the underground storage areas. But what finally finished LHP was the closure of the London Docks in the 1970s, and when the company was wound up in 1977, it had only five hundred customers left, all of whose equipment was converted to electricity. The network was bought in the early 1980s by Mercury Communications; that was subsequently acquired by Cable & Wireless, which used the mains as conduits for laying fibre-optic cables in central London. In the arteries of one defunct infrastructure another now thrives.

# House Environment
## Eszter Steierhoffer

First presented at the seminal exhibition 'Italy: The New Domestic Landscape' at the Museum of Modern Art, New York, in 1972, Ettore Sottsass's 'House Environment' proposed a new modular system for the optimization of domestic space. The exhibition set out to introduce a panorama of the latest Italian industrial design, based on two thematic sections: objects and environments. For the latter, the exhibition's curator, Emilio Ambasz, invited designers to create spaces and artefacts that gave structure to everyday domestic rituals, and to propose radical micro-environments and micro-events that would define contemporary rituals and challenge conventional ideals of living.[1] Responding to this brief, Sottsass presented a full-scale environment that sought to make prevailing ways of life extinct or, failing that, to expose the irrelevance of the existing material and social framework of domestic living.

Merging furniture with infrastructure, Sottsass's installation consisted of a set of grey cuboid units on wheels. The units were designed to be easily transportable, and, each fitting a modular grid of 480 × 480 cm, were meant to offer maximum freedom to organize domestic life. In detailed drawings that accompanied the installation, Sottsass illustrated the potential of his connecting units, which contained a mobile kitchen, shower, toilet, bookshelves, wardrobe, jukebox and other elements of living spaces and bedrooms. According to his black-and-white cartoon-like storyboards, these units – which he often called '*mobile serpente*' (snake furniture) – could easily be moved around and reconfigured as new domestic demands and needs arose. Unlike existing bounded and contained interiors, the endless possible combinations of the modules implied spatial fluidity and infinity.

Sottsass's design drew together two of his long-standing preoccupations: modularity and computing. The first stemmed from the formative time he had spent in the USA in collaboration with the designer George Nelson in 1956 and 1957. The modular prefabricated system deployed by Nelson in his Experimental House (1951–7) became an important reference for Sottsass, who in the late 1960s returned to this interest in temporary architectural structures. Whereas portability was a feature of many of the industrial products he designed for corporate clients such as the Italian furniture producer Poltronova and Olivetti (consider his Valentine typewriter), it was the 'House Environment' that enabled him to explore modularity at full scale.

Mobile and Flexible Environment Module, Ettore Sottsass Jr, installation exhibited at Museum of Modern Art, New York, *Italy: The New Domestic Landscape*, 1972.

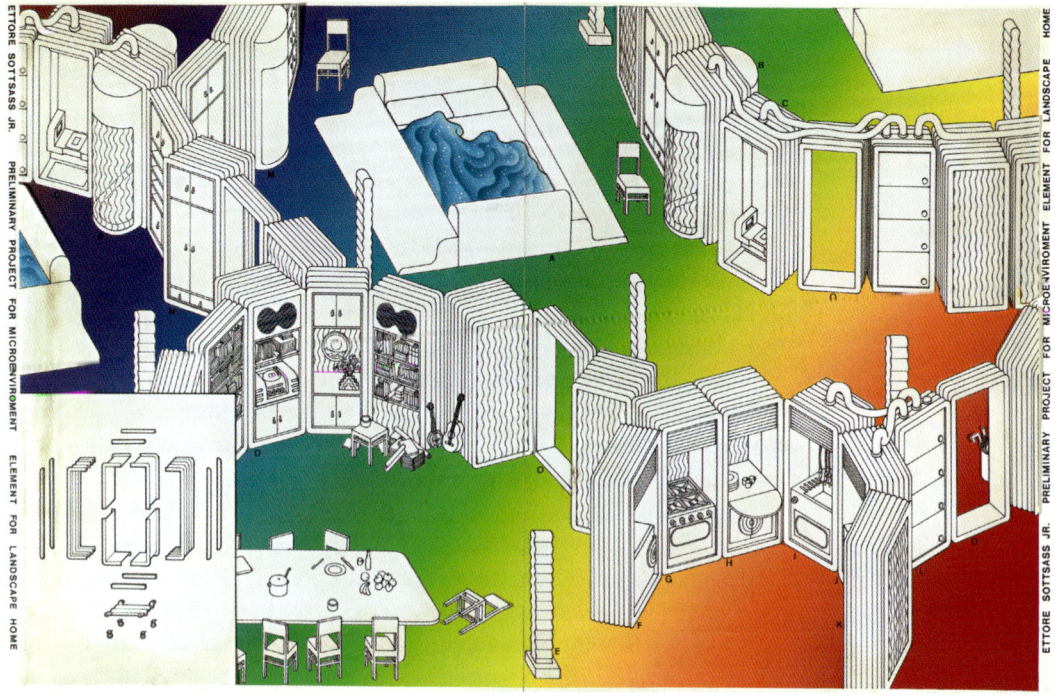

Ettore Sottsass Jr,
*Preliminary Project
for Microenvironment,
Element for Landscape
Home, c.* 1971, gouache
and ink on print with
acrylic.

Modularity was inherently well suited to Sottsass's vision of information-age domestic living. He had studied information systems in relation to his designs for Olivetti's first mainframe computer, Elea 9003, which was produced between 1957 and 1959. He subsequently conceptualized the interior architecture of the computer as the ideal home space; open and indefinite, it could free the domestic environment from convention, habit and existing moral and social norms. Drawing directly on the analogy of the computer, Sottsass imagined his 'House Environment' as a contemporary electronic and digital landscape, where the experience would be determined by relationships – a complex mix of motions, spaces and information – rather than fixed by pre-placed objects. This called for a more nomadic, flexible approach.

Lastly, Sottsass used the 'House Environment' to launch a critique of consumerism and the commercial design industry. As such, the MOMA installation marked both a new era in Sottsass's career and the end of a decade of productive collaboration with Poltronova and Olivetti. However, he still relied on some of the ideas and aesthetics developed during his time with these companies. In particular, the 'House Environment' bore a strong relationship to his Mobili Grigi (grey furniture) series, launched in 1969 at the Palazzo dell'Arte in Milan. Consisting of a double bed, chair, wall mirror and reading lamp in dark grey fibreglass, Sottsass described this domestic ensemble as a 'sad' environment, which strove to create an unreal atmosphere.[2] The melancholic promotional photographs of the furniture series shot by Alberto Fioravanti for the Poltronova catalogue were more reminiscent of an existential disaster than a commercial event.[3] Poltronova put only a few elements of this series on the market, partially because of the experimental use of fibreglass, which rendered production costs very high, but possibly also owing to its pointed lack of consumer appeal.

Sottsass subsequently decided that he would no longer develop specific commercial products. Reflecting some of the wider societal changes and a general sense of political disillusionment after 1968 in Italy, he adopted a form of 'cultural strike', working only on developing more conceptual and experimental propositions. Thus, by the time of the MOMA show, he was able to push the perverse strategy of Mobili Grigi yet further. Through simple forms, muted colours and affordable materials, he deliberately presented furniture as a tool rather than a consumer object or possession. The grey plastic material was chosen as a neutral medium that would prevent attachment to the object itself, instead seeking actively to alienate inhabitants by creating an abstract, estranged environment that resisted fashionable trends.

'House Environment' then was something of a paradox. Even as it boldly proclaimed the obsolescence of existing patterns of domestic life and offered a creative networked alternative, its critique of commercial culture meant that, as an actual design, it was itself extinct on arrival. It was certainly never intended for mass production; its technical details remained unresolved and the units were purely illustrative of their proposed function. Not surprisingly, the design was never implemented in its physical form outside the exhibition. And, notwithstanding the gradual incursions of so-called smart networked devices into the contemporary home, the traditional patterns, experiences and arrangements of domestic living environments carry on largely uninterrupted today.

# 'Hummingbird' Taxi
## Lucinda Hawksley

In the summer of 1897 a new type of vehicle made its debut on the streets of London. The Bersey Electric Cab was known initially as a 'Bersey', but was soon nicknamed the 'Hummingbird' because of its striking yellow-and-black livery and the humming noise made by its battery.

The Hummingbird was designed by Walter Charles Bersey, general manager of the London Electrical Cab Company, and initially seemed set for great success. The new cab company had some impressive backers, including the Hon. Evelyn Ellis, one of the people who had successfully called for an end to the Red Flag Act limiting road vehicles to a maximum speed of 4 mph (about 6.5 kph). In 1896 the repealing of the Act meant the road speed limit was raised to 14 mph (22.5 km), just in time for the Hummingbird's arrival. Another of the company's backers was the Hon. Reginald Brougham, after whose family the iconic Brougham horse-drawn carriage had been named.

The Hummingbird was revealed to the public in 1896, at the first Horseless Carriage Exhibition, held at the Imperial Institute in South Kensington. The new vehicle, which was reportedly capable of a top speed of 12 mph (just over 19 kph), prompted instant excitement among the press, who waited eagerly for its arrival. On 20 August the following year, *The Engineer* reported: 'a service of electrical cabs . . . are to ply for hire in the streets of London in competition with the ordinary hackney carriages . . . Thirteen of these cabs are now ready for work, and a staff of drivers have been instructed in the use of them . . . The "cabbies" are, we are informed, quite enthusiastic about the new vehicle.' On the same day, a journalist for the *London Evening Standard* enthused:

> They are upholstered in leather, and possess many more small luxuries than the public vehicles now in use . . . The steering and driving of the cab are simplicity itself, and require no skilled knowledge whatever . . . A special plug or key is in the possession of each driver; without this key it is impossible for anyone to move the carriage. When leaving his seat, the driver simply places the key in his pocket.

The *Daily Telegraph* of the previous day heralded the Hummingbird as a welcome, environmentally friendly change:

Hummingbird electric taxi, designed by Walter Charles Bersey, manufactured by London Electrical Cab Company, UK, c. 1897.

The new cabs will be undoubtedly a vast improvement from every point of view, as compared with those drawn by the insanitary horse. There is no animal more subject to disease, and his presence on the wooden pavements of the City is responsible for most of the disease germs which every breeze sweeps up in myriads from the filth-sodden streets.

In an interview for the *St James Gazette*, Bersey explained that it was possible for 'an intelligent man' to become a proficient Hummingbird driver within two days; he also spoke about the money that his cabbies were already earning, allegedly outstripping their horse-drawn counterparts. When asked if he thought the Hummingbird signalled the end of the horse-drawn hansom, Bersey replied, 'I should be sorry to see it disappear. There is room for both kinds of vehicles in a great city like London.'

Drivers might have been enthusiastic, but the London Electrical Cab Company had more difficulty in persuading the owners of taxi companies to invest in its new vehicle. The concept of electricity was something few people in Britain really understood. The first electric street lighting in the country had appeared less than two decades earlier, and electricity in the home was still almost unheard of. Luckily for Bersey, London's Commissioner of Police, Edward Bradford, was a fan of the new horseless sensation – and it was Bradford who was responsible for issuing cab licences.

It cost the same to ride in a Bersey cab as to take a horse-drawn hansom, and the Hummingbirds were much brighter than their rivals, because they were lit by electric light, which ran off the cab's battery. On one battery charge, a Hummingbird was able to travel for about 56 km (35 mi.), although drivers needed to work out their routes carefully, since there was only one battery-charging station in London, at the company's premises in Lambeth.

Despite the initial excited hype, the press soon began looking for flaws, and Bersey and the London Electrical Cab Company were furious when, just a few weeks after the cabs' introduction, a cabbie was arrested for drink-driving. George Smith drunkenly steered his Hummingbird along fashionable Bond Street, for which he was fined £1. The cabs themselves were also experiencing problems. The cost of electricity made them more expensive to run than horse-drawn cabs, although it had been hoped that they would prove much cheaper. In addition, the worlds of technology and engineering were not yet able to cope with the finesse of Bersey's design; the cabs' tyres proved much too flimsy for the poor state of London's roads, and the elegant glazed windows sometimes shattered as the cabs rattled over cobbles and other obstacles. The Hummingbirds also became infamous for frequent breakdowns.

The worst news came, however, in the autumn of 1897, when a little boy named Stephen Kempton became the first child in Britain to be killed by a motor car. He had jumped on to the back of a moving cab, on the Stockmar Road in Hackney. Tragically, his coat became caught in the mechanism and he was crushed to death – and the cab that killed him was one of Bersey's Hummingbirds.

In 1898 the company unveiled a new Hummingbird, but it was considered even less reliable than the original model, and both public and media were losing interest. The following year the Hummingbird's death warrant was signed when a cabbie lost control of his vehicle near Hyde Park Gate and crashed – very publicly. The Hummingbird was retired in 1899, after only two years in service.

Walter Bersey was a passionate exponent of what he saw as the grand future of electric vehicles, famously saying, 'There is no apparent limit to the hopes and expectations of the electric artisans . . . in short [it] is the natural power which shall be the most intimate and effective of all man's assets.' Sadly, Bersey was more than a century ahead of his time.

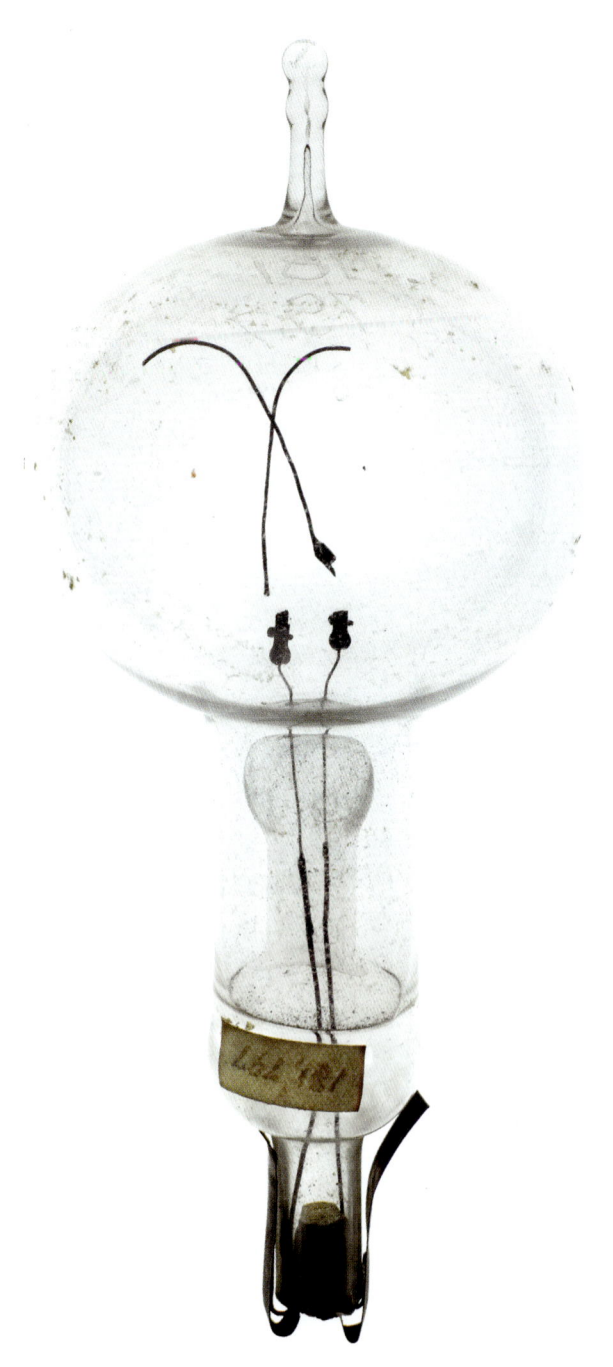

# Incandescent Light Bulb
## Mari Hvattum

The light bulb was *the* quintessentially bright idea – so much so that it became the symbol of one. Gyro Gearloose – Disney's inventor genius of the fictional city of Duckburg – has a light bulb on legs as his muse and helper, and a lit bulb above a head has become a universal symbol of sudden insight. The expression 'a light-bulb moment' has even made its way into the *Cambridge Dictionary* as 'a moment when you suddenly realize something or have a good idea'. Yet the light bulb's emergence was anything but sudden; it was a product of trial and error over almost a century. And, despite being an enduring symbol of genius and invention, the incandescent bulb's own endurance is hardly assured.

Experimentation with electric light began in about 1800, leading to a series of more or less impracticable lamps and bulbs, of which few were commercially successful. In 1802 the British chemist Humphry Davy connected a platinum strip to a battery, producing light as the platinum filament heated. Although the platinum burned up quickly, he had essentially made the prototype of an incandescent lamp. In 1809 he developed the electric arc lamp, which emits light when an electric current is made to 'leap' between two carbon rods, creating a spark. In various modifications, the arc lamp would become the technology of choice for decades to come. The first electric street lights, installed in Paris in 1878, were arc lights: so-called Yablochkov candles, named after their inventor, the Russian engineer Paul Jablochkoff, lining Avenue de l'Opéra. Contemporary visitors recount how the intense bluish light made the new opera house loom like a menacing black mass over the brightly lit avenue.[1]

While the arc lamp was being installed in Paris, several inventors were at work trying to develop a softer and more reliable lighting system. The incandescent bulb, which produced light by heating a filament until it glowed, seemed the most viable alternative. The English chemist Joseph Swan demonstrated a filament lamp to the Newcastle Chemical Society in December 1878, but could not make it practically or commercially viable until a few years later. In the meantime, engineers such as Hiram S. Maxim, St George Lane Fox-Pitt and others had patented bits and pieces of the filament lamp. Swan's most ardent competitor, however, was the American inventor Thomas Edison, whose incandescent lamp was demonstrated on New Year's Eve 1879 in Menlo Park, New Jersey, and put into production a few months later. Edison won the competition for the most efficient

'New Year's Eve' incandescent lamp, Thomas Edison, USA, 1879.

incandescent lamp at the Paris Electrical Exhibition in 1881, beating Fox-Pitt, Maxim and Swan by a few 'candlepowers' (the unit of measurement for luminous intensity, defined by the UK's Metropolitan Gas Act of 1860).

Edison's bulb consisted of a pear-shaped glass vessel with two metal rods entering at one end, and a 'pip' from which air was extracted at the other. The incandescent bulb relies on vacuum, and Edison's most significant improvement on earlier solutions was new mechanisms for air extraction and sealing. The metal rods were connected by a filament made from charred bamboo fibre, the result of a persistent search for a stable and durable filament material. Later, most manufacturers would change to tungsten, a metal with a very high melting point.

The incandescent bulb became a huge commercial success, and soon outcompeted the arc light. 'The light, though very strong, is at the same time mild and in no manner unpleasant or injurious to the eye; it enables my workmen to clearly distinguish the most delicate shadings in the color of their material,' wrote a New York piano manufacturer in February 1882 in an enthusiastic testimony to the merits of Edison's system.[2] Compared to the glaring arc lamp, the incandescent bulb gave a soft and even light, suitable for the domestic interior. With the development of local (and later national) electricity grids, the incandescent light bulb became part of every household, to the extent that a General Electric advertisement proclaimed in 1910 that the Sun is the incandescent bulb's 'Only Rival'.

But the incandescent bulb turned out to have more rivals than the Sun. The filament system is based on heating the filament until it glows. As well as light, therefore, the bulb produces heat, the latter more than the former; 90 per cent of the energy fed into a standard incandescent light bulb comes out as heat. This is obviously wasteful, and since the turn of the millennium more and more countries have banned or restricted the sale of incandescent bulbs. In 2007 the UK government pledged to phase out incandescent bulbs within five years, a move that was quickly followed by the European Union. Despite some delays and setbacks (including a failed attempt by a German company to market the traditional light bulb as a 'mini-heater'), the phasing-out process is going ahead. The USA has not banned the bulb outright, but demands 25 per cent greater efficiency for household light bulbs between 40 and 100 watts. China banned the import and sale of certain incandescent light bulbs in 2012, and will gradually expand the ban. With a few exceptions, most countries in the world have taken measures to quit the incandescent bulb, promoting instead light-emitting diode (LED) lamps, a system based on semiconductors that produce light when connected to an electric current. LED lamps last for a long time, produce very little heat and do not contain mercury, giving them an advantage over not only the incandescent bulb, but its other rival, the compact fluorescent tube.

A trickier point to beat has been the incandescent bulb's unmatched ability to render colour – the quality that so enthused our New York piano manufacturer. Compact fluorescent tubes and early LED bulbs have a narrower colour range, leading to more distortion; in fact, Donald Trump's claim that the 'new light bulbs' turn him orange may have had a grain of truth.[3] But only a grain: more recent LED bulbs can generate the entire spectrum of visible light colours, and the cold (or orange) glare of the early LED lamps is largely eliminated. The incandescent light bulb will probably remain the symbol of a good idea for a long time, but even a good idea doesn't necessarily last forever.

# Integrated Radio/TV Cabinet

## Anders V. Munch

The television set has been a dominant feature in domestic interior design since the middle of the twentieth century. The heyday of its dominion over the living room was the 1960s, when television sets were pieces of furniture in their own right, often combined with radio, gramophone, record storage and speakers in a conspicuous piece of wooden furniture – such as the BeoVision 2000, produced by Bang & Olufsen from 1965. This was to be the last of this kind of integration by the Danish manufacturer, as well as the final design for the firm by the architect Ib Fabiansen. He had been B&O's first designer, and the task he had been assigned was to match modern devices with modern furniture.

The formula of Danish Modern furniture that made it so successful was the pairing of an architect with cabinetmakers, as many showpieces at the Copenhagen Cabinetmakers' Exhibitions demonstrated, and B&O followed this pattern by hiring a furniture architect. Fabiansen's first designs were shown at the Danish Society of Arts & Crafts' exhibition in 1958 at Charlottenborg, Copenhagen: a series of flexible module furnishings to encase B&O products following Functionalist ideas. But the goal was the total integration of device and furniture. Fabiansen continued by combining slim wooden cabinets with devices designed in accordance with the sober minimalism of the German Ulm School, known best through the products of the electronics company Braun. Fabiansen's work laid the foundation of a product style for B&O.

The BeoVision 2000 was less spectacular than some of its American or German competitors, although it exemplifies many of the typical features of integrated radio/TV cabinets. It is long and low, in the same form as Mid-century Modern sideboards (another extinct object), with speakers set some distance apart at the ends, enabling a fine stereo effect. Behind sliding doors, the television is to the left; to the right, on two shelves, are a gramophone above and a reel-to-reel tape recorder below, with a radio running down the frame beside them. While other manufacturers designed models that were opened from the top, in the BeoVision 2000, the gramophone and tape recorder slid smoothly out. B&O also avoided the stylistic proliferation found on the international market, where consumers could choose a cabinet to suit their interior decor (in the USA, the Magnavox Astro-Sonic series offered more than forty different styles, from 'Aegean Classic' to 'Early American' to 'Contemporary' – or 'Scandinavian Modern'). With B&O, by contrast, stylistic choice was not an option.

Bang & Olufsen 'Beovision 2000', designed by Ib Fabiansen, Denmark, 1965–8; gramophone, tape recorder and radio are to the right behind the sliding door, stereo loudspeakers at the ends.

The naming of this hybrid device-furniture object was confusing. Today's collectors seem to agree on 'TV/stereo console', but the marketing of the 1950s and '60s used many different names. The American Magnavox was a 'stereo theatre', and the most spectacular European design, the German Komet (1957) from the manufacturer Kuba, was a *'Fernseh-Stereo-Konzerttruhe'*, basically, a 'concert cabinet'. These and other examples were presented as one-off marvels, rather than a new generic type. 'TV-radiogramophone' was the most sober of the descriptive names used in the period, an extension of the radio gramophone of the interwar years: their earlier appearance in the USA followed the market introduction of the television set just after the Second World War. Du Mont's 'Hampshire' TV-radiogramophone, for example, introduced in 1947, was the company's deluxe model, an aspirational object in the new consumer society.

Rapid technological development and changing lifestyles in the late 1960s, however, put an end to this monolith of home entertainment. In the case of the BeoVision 2000, we can explain its obsolescence in different trajectories, in relation to the change of taste in home decoration and lifestyles, and to new trends in hi-fi. In 1970 Jens Bang, head of B&O, justified the dominance of wood in the company's designs: 'The rosewood spell was forced upon us by the furniture industry, for it has a certain image all over the world, and that means that we have to stick to those symbols to be in conformity with the accepted concepts of Scandinavian design.'[1] By that date, though, the popularity of Scandinavian Modern design was in decline, and the consumer electronics market diverged into a high-tech style.

The integration achieved by Fabiansen at B&O was quickly followed by disaggregation. The individual components and their design were separated; television sets were taken over by Henning Moldenhawer, and radios and hi-fi equipment by Jacob Jensen.[2] The hi-fi trend made the integrated, built-in devices obsolete from about 1970, as the speakers and the many control buttons were now favoured as visible features and spread over the living room. The focus changed to a 'system' distributed throughout the home – instead of a 'central showpiece' of the home – and an important component became the remote control, or 'commander', as B&O called it when David Lewis designed the company's first in 1974. At the same time, a shifting attitude to domestic spaces brought about a more fluid way of inhabiting them, and 'interior decoration' became 'interior landscape'.[3] Radios and tape recorders grew smaller, and television screens and speakers grew larger for a more immersive experience. The miniaturization of electronic devices changed media consumption, and the soundscapes, no longer contained in the living room, were distributed across kitchens and teenagers' rooms and extended into public spaces with wireless devices such as portable television sets and radios. All these developments spelled the end of the

Advertisement for Philips integrated TV cabinet 'Leonardo' 21RD153A, with integral radio and record player, 1957.

integrated radio/TV cabinet, its extinction the result of changes in domestic inhabitation, and in listening and viewing habits, as much as of technical obsolescence.

The integration of technical functions remains a driving force in electronic devices, along with the continuous miniaturization of parts and the expansion of media experiences. In the 1990s audio-visual centres combining television, radio and CD player were introduced as a solution for smaller dwellings, but combined devices of this sort have never again been encased in wooden furniture. Today, most of the technical functions once contained in the cabinet are, along with many other functions, built into the smartphone, which serves as media hub. Yet the original television/stereo consoles of the 1950s and '60s – both obsolete technologically and, in home culture, manifestations of an epoch long gone – are cherished by vintage hi-fi enthusiasts and retro aficionados. Hi-fi nerds savour the challenge of getting the technology to work, especially the picture tube – but mostly they live on as elegant drinks cabinets.

# Invacar: The 'Invalid Carriage'
## Elizabeth Guffey

*The invalid carriage. The motorized trike. The spaz chariot. The Noddy car. The wee Bluey. The cripple cruiser.* All these monikers refer to the Invacar, the three-wheeled, single-user automobile that was a common sight on British roads in the decades after the Second World War. Designed specifically for use by people with disabilities in order to help them integrate better into society, the Invacar might best be described as a road-going wheelchair. The British government provided these cars free of charge to disabled people from the 1950s to the 1970s, and the last was officially retired in 2003. Although it is hard to say how many were manufactured in all, at the peak of their use, in 1976, an estimated 21,000 were on the road. The diminutive vehicle's trajectory from conception to extinction also describes the changing premises of the British welfare state, a story in which shifting attitudes to disability and design collided with human rights and hard political fact.

Born out of the confident belief that a government-run medical programme might counteract or compensate for a physical impairment, the 'invalid tricycle' featured a single seat and adaptable controls built to the specifications of the individual driver. Two versions were designed for different purposes. The smaller, single-headlight 'invalid carriage' crept along village pathways and local streets, its engine unable to move it at more than 32 kph (20 mph). Snug little models of the carriage were designed to such narrow widths that they could pass through garden gates and even a standard domestic doorway (where they could be parked in the hallway). The 'invalid tricycle', meanwhile, reportedly reached a top speed of just under 130 kph (80 mph) and was allowed to join traffic on highways and city streets.[1] The distinctively tapered nose and icy-blue colour of both versions cut a distinctive appearance on the road, while the rear-mounted engine allowed a flat floor, left unexpected room under the dashboard and also simplified its manufacture.

The minimalist simplicity of the Invacar emerges from a venerable – though largely unrecognized – tradition of disabled people 'making do'. For example, a hand-cranked tricycle constructed in 1685 by Stephan Farffler, a partially paralysed clockmaker from Altdorf, Germany, anticipates the more recent domestic 'invalid carriage'. Over the next two centuries, this and other self-propelled vehicles were often confused with the various litters and carts that culminated in the late nineteenth-century bath chair, a three-wheeled vehicle that haunted Victorian spas and resorts. Crucially, the bath chair

AC Invacar Model 70 invalid car, UK, 1976.

and its brethren required a servant or 'push-man' for power. The true descendants of Farffler's vehicle were the self-contained, hand-propelled tricycles distributed to disabled war veterans across Europe in the devastating aftermath of the First World War. Perhaps unsurprisingly, numerous enterprising individuals across the Continent attached small motors to these vehicles. Seeking to help a paralysed cousin, an engineer and motorcycle buff named Bert Greeves modified one using an engine from a lawnmower. In 1946 he founded Invacar, the company that would dominate the British market. His timing could not have been better. Not only was Greeves's poky little car well suited to accommodate the needs of thousands of wounded soldiers returning from the front, but also it was poised to ride a wave of national enthusiasm for the new welfare state ushered in by the new Labour government in 1945.

The great optimism that surrounded the early years of Britain's anticipated cradle-to-grave welfare state profoundly shaped both the design and the use of the Invacar. Britain's National Health Service (NHS) promised a future in which society would take care of all its citizens. Within the administrative culture of the NHS, these motorized tricycles were classified as an 'object of care', similar to a cane or prosthetic device; bought and owned by the government, they were leased free to anyone with a mobility disability. Not only did the NHS freely supply these cars to disabled people, but it offered more than fifty variations of controls, outfitting individual cars with hand-operated brakes or steering wheels shaped like tillers, allowing each user to tailor a vehicle to meet their specific needs. Once the cars hit the road, maintenance and repairs were also provided free of charge. Some affectionate users conflated the colour of the car with the government agency that oversaw its production, distribution and maintenance, dubbing the NHS-issued cars 'Ministry Blue'. For many more, the unprecedented freedom and mobility the car provided embodied the most compelling promises of the welfare state.

By the early 1970s, however, stubborn realities began to overshadow the Invacar's substantial promise. Britain was in recession, and inflation and unemployment rose precipitously. A new nickname, the 'mobile roadblock', emerged as people increasingly found the car underpowered, unreliable and even downright dangerous. Later models constructed from fibreglass proved so lightweight that they toppled over easily in a crosswind. Perhaps even more alarmingly, the engine proved prone to catching fire. But complaints also arose about the solitary design itself; while the cars might seem to offer more independence to disabled people, many drivers decried their feeling of loneliness and isolation while driving the car. In the welfare state, they were seen as antisocial. It was, in fact, illegal for drivers of the Invacar to carry a passenger, even a carer or family member, and at one point a minor

AC Invacar Model 70 cars on a street in Portsmouth, UK, 1993.

scandal arose when a disabled mother was sent to court for using her Invacar to take her children to school.[2] Invacar's distinctive shape and colour, once points of pride, were increasingly seen as stigmatizing.

But the internal politics of disability policy also played a role in the Invacar's demise. While some disabled people embraced the subsidized car, others argued against the government providing such personal transportation devices. Against a backdrop of economic frustration, anger and decline, a series of both Conservative and Labour governments reformed the country's social-security system in the 1970s, eventually allowing government subvention of car purchases as well. Under increasing pressure to begin winding down the programme, the Invacar ceased production in 1976. In the wake of privatization efforts under Margaret Thatcher's Conservative government in the late 1970s, the NHS-issued Invacars were gradually replaced by a government-led programme that helped disabled people to purchase and drive modified commercially produced cars. The newer programme marks a shift in thinking about people with disabilities, and their place in society now and in the years to come. This approach suggests that disabled people are but one more type of consumer in a late capitalist economy, rather than citizens of a caring state.

Singled out for extinction, the Invacar was not officially banned from British roads until 31 March 2003. The biggest collector today is not the NHS – which began destroying the cars wholesale shortly after the turn of the millennium – but an American millionaire. Attracted by the car's eccentric shape, he bought his first one online and has continued to accumulate them ever since. Perhaps this is the surest sign of the car's extinction; it is valued now for its cuteness, while its radical social agenda is almost completely forgotten.

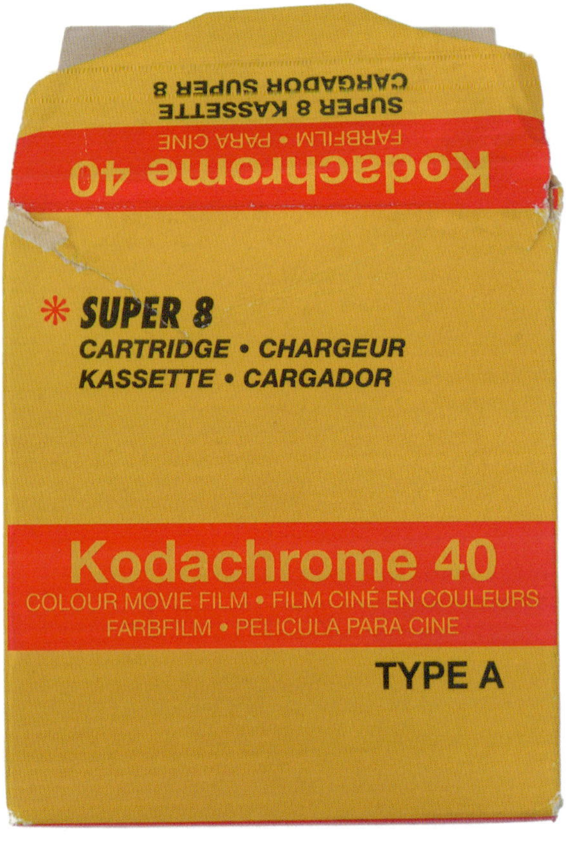
CARGADOR SUPER 8
SUPER 8 KASSETTE
FARBFILM • PARA CINE
Kodachrome 40
SUPER 8
CARTRIDGE • CHARGEUR
KASSETTE • CARGADOR
Kodachrome 40
COLOUR MOVIE FILM • FILM CINÉ EN COULEURS
FARBFILM • PELICULA PARA CINE
TYPE A

# Kodachrome
## Tacita Dean

In his 1973 song 'Kodachrome', Paul Simon beseeched his mama not to take his beloved film stock away. But Kodak did, in 2009, when the company finally stopped its manufacture, and ceased processing a year later. Within a decade, new management wished to bring it back and began investigating ways to create a non-toxic formula adapted to the twenty-first century. But, with the infrastructure gone, this proved impossible. A monumental force of will and sustained investment in synthetic chemistry would be needed to produce Kodachrome again, and this won't happen. With no product, there can be no market viability – no investment, no gain; and no gain, no investment. Sadly, Kodak's eponymous film stock appears well and truly extinct.

Kodachrome was colour-reversal film. Reversal means that there is no negative, only a positive image exposed directly onto the film. It was first produced by Kodak in 1935 as 16mm film and later marketed as 35mm slide film for still photography. It is this – the roll of 36 exposures of Kodachrome – that gave the brand its household name, making the word Kodachrome synonymous with colour photography. It was also manufactured as Super 8 film, and both were sold 'process paid', which meant that Kodak and its licensed laboratories were responsible for processing the film. Consumers would post their canisters and Super 8 cassettes to locations worldwide and then wait for them to be returned as a box of slides or a roll of film. Kodachrome slides and Super 8 film were responsible for creating an amateur market that emboldened every family to own a camera, take photographs and shoot home movies. Kodachrome was only ever projected; families looked at their holiday snaps or films gathered around a slide or Super 8 projector.

My first film was made on 16mm Kodachrome. I bought the rolls in their distinctive yellow cardboard box from a photo shop and sent them off to be processed one by one; it was easy and convenient. I was in Cornwall in the mid-1980s, making an animation in brightly coloured pastel, changing the drawing for each exposure to create the illusion of movement. It was laborious and time-consuming, performing for that three-minute roll of film frame by frame – then waiting a week or more only to discover that I had got the exposure wrong and had to start the whole process again. Such was the emotional trajectory of photography at that time: patience and anticipation, then excitement and delight, or disappointment and dismay. In behaviour alone, there has been a huge cultural shift in society's habits in photography.

Kodachrome 40 Super 8 colour movie film cartridge packaging.

I later started using Kodachrome Super 8 film: the perfect short form, like a celluloid haiku, that was a synthesis of time, place and experience distilled into three minutes of evanescent magic. Kodachrome Super 8 film was the medium of introduction for a new breed of film director that came of age in the 1970s.

The exuberance of Kodachrome colour was famous: its hues rich, distinctive and pure analogue – not a pitch-perfect chromatic verisimilitude but something else, almost an abstraction of it, or a summation. The purity, depth and intensity of colour, as well as the film's distinctive concentrated sharpness and greater contrast, are chemically embedded in the complex coatings of the emulsion and esoteric processing technique. In fact, it is precisely this relationship between emulsion and development that sets Kodachrome apart, with an essential part of the chemistry taking place in the processing and not the exposure. Negative film never achieved the focused chromatic range of Kodachrome. Nor has the digital image, which is a whole other category of flat and normal, come close to rivalling it. From an archival perspective, Kodachrome continues to retain its dark-storage colour integrity over time as no other emulsion has yet managed to achieve.

That colour is reflected light is one of the strangest lessons one learns as a child: that no object really *has* a colour, but merely *reflects* a colour. The way a surface physically receives a ray's wavelength is what gives the appearance of one colour as opposed to another. Colour is therefore perceptive and illusory, so that with the invention of photography, few believed it would ever be possible to catch it photochemically. Those who pursued the idea doggedly were relegated to the status of medieval alchemists. Understanding how colour photography evolved is to appreciate the wonder behind half a century of science, invention and competing patents.

The journey to find photographic colour is a story of the perception of light through chemistry, and of subtracting prismatic rays from white light using their colour complements. First forays involved adding colour, known as 'additive colour', as an obvious next step from colour tinting. It was known from the use of prisms that white light was made up of the seven colours of the rainbow, so initially some semblance of colour was achieved by shooting and then exposing a black-and-white negative through mixing red and green filters. Colour was evident, but it was dull and incomplete, and also dark because the filters absorbed a great deal of light. Moving images also had the halo that we recognize today from badly registered video projection that still adopts a similar RGB ('red green blue') additive method.

It was only when a system of 'subtractive colour' was developed that true experimentation began. In this method, colour was filtered out or subtracted from white light using its colour complement, so a ray of light filtered through red made cyan (as the blue/green came to be identified in the developing colour theory known now as CMYK, 'cyan magenta yellow black'), light filtered through magenta made green and through a blue filter made yellow. Using pigments and dyes to absorb light accordingly, three separate dyed images could be created, which, when superimposed on each other with a black-and-white image, created full colour.

Suffice to say that such a level of blind faith experimentation was mostly the preserve of the dedicated amateur chemist and intrepid photo enthusiast. Exhausting, exhaustive and exhilarating in equal measure, it ultimately became a race to fix three-colour separation within several layers of light-sensitive emulsion on a single surface. That was eventually realized in 1935 by a pair of scientifically educated musicians, Leopold Godowsky Jr and Leopold Mannes, who were under grace-and-favour laboratory-time patronage to Kodak. It was a twentieth-century miracle of chemistry and endeavour that converted the world's memory into colour.

When Kodak stopped producing Kodachrome, it lost its oldest brand and something unique that a visionary company with its own in-house synthetic chemistry laboratory could have strived to reformulate and have kept available. But the times were against it. Kodak had already been diminishing Kodachrome's market by manufacturing Ektachrome, a slide film that was more accessible because it did not rely on the company's licensed processing techniques. But the real threat came with the rapid rise of digital photography. The company was already haemorrhaging money and appeared more than happy to dispense quickly with its photochemical legacy, content to all but wipe the slate clean of more than a century of its own image-making invention. With Kodachrome's processing chemicals so toxic, it was quick to give up on its beloved signature product. Photographers across the globe objected, but the world was in the midst of a monumental shift away from analogue technology, and resistance was hopeless.

But Kodachrome has never been entirely forgotten. In 2017, under more enlightened management, and with a greater intelligence and discussion of what constitutes the difference – and the importance of that difference – between a photochemical and a digital image, Kodak looked into bringing Kodachrome back. There was an immediate ripple of excitement at the announcement, but it came too late. A decade earlier, with the infrastructure still in place, a solution might have been found – but at that time the aggressive marketing of digital media eclipsed all hope for the future of photochemical film. Thankfully, this is now changing and the use of film is on the increase again. But it would take pioneer energy and persistence against the market naysayers to get Kodachrome into production again. For now, it must endure in memory, both actual and in how we see the twentieth century in our mind's eye.

# B C
# CDDD EEE
# EE EFF .
# ,
# HHH IIJK '
# ,
# KLLL M :
# ,
# N NN;

Letraset products are protected world-wide by patents and patent applications throughout the world.
Letraset, Instant Lettering and Spacematic are Trade Marks.
©LETRASET INTERNATIONAL LTD. 1972.

B8559

# Letraset
Robin Kinross

Printing from metal type was always withheld from popular use, for reasons that include technical difficulty, cost of equipment, and guild or trade-union protection. The idea of printing as a 'black art' – an inky, secretive business in which mirror-imaged letters magically became readable text – accompanied its start in fifteenth-century Germany, but was carried through into the modern age. In the 1960s, in Western societies, metal type and its associated method of printing – letterpress – began to decline, to be replaced by photocomposition of type and offset lithographic printing. The processes we use now, of text composition on small computers to generate plates for offset lithography, came into use in the late 1980s. In this gap between the fall of metal typesetting and its associated process of letterpress printing and the rise of the personal computer, some simple techniques of composing words for printed reproduction were able to flourish. Metal composition and letterpress printing live on in small enclaves with no connection to mainline produc-tion, but the machines and processes of this intermediate period, 1960–90, have mostly become extinct.

Dry-transfer lettering is one of these extinct techniques. Its leading manufacturer was the British company Letraset International Ltd.[1] There were rivals, notably the French company Mecanorma, but it was Letraset that spread its products most widely. Its name became synonymous with the technique, sometimes becoming the verb that lay within the word ('can you letraset that headline for me?').

The Letraset company was formed in Britain in 1959. In its first years, letters held on a backing sheet were dampened with a wet brush, then slid off as 'transfers': a slow business and only easily workable for larger letters. Dry-transfer lettering was introduced in 1961, and this provided the break-through. The product was simple: letter shapes imprinted on to the back of a polymer-coated carrier sheet, and protected by a loose siliconized blank sheet. The letters could be rubbed down with a blunt tool (the company's own burnisher, or a ballpoint pen or soft pencil) on to a receiving surface, to form words and whole lines of text. As well as letters, a large variety of other marks, patterns, plain colours and images were produced and marketed.

The revolution of Letraset lay in its popular dispersion. Suddenly type could be bought for not much money from shops – the art materials shops or stationers that even small towns possessed. Almost no instruction was needed for use. Respectable artwork for cheap offset printing could be produced

Sheet of Letraset,
38.2 × 25.6 cm
(15 × 10 in.).

very directly, perhaps incorporating smaller text from an electric typewriter and photographic prints for images. The Letraset company was generous in its publicity materials. Its carefully designed catalogues could be picked up at little cost from local shops, and opened up a vista of sophisticated typography. Users ranged from highly sensitive typographers doing good work for low-budget clients, to church or parish council notice-makers, to untutored kids with urgent messages to get across in ransom-note fashion.

The company's success – it boasted two Queen's Awards for exports – followed partly from its commitment to good standards in typeface design. Letraset's production team developed great skill in cutting the stencils that generated the final letterforms. The company paid proper licensing fees to type foundries for the use of their typefaces, and it began to issue its own well-made types. The first of these, Compacta – a strong, space-saving headline type – came out in 1963, and exactly caught the modernizing spirit of the times. From 1970, the Letragraphica series of new typefaces was issued, chosen by a panel of celebrated designers from Europe and the USA.

Despite popular success and good standards of design, Letraset never quite lost its aura of being a cheap substitute for something better. If budgets ran to it, one would use high-quality prints from metal type ('repro pulls'). As photocomposition improved in quality and became cheaper throughout this period, this newer technology became the best source for typesetting for offset litho printing. Making words with Letraset was always a rather awkward and fiddly process, for which one could develop definite skills. The larger the letter, the harder it was to rub down in one piece, and artwork that showed letters with cracks and missing parts – perhaps inked in later – betrayed an inexpert hand. The carrier sheets hardened physically and became more difficult to manage as they aged, and some might pick up hairs from the homeworker's cat. But the most familiar defect of the sheets was the finite number of letters; one would run out of characters and have to dash back to the shop in the hope that they had that sheet in stock, just for those extra Es or Bs.

The most essential limitation of dry-transfer lettering lay in its material constitution. The letters on the sheet were precisely arrayed, but the Letraset sheet was without fixed relation to the plane of application. So letters might be rubbed down at slightly different angles to each other, and with visibly inconsistent spaces between them. In this sense Letraset letters belonged to the category of *lettering*, as opposed to that of *type* or *writing*. They might look like 'type', but they were without the regulated system of space that comes with a type that sits on a physical body (of metal) or which is built into the unitized system of a composing machine (such as a photocomposing machine). Perhaps the closest comparison of dry-transfer letters is with stencil lettering, in which the letter image is marked on to a baseline drawn on the receiving plane, but in which there are no physical constraints in placing the image; it is simply up to the skill and good visual sense of the operator.

The Letraset company was clearly aware of this limitation early on. In 1968 it introduced its Spacematic system, in which every letter was provided with guidelines underneath it, and to right and left. One rubbed down the right guideline with the letter and then, with the following letter, rubbed down the left guideline of that letter – exactly abutting the right guideline of the preceding letter. When the line was complete, all the guidelines would be taken off with a rubber or masking tape. But even if the guidelines were used, there was no indication for the spaces between words. Confident users – probably most users – did not bother with the spacing guidelines, and did it by eye instead. The Spacematic system was there as a hint, a suggestion that space does indeed make all the difference in typography.

Looking through the patent applications of Letraset International, one sees that the company continually addressed the problem of regulated spacing. As late as 1979 a patent was sought for an 'apparatus for applying indicia to a receptor': the carrier sheet is held in a parallel motion device that can move up and down or along a drawing board; the vertical arm is stepped, achieving constant baselines. The device was never put into manufacture. But this is the closest Letraset could have come to a proper method of typesetting, in which the individual types are set down in good order, with a system of spacing that is built into their generation. It would have been a hefty and expensive piece of equipment, contradicting the instant, do-it-yourself character of dry-transfer lettering.

The end of Letraset came quite quickly. In 1981 Letraset International was acquired by the Swedish office-equipment company Esselte. In 1984 the first Apple computer was launched, and towards the end of the 1980s Apple Macintoshes were beginning to be used by graphic designers. The laser printers associated with these small computers were made with fonts as part of the package. Designers quickly realized that personal computers could be used for drawing and producing new typefaces, as well as for typesetting with the 'resident' fonts installed in laser printers. In 1990 Esselte Letraset acquired the German digital type company URW. But by then dry-transfer lettering was on its way out. It was carried on, in obscure contexts, into the new millennium, but in 2012 the Letraset company was sold again, and in 2016 the business and the company name were ended.

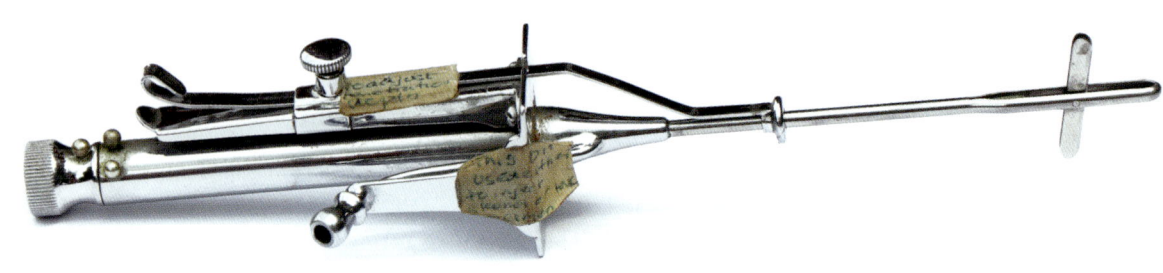

# Leucotome
## Carsten Timmermann

What happens to medical instruments when they fall out of use? They may
end up in a collection such as that of the Museum of Medicine and Health
at the University of Manchester, where they are occasionally displayed as
witnesses of past medical practices. Our museum holds several specimens
of obsolete instruments, such as stomach pumps, endoscopes and obstetric
forceps, which are arranged much as stuffed animals are in natural-history
museums, each classified and labelled with the name of the surgeon or physician who invented it. But the descendants of stomach pumps, endoscopes
and obstetric forceps are still used today. This is different in the case of the
museum's leucotome, an instrument associated with a surgical procedure
that is truly extinct: leucotomy, a form of psychosurgery – brain surgery
with the objective of treating mental illness.

Leucotomy was invented in the 1930s by a Portuguese neurologist,
António Egas Moniz. Egas Moniz believed mental illnesses were caused
by organic malfunctions and should be treated accordingly, with physical
interventions or drugs. He had little respect for psychological explanations
and psychoanalytical approaches, such as those developed by Sigmund Freud.
Egas Moniz suggested that mental disorders were caused by 'fixed ideas'
dependent on 'abnormally stabilized' nerve connections in the prefrontal area
of the brain. He thus concluded that patients could be treated by destroying
nerve cells in this area. He and the surgeon Pedro Almeida Lima experimented with a procedure on asylum patients, first injecting alcohol into the
brain tissue through holes they drilled in patients' skulls. Later Egas Moniz
modified an instrument he had ordered from Paris: the first leucotome. The
thin end of the instrument was inserted into the brain of a patient through
a hole in the skull, and rotated. In 1936 he reported on his experiences of
twenty cases in a monograph, claiming that seven patients had been cured,
seven had improved and six were unchanged.[1]

Egas Moniz's publication was discussed at conferences and reviewed in
medical journals in many languages. Neurologists and psychiatrists had long
been interested in the biological foundations of mental illness, and in the
interwar period confidence grew that interventions were possible that helped
to ameliorate if not cure madness, and thus reduce the growing numbers of
asylum inmates. It had become clear that at least one frequently diagnosed
condition, 'general paralysis of the insane' (GPI), was caused by syphilis,
which could now be treated with a new drug, Salvarsan (Arsphenamine),

Leucotome, Down Bros,
UK, 1940s.

and there was no reason to assume that biological cures were not viable for other disorders. Psychiatrists experimented with new interventions, for example injecting patients with insulin to trigger a coma, or administering electricity to trigger convulsions, which were assumed to reset the brain and thus alleviate mental illness. The growing attention to psychosurgery was part of this wider enthusiasm for radical therapies.

Within a few years of Egas Moniz's publication, teams of neurologists and surgeons in various countries had developed variations of the procedure, along with their own instruments. In the USA, a modified version of leucotomy operation, the transorbital lobotomy, was developed and enthusiastically popularized by the neurologist Walter Freeman and the neurosurgeon James Watts. Egas Moniz was awarded a Nobel Prize in 1949, which further enhanced the credibility of leucotomy. While instruments and procedures were not evaluated to today's standards (the randomized controlled trial had not been invented, for a start), the publication of case series was a common way of reporting results in surgery. The side effects of such operations could be drastic and irreversible; if too much brain tissue was destroyed, the patient could be transformed into a human vegetable or killed. Such risks, however, were viewed as an acceptable price for getting people out of the asylum. The belief in the effectiveness of the procedure was an expression of an engineering mindset in medicine, of enthusiastic modernism; cutting a few nerve connections was equated to flicking a switch, turning patients tormented by their conditions into useful members of society, who might no longer be capable of producing great art, but were able to live contented lives and, above all, work.

In Britain in the 1940s and '50s leucotomies were performed frequently as treatments of last resort, for patients with particularly serious or persistent forms of mental illness. This was a time of great change in psychiatry. Mental hospitals were now part of the National Health Service, and the Ministry of Health was encouraging targeted interventions based on scientific research, which included leucotomies. The historian David Crossley estimates that by 1954 more than 12,000 Britons had been leucotomized.[2] A BBC television programme from 1957, *The Hurt Mind*, featured a leucotomy patient and his wife in a relaxed conversation with the presenter and the psychiatrist William Sargant, an enthusiastic supporter of biological and surgical therapy for mental conditions. The message was clear: this medical procedure had liberated the patient from his suffering by removing the cause of that suffering, a disturbance in the brain.

The leucotome in our collection is of a type designed in the early 1940s by James McGregor, Senior Psychiatrist at Warlingham Park Hospital in Croydon (the former Surrey county asylum), who was also a trained engineer. The prototype was constructed by the hospital's assistant clerk of works, J. W. Pearson, and the instrument was tested in a series of operations by John Crumbie, a visiting surgeon from nearby Croydon General Hospital. The leucotome was meant to make the operation more precise and controllable. Yet it caused two deaths from bleeding in the brain as the rotating blade caught and severed small blood vessels, leading one contemporary to compare the instrument to an egg whisk. Reportedly cumbersome to use, the Warlingham leucotome had become extinct by the mid-1950s and disappeared from trade catalogues.[3]

The extinction of the leucotomy itself was not sudden, however, but a complex, drawn-out process, as the ecosystem of psychiatry changed. In the 1960s Freeman was still criss-crossing North America in his camper van, visiting mental hospitals and former patients, and promoting modified lobotomies. But the number of leucotomy operations overall was declining. Leucotomy had always had its critics, and in this era critical voices grew louder. Intellectual culture in Britain and the USA increasingly celebrated anti-authoritarianism, which also influenced the rise of the 'anti-psychiatry' movement. Reinforced by representations in popular culture such as Ken Kesey's novel *One Flew over the Cuckoo's Nest* (1962; turned into a successful film by Miloš Forman in 1975), leucotomy started to look like the ultimate punishment for unruly behaviour, destroying personalities and turning patients into docile cogs in the machine. New pharmaceuticals, such as chlorpromazine, replaced surgical and physical interventions as the treatment of choice, despite significant side effects and scant evidence that they were more effective. But drugs were central to a new paradigm in brain science that focused on the role of receptors, and were a better fit for the clinical trials that came to dominate clinical research, often funded by pharmaceutical companies.

As a small medical museum, how do we deal with an instrument that at the time of its invention represented medical progress and hope for a cure, but that is associated with a procedure today viewed as inhumane, unethical and totally unacceptable? This is not trivial. Almost certainly, some visitors to our exhibitions will have been diagnosed with mental conditions that in extreme cases would have been treated by leucotomy in the 1940s and '50s. The leucotome, however, is a symbol of the situatedness of medical technology and techniques, and the necessity of understanding their rise and fall contextually. It is understandable if we feel relieved to live in a time when leucotomies are not routinely performed. But the extinction of this instrument reminds us to approach today's routine therapeutic interventions carefully and critically. They may yet turn out to be the leucotomies of future pasts.

# Manchester Pail System: 'Dolly Vardens'
## Barbara Penner

The photograph here shows a brightly painted cart, filled with pails, sitting in a vitrine at Manchester's Museum of Science and Industry. The caption reveals this object to be a purpose-built night-soil cart, just one of dozens that would have filled the streets of late Victorian Manchester. These jaunty carts were dubbed 'Dolly Vardens' after the fashionable coquette in Charles Dickens's *Barnaby Rudge* (1841). Its picturesqueness makes it hard now to see this cart for what it was: an element of a modern municipally run sanitation system. And yet, between the late 1860s and the 1890s Manchester's 'pail system' was not only tolerated but even touted as an effective solution to one of the greatest problems of industrial urbanization: how to manage water and waste safely.

It was inevitable that this problem surfaced early on in Manchester. As the world's leading manufacturing city by the 1830s, it was also the first to demonstrate the price of capitalist progress. Even even as new factories such as cotton mills created tremendous wealth, they destroyed the environment, creating a landscape of blackened skies and choked rivers. Contemplating the apocalyptic scene in 1835, the French historian Alexis de Tocqueville declared: 'From this foul drain, the greatest stream of human industry flows out to fertilize the whole world.'[1] As thousands of workers crowded into dense inner-city slums, piles of uncollected refuse and overflowing cesspits leached into the soil and polluted wells. Tocqueville's 'foul drain' became reality, as infant mortality and epidemics of cholera and other diseases spiked.

The public-health crisis forced the municipal government to accept at last that environmental conditions would be alleviated only by coordinated action on its part. Faced with equally disastrous conditions, London authorities in the 1850s had concluded that the best solution was to flush human waste away, using water closets connected to sewerage systems. In order to manage the resulting increase in waterborne effluent, the city authorized the construction of a vast new main drainage system that intercepted sewage before it reached the River Thames and bore it downstream, thus improving water quality and reducing cholera outbreaks in the city centre. Although well aware of London's example, when Manchester acted a decade later to address its environmental crisis, it chose a different 'dry' route, implementing a pail system from 1869.

Even if it did not have the engineering sophistication of London's scheme, Manchester's pail system was also a centralized technological

Night-soil cart, Manchester City Council Cleansing Department, *c.* 1910.

solution to the problem of waste, requiring specialized equipment, spaces and manpower to operate. In 1877 the system covered 6,000 houses, roughly a tenth of the city, and was being rolled out to thousands more. As part of the scheme, the municipality installed a dry closet in the yard of each house and provided residents with two galvanized iron pails: one for the collection of excrement, another for dry refuse. Once a week, municipal workers sealed up full pails with rubber-lined lids, loaded them on to their carts and replaced them with fresh ones. Once back at the vast depot on Water Street, the pails' contents became the base ingredients for commercial-grade fertilizer. Made of excrement, slaughterhouse refuse, putrid fish and fine ash, it was sold to apparently enthusiastic farmers at 12s. 6d. per ton.

Now that we have become so accustomed to flushing waste away automatically, the very idea of a municipality laboriously collecting residents' waste to mix into fertilizer seems almost fantastical. And it certainly wasn't done quietly: unlike the more traditional practice of scavenging, which took place at night, pails were collected by day, ensuring that Manchester's streets were filled with a constant, odoriferous parade of horse-drawn Dolly Vardens rumbling off to the depot. Why exactly they were given that moniker is a mystery. By this period, however, Dickens's creation had a startling range of articles named after her, including a hat, a perfume, a polka and a species of trout. Whether meant as an ironic tribute to the stench (the Dolly Varden 'perfume') or a visual pun (the Dolly Varden hat resembled a pail lid), the nickname implies a degree of sardonic acceptance on the part of Mancunians and confirms the pail system as a highly visible feature of daily life.[2]

In fact, it is not hard to explain why the pail system was first accepted as a viable solution to waste disposal. Despite its success in combating cholera, London's system continued to generate fierce opposition on the grounds that it diluted human manure, rendering it worthless as fertilizer and increasing English reliance on 'foreign' manures such as guano. By capturing a free local source of manure, Manchester's pail system promised to maintain the working organic economy, and to make money so doing. Moreover, in the era before biological methods of sewage treatment, keeping excrement out of rivers entirely remained the most foolproof defence against pollution and disease. A report on the pail system crowed in 1877: 'during the past year about 3,000,000 gallons [11,300,000 litres] of urine, with accompanying faecal matter, has been diverted from the drains and sewers of the city; and the problem connected with the pollution of rivers is said to be partially receiving its solution at Manchester.'[3]

Behind the scenes, however, there was less cause for self-congratulation. The obvious weakness of the pail system was that it served only those in inner-city districts. Large-scale polluters, such as factory owners and suburban middle-class residents with indoor plumbing, were allowed to continue

William Powell Frith,
*Dolly Varden*, 1863,
oil on canvas.

dumping effluent straight into Manchester's rivers. As the historian Harold Platt has documented, the ruling classes – the 20 per cent of the male population who had the franchise – were deeply committed to this two-tier model of water management, in which not only access to public resources but the right to pollute were determined by class. This is no doubt why, even when faced with evidence that the pail system was badly managed and the fertilizer would never turn a profit, city officials refused to give it up. Platt argues convincingly that the council was driven by a desire to hoard water for use in mills and suburban villas, even though the working classes had also paid their share for the waterworks through taxes and had an equal right to indoor plumbing.[4]

What finally gave the game away was an old-fashioned scandal. In 1889 it was revealed that instead of being processed, the raw contents of pails were regularly dumped straight into the Medlock River at night, making a mockery of the city's virtuous boasts of conservancy and of public-health safeguarding. This scandal, along with political reforms enfranchising greater numbers of working-class voters and the vocal criticisms of health reformers, ensured that the now 'accursed' Dolly Varden system could not endure.[5] In a dramatic reversal, in 1890 the city council decreed that all new housing should include indoor plumbing, and approved the construction of an intercepting sewer system to keep the greater volumes of waterborne waste out of urban rivers – the same system that had been implemented in London more than three decades before. Dolly Vardens were fully decommissioned by 1917.

Although it was a positive development in terms of social justice, we should resist simply celebrating the Dolly Vardens' extinction or regarding it as a tale of a superior infrastructural solution 'triumphing' over an obviously inferior one. Indeed, when seen in light of today's environmental crisis, not only is the principle of conservancy hard to fault, but also it shows up the perversity of our now-pervasive waterborne systems, in which one of our most precious resources (clean water) is polluted and then flushed away. Yet, however laudable in theory, Manchester's pail system was a dismal failure in practice, poorly implemented and managed, and indelibly marked by the classism of city officials. Reviewing the whole sorry saga, the eminent politican and feminist Shena Simon concluded: 'That property owners and ratepayers' associations, in their short-sighted zeal for saving the rates and their own pockets, may not always be the best judges even of their own interests and certainly not of those of the community as a whole, is clearly proved in this instance.'[6] As we enter an age of growing scarcity of resources, the Dolly Vardens serve as a timely reminder that any infrastructural system is only ever as sustainable and just as the bodies that design and run it.

# Mechanical Polygraph

Danielle S. Willkens

Painfully, over the course of a week, Thomas Jefferson wrote his famous 'head and heart' letter to his cherished Maria Hadfield Cosway in early October 1786. Although the twelve-page letter featured a lively conversation in which a logical head chastises an impassioned heart, it was written with one pen, held awkwardly in Jefferson's left hand. His right hand had been incapacitated by a fall that dislocated his wrist, and the injury would cause him discomfort for the rest of his life.

As a prolific correspondent, litigator and statesman, Jefferson noted that he was often 'chained to' or 'drudging at a writing table'.[1] His dedication to his writing, and its preservation, was single-minded, resulting in the conservation of nearly 19,000 of his letters. The loss in 1770 of 'every paper I had in the world' to a fire at his family's home explains his early adoption of copying technology.[2] In 1784 he received his first copying press, which had been invented by James Watt just four years before. It was a gift from another technophile, Benjamin Franklin, who, passing the baton as America's Minister to France, recognized the value of the press as a diplomatic tool for record-keeping. Watt's invention relied on several proprietary elements: special ink for the composition of the original letter, unsized and wafer-thin paper, and a pin-screw or rolling press. Once the original letter was dry, the author would dampen the unsized paper and press it directly to the original. The ink would bleed through the wet transfer to create a duplicate. Although this process compromised the original letter by lifting a portion of the ink, it provided the author with a record copy, albeit fragile and often with feathered or blurry script. It was probably with Watt's copying press that Jefferson duplicated his 'head and heart' letter; and without Jefferson's fastidious replication of even his most personal epistles, this uncharacteristically intimate letter would have been lost to history. Cosway's original has never been located.

Convenient and relatively portable, copying presses still left much to be desired for Jefferson and other meticulous record-keepers. Instead of relying on a method that could damage the original documents, 'polygraphic' writing machines seemed more promising, because they created instant copies. First recorded in 1794, and true to the Greek root *polygraphos*, the polygraph was initially a mechanical device for making several facsimiles. Building on seventeenth-century pantagraphic machines, inventors in the eighteenth and nineteenth centuries, such as Marc Isambard Brunel, developed writing

Hawkins & Peale's Patent Polygraph No. 57, used by Thomas Jefferson from 1806 until his death.

machines that used a series of parallel bars and brass fittings with special reservoir quills to mirror an author's script on a second, third or even fifth piece of paper.

Of myriad inventions for mechanical copying, it was the English-born John Isaac Hawkins's polygraph, granted u.s. and British patents in 1803, that held the most promise. From 1804 the American painter-cum-inventor Charles Willson Peale collaborated with Hawkins and managed sales of 'Hawkins & Peale's Patent Polygraph' in America. Peale would pursue modifications and improvements with a clockmaker's precision, and even manufactured the first steel pens in the USA to accompany the device. He was a true believer in the polygraph's potential, and was convinced that it could serve as an essential, extended appendage of the human body and increase natural production and efficiency for law and governmental offices across the country. Keen to advertise the device, Peale placed the polygraph in the same room in his famed Philadelphia museum as his excavated mastodon skeleton, almost foreshadowing the device's imminent extinction.

At the start, Peale's polygraph seemed destined to do well, having the enthusiastic support of key figures in the American Early Republic. Jefferson was an eager client; a day after his election to a second term as President, *Poulson's American Daily Advertiser* featured his fervent testimonial for the polygraph – this 'most precious invention' – and the President lamented that it had not been invented '30 years sooner'. He found the polygraph so indispensable to the Executive Office that he kept one at the President's House (renamed the White House after the war of 1812) and another at Monticello, so that he could maintain record-keeping while supervising his plantation and the construction of his house. The polygraph was also taken up by other innovators, such as the architect-engineer Benjamin Henry Latrobe. In a letter to Jefferson, Latrobe said that the polygraph took a mere hour to master; he endorsed the device in publications in 1804 and 1805, and relied heavily on it for his practice's bookkeeping until his bankruptcy in 1817.[3] It is thanks to the polygraph that Latrobe's correspondence has been preserved.

Notwithstanding such endorsements, Peale's ambitions for the mechanical copier faltered in the first decades of the nineteenth century. His device was more reliable than its predecessors, but still delicate and temperamental. It was also expensive; Peale's clients were willing to pay only $50–60 for a device that cost over $100 to fabricate because of its components made of music wire that required assembly with customized tools. Even his most solid support base – those technically inclined autodidacts willing to experiment with different pens and writing surfaces – ultimately undermined him. Jefferson and Latrobe commissioned several devices for their peers, but they also requested alterations and replacements at Peale's cost. In the end, Peale's small workforce completed only eighty devices. For Hawkins, the market was slightly more favourable in both England and continental Europe, where three times the number of machines were produced – although not enough to sustain the business. Production had largely ceased by 1807 owing to the introduction of Ralph Wedgwood's 'highly improved' carbonic paper.

Although not a commercial success, Hawkins & Peale's Patent Polygraph is still important to acknowledge since it imagined a new, transatlantic world with accessible, duplicate records, and facilitated the development of countless other mechanical copying machines, such as Wedgwood's Penna-Polygraph (*c.* 1808). The early twentieth-century interest in rationalizing work brought the invention of signature machines, such as the Signagraph, that used up to ten pens for expedient cheque endorsements at the U.S. Department of Treasury. Whether they were used to sign one or ten documents, copies produced by these machines remained intimately tied to the hand of the author, since original and duplicate were created at the same time and were composed physically adjacent to each other. This proximity was a guarantee of the copy's authenticity.

This propinquity of author and duplicate would be disrupted by later models of mechanical copying. With David Gestetner's 'cyclostyle' file plate process of 1881, the mimeograph of the late 1880s and Xerox's commercial 'dry copying' patented in the 1950s, copies could be made at any time, and without the physical presence of the original author. Not surprisingly given the greater convenience and flexibility of such machines, this was the model that came to dominate copying, and polygraphic copying fell out of use. Instead, the polygraph has today come to be most commonly associated with devices for detecting the truth by monitoring a subject's blood pressure, breathing rate, pulse and perspiration. Ironically, such machines continue to insist on the body as an index of veracity and accuracy, even as they seek to ferret out lies.

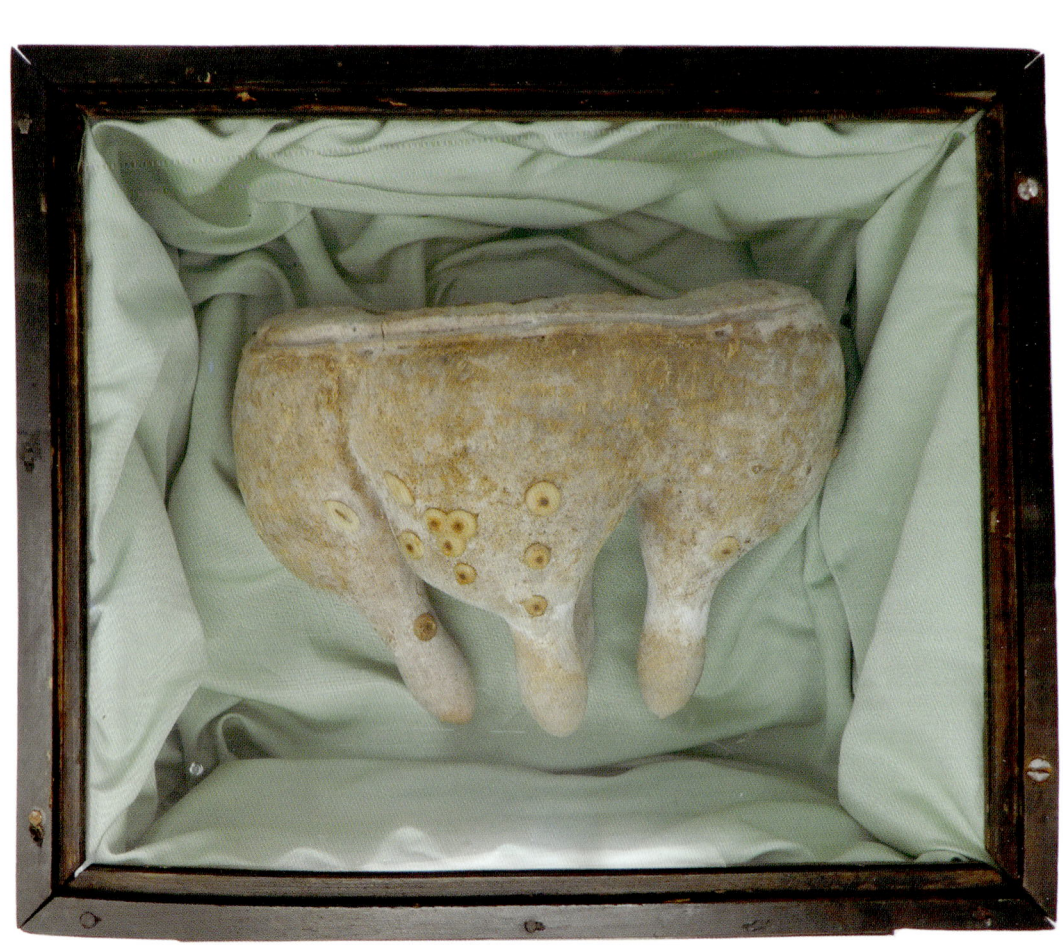

# Medical Wax Model
## Thomas Kador

One of the great powers of objects is that they act as conduits for personal meaning making. They can conjure historical events, but also trigger more intimate memories. The object illustrated here reminds me of the many summer holidays I spent, as a child, on a small farm. On occasion I got to try my hands, quite literally, at milking cows. Although now this seems a very distant memory, I still have a haptic recollection of what the cow's udder felt like and the combination of massaging, tugging and squeezing that was required to get it to produce a fine jet of milk into my bucket.

The udder represented here is highly unlikely to be touched by any hand, especially ungloved. It is a wax model that was designed as a medical teaching aid, probably in the early nineteenth century. During this period, wax models (*moulages*) were a staple part of medical teaching across Europe, depicting important diseases and conditions at life-size and in a lifelike fashion, to help aspiring medical practitioners diagnose their patients. While on the whole, the model depicted here has the appearance of a rather 'average' udder of a milch cow, it shows a distinctive rash that represents the appearance of a condition known as cowpox (*Orthopoxvirus vaccinia*).

Medical wax models originated in post-medieval Italy. Italian sculptors recognized the versatility of wax in the fifteenth century, and in the seventeenth century the first anatomical models for use in medical teaching appeared. Their great advantage over dissected human bodies was that they could be reused over a long period. From the early nineteenth century the use of wax models spread across Europe and became ever more specialized, demonstrating specific medical conditions (pathology) rather than only general anatomy. Until the early twentieth century, there was a lucrative and highly skilled cottage industry producing wax models for medical and educational purposes. Many hospitals employed their own wax modellers directly, such as the Vienna-based father-and-son team Carl and Theodor Henning (active between 1893 and 1946) and Joseph Towne (1806–1879), the official wax modeller for Guy's Hospital in London for more than fifty years.

The example illustrated here shows just how important this technology was before accurate and cheap reproductions could be achieved in print, because it played a role in one of the most successful human-led extinctions of all time, the only recorded eradication of an infectious disease transmitted by humans: smallpox (*Variola vera*).

Wax model of cow's udder infected with cowpox, *c.* 1800.

In Gloucestershire, western England, in the 1790s, the local physician Edward Jenner (1749–1823) discovered that dairymaids, who had contracted cowpox through hand-milking cows carrying the virus, appeared to be immune to the deadly effects of the much more aggressive human smallpox virus. Consequently, Jenner managed to isolate cowpox and infect patients with a low dose of it – vaccinate, from *vacca*, the Latin for milch cow – before exposing them to smallpox, an almost certain death sentence up to that point. Astonishingly, the individuals did not develop smallpox, and the rash from their cowpox infection disappeared relatively swiftly.

This wax model of an udder was clearly intended to help doctors to identify cowpox cases from which to source vaccines. Such models were in wide use, and near-identical copies can be found in medical museums in Padua, Milan, Pavia and Bologna. They appear to have been derived from illustrations in a treatise of 1809 by Luigi Sacco (1769–1836), who spearheaded the smallpox vaccination campaign in northern Italy. Jenner, Sacco and their counterparts elsewhere saved many lives with these small-scale local campaigns. However, the successful eradication of the disease did not occur until the second half of the twentieth century. It took a major concerted and nearly global immunization effort to reach the required 86 per cent

'L'Origine de la Vaccine', engraving of a physician inspecting a dairymaid's cowpoxed hand, while a farmer or surgeon offers to a dandy inoculation with cowpox that he has taken from a cow, Paris, *c.* 1800.

vaccination rate to pass the herd-immunity threshold that would ensure the virus's elimination. The World Health Organization officially declared smallpox extinct in 1980.

Thus, we might say that the wax model of the cow's udder has been subject to a double extinction. Like smallpox, wax models themselves became redundant with the advent of affordable colour photography and developments in printing technology for textbooks, in the first half of the twentieth century, although the onset of digital technology has encouraged a revival in model-making. While making carefully handcrafted and painted models of patients' body parts in wax may be a thing of the past, producing three-dimensional models for medical teaching and research has re-emerged with ongoing advances in three-dimensional imaging and printing technology. This includes 3D printing of specific patients' hearts, allowing surgeons to study them in detail before performing surgery on the real organ. There is no longer any need to 3D-print a pox-infected cow's udder, however, since – even had the variola virus not been eradicated – doctors nowadays would use synthetic sera rather than harvesting vaccine from the udders of cows. Yet even if epidemiologists in the twenty-first century have little use for a two-hundred-year-old model of a cow's udder, the principle discovered by Jenner and popularized by him, Sacco and others, with the help of this object, remains the foundation of the fight against infectious diseases to this day.

To: 1.       From:
  2.       Date:
  3.       B.F. to    on
  4.       P.A.

| FILE REF : | SUBJECT. |
|---|---|
| | |

Wt. P1954/29 1000 pads 2/53 K.H./6797/8 Gp. 860/2

# Memo
Adrian Forty

Anyone who worked in an office before 1990 will remember the memo. A form printed on a standard paper size, with headings 'To', 'From', 'Date' and 'Subject', and sometimes more, it was, until the arrival of email, the most common form of written communication within businesses and organizations of all kinds. The example shown here, from the British Civil Service in the 1950s, reflects the hierarchical nature of its bureaucracy, and the importance attached to record-keeping. From a dictated or handwritten draft, it was typed by a typist, addressees were named in a numbered list and the memo, attached to the file of documents to which it related, was circulated, each recipient in turn initialling it once they had read it, and sometimes adding a handwritten comment, before circulating it to the next. Once all recipients had seen it, it would return to the records office, until it was, if required, 'brought forward' ('B/F') to a meeting on the date indicated. Finally, it would return to the filing department, whence it might be retrieved at any time in the future. Other organizations used different styles of memo, and so, for example, in flatter organizations, rather than a rank order of named recipients, the use of a carbon copy (c.c.) allowed the memo's simultaneous delivery to several different people. This is the format that was adopted by email, with its 'c.c.' now a skeuomorphic anachronism, while 'b.c.c.', 'blind carbon copy', was an addition to the arsenal of communication unknown before the digital revolution.

Although the memo might at first sight look to be a long-existing mode of communication, it goes back no further than the last years of the nineteenth century, when it appeared as one of a range of innovations in business and government practice, features of a new movement that has become known as 'systematic management'. Previously, a memorandum, 'a thing to be remembered', was addressed solely to oneself. Only at the turn of the century did it become attached to a newly developed mode of internal office communication with its own format and literary style. Fundamental to the memo was its distinction from the letter. Dispensing with the customary forms of address and courtesies of the letter, economical in language and restricted in content to a single subject, it was a novelty. And, whereas letters were primarily for communication outside the organization, the memo was exclusively for communication within it. The impetus for the innovation came from the growth in the scale of business and government, and the difficulties in keeping track of the corresponding increase in the flow of

Printed memo form,
British Civil Service, 1950s.

information, often across large distances. At the same time, organizations of all kinds were increasingly concerned that knowledge of their operations, which hitherto resided largely in individuals, should be transferred to a corporate memory, accessible to others and to posterity – a change akin to scientific management's aim of wresting knowledge of manual labour practices away from tradesmen and into management (see Cyclegraph).

The memo supplanted various features of the letter, in particular its length, and its encumbrance with salutations and courtesies: the 'I am in receipt of your esteemed communication of the 14th instant to hand and beg to advise you . . .' style that was normal in nineteenth-century business. While this might continue to have been expected for external letters, and omitted at the risk of offending the recipient, for internal communications all such formalities could be dispensed with. Advocates of the memo were aware that there might be some resistance to this. In the Du Pont Company, the Efficiency Division commented in 1913 on the new style that they wished to introduce:

> It may be difficult . . . to overcome the inertia of our long-standing habit of clinging to traditional forms and usage, but after the right attitude is attained there should be little difficulty, and a constantly diminishing tendency on the part of the recipient of a letter to 'get sore' at the terseness or bluntness of the communication.[1]

The invention of the memo was contingent on two other more or less contemporary innovations in corporate and institutional practice: filing; and changes in document reproduction following the adoption of the typewriter. Hitherto, record-keeping in offices had been separated into two procedures. Copies of outgoing letters were kept in bound, chronologically arranged letter books. In larger organizations, letter books might be allocated to particular correspondents, but in smaller ones, all copies were bound together in a single volume, ordered by date. Incoming letters were generally kept by the recipient, either in bundles or in boxes. With this system, it was difficult to find a copy of a particular letter unless one knew the date it had been sent, and even more difficult to match it with related incoming correspondence. Systematic management aimed to combine incoming and outgoing letters, in either a box file or a vertical filing system, arranged by subject or correspondent (see Vertical Filing Cabinet). The success of the system relied on each communication being, so far as was possible, confined to a single topic; and while this might prove insuperable with external correspondence, it was possible with internal correspondence. For the memo to attain its full potential as a document that could be filed, it had to be confined to one topic – hence the logic of the printed 'subject' heading.

For most of the nineteenth century, letters were copied by pressing a sheet of dampened tissue paper down on the face of the manuscript original, which would have been written with an ink containing an aniline dye, enough of which transferred to the tissue paper to leave a legible facsimile. The standard letter copy books were leaves of tissue paper, and copies were made directly into the books, the adjacent pages protected by oilcloth sheets to prevent moisture from the dampened page from seeping through to other pages. This laborious process was revolutionized by the typewriter, and the use of carbon paper, which made possible the reproduction of up to ten legible copies on onion-skin paper. For the first time it was possible to produce multiple copies of a document at the same time as the creation of the original, on individual sheets that could be circulated and filed independently (although, in practice, bound letter books continued for much longer, as aniline dyes in the typewriter ribbon ink enabled impressions from the original to be taken on the dampened pages). The two innovations together – filing and typewritten letters with carbon copies – were necessary preconditions for the memo, although not its cause, which lay in a desire to change the style of internal communications.

With the memo came a new literary style. Recent experience with email, SMS and social media has accustomed us to new communications technology bringing about changes in language and forms of address. There is nothing new about this; the telegraph and the telegram had had similar effects even before the invention of the memo. But the memo was different, because the result was deliberate, whereas with these other media it was not intended as an outcome. Part of the purpose of the memo was to reform the language of business and bureaucracy, simplifying and abbreviating it. The dimensions of the memo itself provided a constraint on verbosity. Along with other developments, from the decline of classical learning to the rise of specialized scientific languages, the memo helped to bring to a close the 'rhetorical empire' that had hitherto dominated language. While language retains its traditional function of persuasion, this is no longer effected, as was once the case, so much by rhetoric as by information. The linguistic style of the memo was symptomatic of a new literary form, sometimes described as the 'information genre', which has become the norm of our times. Neither rhetorical nor scientific, it is the language of forms, instruction manuals, reports, documents and much of the Internet. The memo may have gone, but its legacy survives, not just in the format of the email, but also in the pervasiveness of 'information' as the pre-eminent mode of language.

# Milk Spoon
## Hugo Palmarola

A small white plastic spoon is today the last specimen of thousands, handed out as part of a programme to reduce malnutrition and infant mortality in Chile. In the early 1970s these spoons – used for measuring powdered milk – became a powerful tool in the free distribution of milk to children throughout the country, one of the outstanding episodes of President Salvador Allende's socialist project. Carrying significant political, social and economic implications, the spoons were designed by the Industrial Design Area of the Institute of Technological Research of Chile (INTEC), under the influence of the Hochschule für Gestaltung in Ulm (HfG Ulm).

Among the first forty measures of Allende's government (1970–73) – the first democratically elected Marxist government in the world – was the provision of a ration of 'half a litre of milk per day' for all children under sixteen in the country. Directed especially at the underprivileged, the measure was emblematic of the new administration and was widely publicized. Although the Chilean state had been distributing free milk since 1924, under Allende's 'Popular Unity' government the provision was greatly expanded. In the first year of the new government, 48 million kg (106 million lb) of dried milk were distributed, four times the annual quantity under the previous administration; since national production was not sufficient, additional supplies were imported.[1]

Inadequate nutrition in Chile resulted in one of the highest infant mortality rates in the world, and in 1970 around 20 per cent of children suffered from malnutrition. In this context, the milk distribution project was a matter of life or death. A well-nourished New Child became the first necessity of the Popular Unity project, to make way for the Christian and Marxist messianic concept of the New Man.

The National Complementary Food Programme was set up to promote normal growth in children, prevent malnutrition and raise standards of nutrition and health in the population, especially through the distribution of powdered milk, which provided proteins, vitamins A and B2, calcium and phosphorus. The basic monthly allowance was to be 3 kg (just over 6½ lb) of powdered milk per month for children under six months old, 2 kg (just under 4½ lb) for children from six months to two years old, 1.5 kg (3⅓ lb) for children aged two to six, 1 kg (2¼ lb) for those aged six to fifteen, and 2 kg for expectant mothers. Preparing the milk required mixing a specified weight of powder with a defined volume of boiled water, and initially the

Spoon for the dosage of powdered milk, 5 g.

programme supplied a table giving the quantities for the different age groups, with the powdered milk measured in level medium tablespoons.[2] Soon after the project was launched, the Ministry of Health discovered that, of the 3.6 million beneficiaries of the National Powdered Milk Plan, approximately 12 per cent received an inadequate dose.[3] Because people were using spoons of different shapes and sizes, the mixture was often either too dilute or too concentrated.

The programme was not without controversy. There was a popular belief that milk caused diarrhoea and should therefore be avoided, although in fact part of the problem was that the stomachs of malnourished children were not accustomed to it. To overcome this, the government supplied powdered skimmed milk with 1.9 per cent fat, below the international standard of 3.1 per cent, but since it was skimmed, it tasted unpleasant, which caused some resistance to its consumption. Reports surfaced that the quality of the milk was so bad that it was being used for marking football pitches. Such stories fuelled the right-wing opposition's criticism of the socialist project. The National Complementary Food Programme also placed special emphasis on the hygiene of milk preparation, since poor cleanliness was recognized to be another cause of diarrhoea in children. The improvement of the conditions under which powdered milk was prepared was one of the main concerns of nutrition specialists in Chile, even before the scandal that later broke on the international stage over the sale of formula milk to some developing countries, where poor standards of hygiene placed babies at increased risk of infection and mortality while depriving them of the natural antibodies from breast milk.

In addition to these problems, the success of the programme was threatened in part by the erratic measurement of the ingredients, and the Ministry of Health commissioned the newly created INTEC to find a solution. Work began on the design of a plastic spoon to give a precise measure of powdered milk. Led by the German designer Gui Bonsiepe of HfG Ulm, INTEC, which included foreign graduates from HfG Ulm and a group of Chilean Industrial Design students from the University of Chile, designed some twenty spoons, giving exact measures of 5 and 20 g (0.17 and 0.7 oz) of powdered milk. Some of these prototypes showed an almost obsessive level of precision in the complexity of their mechanical levelling systems. The final outcome was the mass production of two small one-piece white plastic spoons, with an inscription in high relief reading 'Medida rasa 5 g SNS' (Levelled measurement: 5 g; SNS being the National Health System, Sistema Nacional de Salud) on one, and 'Medida rasa 20 g SNS' (Levelled measurement: 20 g) on the other. These two sizes allowed a faster and more accurate dosage for the preparation of small or large amounts of milk, according to the age of the child. The consumption was estimated at 1 litre (just over 2 pints) a day for children under six months, 0.67 litres (just under 1½ pints) for children from six months to two years, 0.5 litres (a little over a pint) for children from two to fifteen years, and 0.67 litres for pregnant women. As a project, it was indicative of the expectations placed on design to bring about the realization of the utopia Chile aspired to be in the early 1970s.

Under the Allende government, INTEC's Industrial Design Area developed about 22 industrial design research projects, with the objective of intervening in the economy, either by import substitution or by rationalizing production in state or mixed-economy industries. Agricultural machinery, metalworking, electrical appliances, school furniture, housing and food were all industries within which schemes for product standardization were developed. Economic instability and political uncertainty, however, meant that almost none of the products was ever manufactured. The milk spoon was one of the few to reach production, but later under the dictatorship of Augusto Pinochet, data about the supply of powdered milk in the Allende period was falsified, and the spoon, along with the rest of the programme, fell into oblivion. Currently, only one 5 g spoon remains in existence, along with renderings of some of the alternative designs.

# MiniDisc
## Priya Khanchandani

Anyone who bought into the promise of the MiniDisc, an audio format launched by Sony in 1992, will remember that it was meant to transform the way we listen to music. Fashionable for its sleek form at a time when the chunky cassette was still in use, and practical because, unlike the CD, it could record as well as play music, it appeared to solve all our music woes in one miraculous, compact package.

Soon after the invention of the CD by Sony and Philips in the early 1980s, Sony had begun to research ideas for a recording and playback device that used a smaller disc. Sales of audio cassettes, with their tangle-prone tape, were already in decline, and the company wanted an alternative to the CD that was as sharp to listen to but which could also record and be rewritten.

In 1991 it was born. Inside a square plastic shell was a 64 mm (2½ in.) disc that could record up to 74 minutes, a quarter of the size of the CD but with the same capacity and high-quality sound. Unlike the CD, it didn't jump when you used it on a portable device, even while playing sport. The new format had the potential not only to replace the cassette, but to seriously rival the CD. Engineers at Sony worked around the clock to prepare the MiniDisc for market in time for the launch of a rival product by Philips that was being developed at around the same time.

In 1992 the MiniDisc was ready for mass production. Its launch came with a fanfare. Adverts of the time show how much it promised: the protagonist of one television advert that aired in Britain in 1997 manages to make the Sun go down while his MiniDisc is playing fast-paced techno, suggesting that the thing could give its user special powers. The phrases 'freedom of religion', 'freedom of speech' and 'freedom of the press' introduce the MiniDisc in another television advert of the time, followed by 'freedom of music'. A rapid montage of images shows people listening to their portable MiniDisc players while running or otherwise on the move.

But such narratives, however aspirational, did not do enough to tempt consumers, who bought fewer than 50,000 MiniDiscs in its launch year. Although the MiniDisc was a useful product, its £500 price tag did not justify its improvements on the CD, which was significantly cheaper. Moreover, it was targeted mainly at teenagers, which was surprising, given that they could not realistically be expected to be able to afford it. For this reason most record labels stuck with CDs and never transferred to the new format, in spite of the hype that resulted from Sony's extensive marketing campaign.

MDR-74 Recordable MiniDisc, 1999.

MiniDisc sales grew to 1.2 million units by 1995, and then to 2.9 million in 1996. Encouraged by the product's relative success in Japan, which is reported to have accounted for 60 per cent of the global MiniDisc market in 1997, Sony launched a set of new MiniDisc players in 1998. These included lower-cost models, the cheapest of which was in the region of £154.

Such less costly versions of the MiniDisc player had potential in the longer term, but stood little chance against another new format that would soon become practically synonymous with music: the MP3. In 2001 Apple introduced the iPod, a portable device that you could hold in your hand, which could play hours of music and did away with the need for a tangible disc altogether. Although it was not the first MP3 player, the iPod was a leap forward in terms of consumer electronics design: smaller and much easier to use than its competitors, with a scroll wheel that allowed you to zoom through your own playlists. As the MP3 gained ground, the MiniDisc, however innovative it had been initially, struggled. In 2013 its fate was sealed when Sony stopped making it altogether.

Unlike the 33 rpm vinyl LP, the 45 rpm single and the cassette tape – other music media that had been victims of product innovation – the MiniDisc was superseded when it had barely established itself, and it never came to define an era of music. It entered the market during a period of fast-paced digital innovation, when products went out of date long before they had stopped working, as continues to be the case today. But, unlike today, the MiniDisc came about when the rapid turnover of household technology was not yet quite taken for granted. Although it would not be the first music medium to have a short life, in the late 1990s consumers who bought into it did so under the assumption that it was there to stay. Being left not only with defunct MiniDiscs, but with defunct MiniDisc players that had been expensive to acquire, was not something they would have factored into their purchase to the degree that today's consumer might.

Today consumers are only too aware of how the acquisition of consumer technology is stimulated by the constant availability of something newer and more innovative than what we already have, leading to today's situation, where millions of phones, music appliances and computers are regularly replaced by ever better models, even though the original products still work. But that was not the case to the same degree for a generation that had used the cassette and LP for decades, collecting shelf upon shelf of them.

Retrospectively, the MiniDisc's limited duration should not be put down to technical shortcomings. The fact that it continued to be manufactured well into the early 2000s shows that it persisted, to some degree at least, even after the MP3 had become the industry norm. Although Sony has not been willing to disclose specifically why it was not discontinued earlier (possibly reflecting its embarrassment that the MiniDisc did not meet expectations), it is likely that it continued to invest in the product, hoping it would gain a foothold even as the popularity of the CD waned.

But the longevity of the MiniDisc is also a reflection of the fact that it continued to have a niche following among those who persisted in valuing a physical album with sound quality that was superior to that of the cassette tape, equivalent to the CD but with recording capacity, in the form of a small, well-protected shell. It stood out as a technological improvement when held up to disc formats that preceded it; and, rather than being down to lack of innovation, its premature end is explained by the fact that stakes were high and changes had begun to come about so swiftly that it was practically impossible for innovators to keep up.

In the end the MiniDisc could not compete with a device such as the iPod that liberated us from the need to acquire discs altogether, saving us both money and storage space, and enabling us to be less dependent on location. Although this may have been the best outcome for the consumer, for musicians the MP3 has meant declining album sales and the beginning of free music downloads from streaming services, which have caused controversy among artists such as Taylor Swift, who in a *Wall Street Journal* article in 2014 famously argued that 'valuable things should be paid for.' Ultimately, the MP3 was so successful that it triumphed over the MiniDisc and the CD. For the time being, it might be first in the music technology race, but now that we know no digital medium is here to stay, only time will tell what will overtake it, and how soon.

# Minitel
## Shahed Saleem

In the 1970s, the French telephone network was in disarray. In fact, with waiting times of three years for the installation of a new line, it was described as the worst telephone system in the industrialized world. Something had to be done, and the French president Valéry Giscard d'Estaing, who was on a mission to modernize the country through new infrastructure projects, commissioned a report into the state of the nation's telecommunications industry.

The resulting document, *L'information de la société* (The Computerisation of Society) by Simon Nora, a French cabinet member, and Alain Minc, a businessman and political advisor, was released in 1978 and became a best-seller. It observed presciently that in an emerging post-industrial society, access to information was a source of power, and that the growth of information networks would lead to a more decentralized society, destabilizing traditional ideologies. Failure to respond would mean that France would lose the ability to control her fate, so they proposed to harness and promote the increasing interconnection between computers and telecommunications by layering interactive services over the telephone network, accessed through a terminal in every house.

The result was Minitel – a pioneering design experiment in the integration of keyboard and screen, as well as in the emergent field of interactive user-led digital design. Early models had black-and-white displays with simple graphics, and featured keyboards that could fold over or beneath the screen (precursors to the later revolutionary designs of early Apple desktop computers). Since the Minitel was intended to be a ubiquitous part of the household, its small size and convenience of use were critical. Later models featured more curvaceous styling, in keeping with contemporary trends, while retaining the principle of the small screen and keyboard in one unit. By the time of its demise, it also included the facility to insert payment cards.

Minitel was first introduced as a free electronic replacement of the telephone directory. Over the course of its thirty-year lifespan, however, it would spread to include terminals in 7 million households across France, hosting thousands of providers offering a variety of services from buying train tickets and checking exam results, to playing games, adult chat and much more. Minitel was, in effect, a French Internet before the Internet as we know it today. But how did it happen and where did it go?

Minitel's ambitious expansion must be situated in the context of the international rivalry in the 1970s to develop national telecom and data

Alcatel Minitel communication terminal, France, 1983.

services. The UK had the videotext services Prestel and Teletext, and France, always sensitive to anything smacking of Americanization, feared that U.S. dominance through IBM and its computer networks would also infiltrate French culture. The French idea, therefore, of placing a data terminal in every house was to be a revolutionary strategy that would give France the upper hand in the telecommunications race.

The initial challenge was to persuade members of the public to install terminals in their homes. As an incentive to encourage take-up, the French telecommunications service had the brilliant idea of replacing printed tele-phone directories with digital versions that could be accessed only through the Minitel system. The first such electronic phone book was introduced in late 1982 in the region of Île-de-France, and was rolled out incrementally across the country between 1983 and 1987. Minitel terminals were handed out free at post offices. By plugging the terminal into a home phone line, users could dial in to the post-office directory straight away and access its data. For many users this was their only direct experience with interactive computing, and they were billed monthly for the amount of time they spent online.[1]

But the telecommunications service needed revenue to pay for the digitization of the network, and Minitel needed to pay its way by increasing use of that network. The electronic phone book had created a customer base for new, privately run videotext services. To capitalize on this, licences were granted to providers who offered services over the network. Each provider's site had a unique four-digit number that was dialled on the terminal to access it. Apart from buying train tickets, booking holidays or using the directory, Minitel also enabled the first online food delivery services, and even facilitated the organization of a student demonstration in 1986. The provider received two-thirds of the income generated through billing for time spent on its site, with the national telecommunications service taking the remaining third.

At its peak in the 1990s the Minitel service had 25 million users, the same level of use as the global Internet at the time. The French rail service, the SNCF, reportedly earned $20 million a year at this time through Minitel. While it is unclear if France Telecom made money from the Minitel system, there certainly were entrepreneurs who made fortunes. The adult chat service and industry, for example, boomed through Minitel and became a major revenue stream for France Telecom.

However, with the rise of the World Wide Web during the 1990s, Minitel's days were numbered. In 1993 the Geneva-based organization CERN, where the World Wide Web was developed, released its software into the public domain and gave it an open licence, thus ensuring its dissemination. Unlike the open platform that the web offered, Minitel was a closed system, where only registered providers could offer services. And although the design of the terminals was adapted over its lifespan, the technology eventually stopped evolving. This, along with a failure to export the system to other countries, meant that Minitel could not compete. By 2011 there were only 810,000 terminals in use, and in June 2012 France Telecom made the decision to switch Minitel off for good.[2]

Throughout the 1980s and in the early 1990s Minitel was probably the most innovative and advanced mass telecommunications system in the world, in terms of both digital and physical design technology. It computerized French society and gave it technological independence. It introduced many of the services and systems that would come to be found on the Internet: banking, weather reports, stock prices and pornography, to name but a few. Through Minitel, new types of service were invented, industries were connected and people explored new kinds of communication. It promised a digitally interconnected future and showed, at the scale of a nation, the potential of the World Wide Web at global level.

Photo/ELLISON
-Austin-
Tex

# Moon Towers
Bryony Quinn

In the late nineteenth century the advent of electric street lighting brought the phenomena of day to night all over the world. The gas lamps and their gloomy fossil brothers had never been brilliant enough; the quality of artificial light was cleaner and clearer, and could cover a greater area. Light has always been synonymous with revelation and knowledge, but in the late years of one of the most technologically advanced centuries the world had seen, electric light now meant productivity and industry 24 hours a day.

In the beginning, there was competition to illuminate the urban world electrically, but by the close of the nineteenth century the incandescent bulb had emerged as the clear victor. For the next one hundred years or so we lived under the stabilized and standardized glow of the modern light bulb and forgot about the obsolete but ingenious solutions to the dark that had been proposed and built, and which lit whole cities – including artificial moons.

The first 'moon tower' or 'moonlight tower' was erected in the 1870s and featured one of the earliest and most popular electric lighting solutions of the time: arc lamps. Invented by Humphry Davy in 1801 and developed for mass manufacture and installation as street lighting by Charles F. Brush from 1877 onwards, the arc lamp was literally one of the most brilliant inventions of the era, with models that could reach an unbeatable glow of 6,000 'candlepower'. Their brightness aside, arc lamps were cheaper to run and insure than gas. It was these savings for the municipalities that adopted them that secured their initial success in the field of artificial lighting, and led to their indiscriminate use in commercial centres, as well as public and suburban spaces. For the private businesses that installed them – P. T. Barnum's circus, for example – it was the novelty of such bright lights that attracted greater patronage to spaces and activities that were previously too shady.

However, the arc lamp was a frightening level of illumination for some humans (and non-humans) compared with gaslights, which produced a glow equivalent to sixteen candles. Such brightness was unfit for domestic spaces and barely tolerated in larger, public indoor environments such as theatres. Arc lamps, the manufacturers proposed, were best enjoyed outdoors, at a distance, with between four and eight lamps mounted at the top of skeleton towers that could reach upwards of 75 m (250 ft). From ground level, this height diffused the intensity of the light, with the additional benefit of casting a wider glow.

Workers repairing or replacing bulbs on the moonlight tower at 9th and Guadalupe streets, Austin, Texas, 1930s.

On the point of their terrifying brilliance, arguments against the installation of arc lamps in urban areas included the disruption of sleep, the unflattering colour the lights gave one's appearance and a phenomenological anxiety rooted in the inability to tell night from day. The novelty of reading a newspaper or watch face in the early hours of the morning – in the middle of the street – was wasted on the lamps' detractors, particularly those with an interest in the continuation of gas as the prime source of light in the city. As was the romance; for the champions of arc lamps, the notion of personal moons for every new town across America held strong poetic appeal.

For moon towers to exist at all, various technology, independently developed, had to converge. The first step was documented in the arc-lamp prototypes by Davy, who, over the years, had tested various materials against the improving capacity of the new electric batteries (after Alessandro Volta made his seminal announcement in 1800). After Davy first witnessed a spark that flared between two carbon electrodes connected to an early power unit, his experiment lay dormant for half a century because the light could not be supported with any regularity, or without great cost. Fifty years later a continuous arc of light was successfully powered and regularized by the new, more reliable – and cheaper to maintain – steam-driven electric generator. Picking up on this development a few years later still, Brush, who in 1877 invented a single light generator using an arc lamp, successfully scaled his operation into a series of towers with several lamps and switches.

Almost immediately, the Brush System competed with a host of other companies and their variation of arc lamps mounted at a great height (many of these companies would later merge). Soon, moon towers as a species stood sentry-like over many major cities in the United States, including Austin, San Francisco, Denver and New York. In Minneapolis an 'electric moon' was reported. In Los Angeles the towers numbered 36. But while these cities supplemented their arc lamps with other lighting competition, the city of Detroit is notable for implementing moon towers exclusively over any other solution at that time.

It is in the construction and dismantling of the Detroit moon towers that the success and ultimate failure of the design can be most clearly marked. In total, 122 were raised across the city, set at 305–365 m (1,000–1,200 ft) intervals in the centre (where they stood 53 m/175 ft tall), and with 762 m (2,500 ft) between those in the residential streets (at 46 m/150 ft high). The city rationalized that 100 towers would be cheaper to service than 1,000 lamps, and that they would look spectacular.

Despite the scale of the deployment, this firmament of artificial lights suffered from a fundamental flaw. Moon towers were effective in open or low-built areas, where the angle of light from each tower would not throw obstructing shadows as it encountered the buildings below. However, at this time Detroit was known as the 'City of Trees' (a styling that disappeared as many trees were felled in the twentieth century), which meant that, during spring and summer, and particularly in residential areas, the light was obscured by the tree canopy. When foliage was not a problem, mist was. Meanwhile, the high-rise architecture that was beginning to alter u.s. cityscapes in the late nineteenth century had reached Detroit's business quarter; such modern proportions blocked the light completely.

Within ten years of its construction, the Detroit system was dismantled. What was dazzling above the rooftops, cloud cover and tree canopy, and in the lofty imagination of city planners, was frequently dark and patchy on the ground – and an assault on tax dollars. What's more, the incandescent light bulb had developed far enough by this time to be easily, affordably and inoffensively mounted at street level. Elsewhere in the United States, any moon tower still standing at the start of the twentieth century remained in place as a curiosity.

This does not mean that the moon towers have left no residue on the field of artificial street lighting. In the 1990s, for example, Russia launched satellites as part of its Znamya experiment with the purpose of deflecting sunlight in targeted beams 'equivalent to that of several full moons'. With great fanfare in 2018, the Chinese city of Chengdu announced a plan to launch an artificial moon in 2020 that can illuminate an area between 10 and 80 km (6 and 50 mi.) wide, 'designed to complement the moon at night' only eight times brighter. While this plan has not yet been realized, it represents the legacy of the moon towers, which apparently did not end with the failure of the arc lamp as technology, or with the obvious flaws of a highly elevated street-lighting system.

It is easy, at this point, to lyricize the astonishing effect of these inventions or laugh with incredulity at all the men who have claimed to harness or compete or improve upon a cosmic power such as moonlight. Arguably, a more appropriate and sobering inheritance left by the moon tower should be the floodlight and, with it, an urban environment that is more glaring than glowing, more invasive than inviting. With the end of night-time as nature had known it before, the arc lamp introduced ceaseless capitalism and the potential for continuous surveillance. And yet, the legacy of the moon tower is perpetuated by the romance – a technological fantasy – of living life in the glow of a personal 'lunar' light source.

# Nikini
## Rachel Siobhán Tyler

When I was a child, the signs in public toilets fodbidding the act of throwing 'sanitary napkins', 'sanitary waste' or 'feminine hygiene products' down the toilet were a mystery to me. My mum elucidated: she explained that they referred to tampons and sanitary towels, which should be disposed of in a 'sanitary bin' next to the toilet, rather than flushed.

In fact, much of the way menstrual-related products were referred to confused me. I was bemused by the television adverts of the 1990s pouring blue water – a prudish, unrealistic stand-in for blood – over various period products. I was baffled as to why roller skating and football appeared so frequently.[1] Possibly my failure to grasp the point of these adverts was in part because I was not of the generation who remembered the sanitary belt, the main period product in use before modern sanitary pads. Wearing this contraption would have indeed made it difficult to undertake sports and other day-to-day banalities comfortably.

The history of period products is even less well documented than the actual experience of menstruation, with its accompanying pain and diverse bodily effects. Even between generations of women, the former tends not to be discussed conversationally. Sometimes unwittingly, but often deliberately, it is a history that has been kept hidden. When I eventually did learn about sanitary belts, it was through fleeting references in literature, such as Judy Blume's novel from 1970, *Are You There God? It's Me, Margaret.*

Nikini was a 'disposable sanitary system' manufactured in England by Robinson & Sons of Chesterfield. Robinson had been manufacturing sanitary towels since 1880, alongside an extensive range of medical dressings. In 1895 it was the first factory in the UK to introduce cellulose wadding, which was quickly adapted for sanitary towels. In 1957 the inventor Valerie Hunter Gordon lodged the first patent for 'waterproof sanitary garments', and the Nikini went into production just two years later. (Gordon, incidentally, was also the creator of another revolutionary but now extinct product, the PADDI, the first disposable nappy system for babies.) In 1963 Robinson reported that its new product was 'already in a short time firmly established and in great demand'.[2] Designed to absorb menstrual blood, it comprised a pair of waterproof knickers and specialized absorbent pads. The reusable knickers were marketed as the 'Nikini' and sold alongside disposable 'Nikini pads'. The Nikini was made from PVC fabric, with wide elastic straps at the sides to keep it tight and snug against the hip, in an attempt to prevent

Nikini menstrual product, PVC knickers with elastic straps, UK, c. 1970.

movement or leakage. The pads, made from cellulose, attached with press studs to two tabs at the front of the crotch.

The design marked a big step forward in the comfort and practicality of industrial period products, which previously had been dominated by the sanitary belt and looped towels. Looped towels were layers of cellulose bound into an open loop at each end, and secured to two lengths of tape, usually with some sort of pin. The tapes attached to a belt around the waist, at the centre front and back. The belt was made from either cotton corded-drawstring or, later, elastic. The traditional sanitary belt made wearing a bikini impossible, not to mention hip-hugging jeans, a miniskirt or, really, any of the modern garments belonging to the 1960s. In all these outfits, the sanitary belt would have been visible, extremely uncomfortable and most probably ineffective.

By contrast, Nikini knickers were small and low-cut. This design feature was reinforced in the name, which phonetically merged the British word 'knicker' with 'bikini'. The advertising stood out both for its directness and for its emphasis on the fact that the product was designed by a woman. Adverts referred directly to 'periods', which was very unusual in the 1950s. (In fact, the use of precise language in adverts is only slightly more common in the UK today; the word 'vagina' was not used in an advertisement campaign until 2010.) With its frank branding, the Nikini can be seen as an innovation that helped to pave the way for the feminist social revolutions of the 1960s and '70s. At a time when single women were routinely rejected from mortgage applications and were openly paid less than their male colleagues, accessible innovations that allowed women to manage their own bodies – such as the bicycle, trousers, the contraceptive pill and, I argue, Nikini – became vital tools in an ongoing battle for an equitable society. The Nikini's advertising was aimed at the 'modern girl', and showed images of women in leotards in various inverted yoga poses, riding bicycles and wearing fitted trousers. It rejected bodily shame and the very idea of the period as a 'curse'. One advert reads: 'Be safe with Nikini and your period won't be a curse anymore.'

Nikini packaging, UK, c. 1970.

Nikini became extinct when it was superseded by self-adhesive sanitary towels in the early 1970s. These eliminated the need for a purpose-made garment to hold the period products, which instead adhered to the users' own underwear. The rise of Tampax tampons dealt the decisive death blow. The compact nature of the new generation of self-adhesive menstrual products, and the reduced stigma attached to the use of tampons, meant that Nikini was no longer the latest word in modernity; instead, it quickly came to seem cumbersome and antiquated – just another form of the sanitary belt.

Nikinis, once so popular, have now been largely forgotten. Or, rather, they are relatively undocumented. When I contacted Robinson, it had no record of when the Nikini ceased production. Moreover, it is a product that, even though it was no doubt revolutionary in its way, is rarely spontaneously recalled. Unlike Bakelite wirelesses, old computers and toilets 'out back', stories of what people used before the dawn of press-on pads, tampons and menstrual cups are not used to regale grandchildren at the dinner table. If discussed, a bevy of euphemisms are invoked – the curse, on the rag, sanitary waste, time of the month, and so on. Furthermore, it is hard to talk about menstrual products without falling back on the narrow and idealized discourse of ultra-femininity created and repeated by advertisers in the twentieth century. Glenda Lewin Hufnagel points to the irony of these adverts, in which 'to be perceived as feminine women must hide the fact that they menstruate'.[3]

However, today, the belief that menstruation should be hidden away is being challenged far more actively. There were shock waves in the international press when the musician Kiran Gandhi was photographed 'free-bleeding' while running the London Marathon; reusable menstrual cups are popular again; more scholars are publishing on the history of menstruation products; and in 2019 the Science Museum in London began collecting menstruation products. Language is becoming more accurate, more inclusive and less gendered.

Importantly, there is growing public debate about what has been dubbed 'period poverty', an all-too-real situation that many people face, where they not only have limited or no access to hygienic, safe period products, but also are not allowed the right to manage their menstruation without shame and without it affecting other aspects of their life. In 2018 the Scottish government passed a bill that made all period products available free of charge to students in schools, colleges and universities, and in 2020 they approved legislation to make them freely available in a variety of designated public places. The liberated future first glimpsed in the Nikini – in which we are unembarrassed by and even embrace our periods – may yet come to pass.

# The 'No Nonsense' Fountain Pen
## Pippo Ciorra

The standard 'No Nonsense' fountain pen, built in plastic and stainless steel in a limited range of colours, was manufactured by Sheaffer of Fort Madison, Iowa, from 1969 until about 1991. Different versions – slightly modified shapes, special editions, ball pen or calligraphy nibs – were in production until the end of the 1990s, but the standard model was cartridge-filled, steel-detailed and 13.05 cm (just over 5 in.) long when capped. Although it has been out of production for a long time, until a few years ago it wasn't impossible to run into surviving exemplars in stationery shops, especially those close to architecture schools. Those specimens slowly disappeared, and now it must be fished for on eBay.

The popularity of the pen came from its appearance and performance. It was a cheap tool, selling when it first appeared on the market for less than $2. It had a modernist/deco look, derived from flat-top fountain pens from the 1920s, which perfectly fitted the early Postmodernist inclination of the time. It also functioned effectively and was comparable to much more expensive products. But the mythology tied to the No Nonsense comes mainly from its misuse. Designed for handwriting, the tool in fact revealed extreme flexibility and surprising versatility for unpredicted uses, such as scoring music or producing architectural sketches. In North America and in Italy especially, architects and architecture students fell in love with the smooth, sharp, agile line pouring out of the two available nib sizes, F(ine) and M(edium). Fountain pens do not generally suit thick drawing paper, because the ink spreads and it is difficult to control the line, but the No Nonsense performed perfectly, working exceedingly well with the materiality and colour of the yellow tracing paper ubiquitous in North American schools and offices. It also suited the post-Corbusian habit of travelling with a small notebook in your pocket. The colourful pen – sometimes two or three of them judiciously matched – would happily sit in the front pocket of the dandy architect's jacket in place of a handkerchief and allow the owner to use a single, satisfying tool for both writing and sketching.

It was only a short time before zeitgeist and *genius loci* came together to turn a fountain pen into an icon. Rome in the 1970s and '80s was the setting for an architectural community extremely (some would say pathologically) devoted to the immaculate beauty of drawing – *disegno* – versus the dangerous corruption of the building process.[1] According to Glenn Adamson and Jane Pavitt, 'for Italy the '70s were the bleakest decades since the war: a

'No Nonsense' fountain pens, Sheaffer, USA, 1969–90s.

time marked by intellectual crisis, domestic terrorism and severe economic recession.'[2] These conditions allowed the country's most influential architectural intellectual, Manfredo Tafuri, to find no salvation for architecture whatsoever. He stated, in fact, that there was no way of producing architecture without surrendering to the profit-making and power-increasing process of capitalism. He would theoretically favour some sophisticated form of 'abstention'. The outcomes of such a nihilistic atmosphere were paradoxical. In Venice (and partly in Milan) it generated a relentless process of research/design-producing projects and texts that became milestones of architectural theory, but much less in terms of buildings.[3] But in Rome, a city less inclined to intellectualize, Tafuri's nihilism manifested itself as a commitment to drawing as a process in and for itself. In contrast to their northern colleagues, the Romans did not draw to produce theory or political philosophy but to seduce the viewer through the power of their 'ideal' urban views; to turn politics into form and desire. Architettura Disegnata, as the trend was later named, became a legitimate presence on the architectural scene, consisting of a group of *architetti disegnatori* whose mentor was Paolo Portoghesi, curator of the first Architecture Biennale and 'Postmodernist' hero.[4] His journal *Eupalino*, an architecture quarterly that ran from 1983 to 1990, was printed in extra-large format expressly to allow the layout of beautiful large-scale drawings, often traced with the No Nonsense fountain pen on sophisticated rice paper. At least one gallery, AAM (Galleria Architettura e Arte Moderna), continually hosted exhibitions by the group mixed with artists' work, and architectural drawings became increasingly marketable over the 1980s and beyond.

The first time I encountered the No Nonsense pen I was still a student, on a studio visit to the office of my professor, Ludovico Quaroni. While talking to him at his desk I saw a large collection of pens lying on the table. There were the original eight available colours plus some special editions. I immediately purchased one (I think my first was white with black ink), and became addicted. Using it, I discovered that there was a kind of magic in the act of sketching with a No Nonsense on a piece of tracing paper. The fluidity of the movement and the powerful sharpness of the thick line gave the feeling of control and the ability to fix an idea quickly in a sketch that had a value of their own. I realized everybody in school – I was a student at La Sapienza in Rome – was using the same pen. Not just Quaroni's and Portoghesi's army of assistants, but everyone else too: architects, teachers, students. Later I noticed how some of the 'trendy' young teachers were extending the use of the pen beyond the sketch, employing it as a tool for final, pictorial, large-scale drawings, often choosing a dark reddish-brown sepia ink that recalled Renaissance drawings or *sinopias*.

The 'official' life of the standard No Nonsense lasted two decades, and it went out of production in 1991 (possibly owing to the marketing problem of a company offering a low-cost product that competed with its own more expensive items). The fact that its life coincided with the Postmodernist movement made sense even beyond the obvious issue of the parallel debut and rise of computer-aided design. The resulting elegy to the fictional architectural image, striving for artistic and political autonomy, is probably among the richest contributions Italy made to Postmodernist architectural culture, establishing a supremacy of the drawing as a reaction both to a long-term crisis of modernism and to the political 'impossibility' of an autonomous architectural ontology in a post-utopian society. Within this narrative, the No Nonsense fountain pen in clear view in the breast pocket was both a manifesto and a weapon, ready to be unsheathed at any moment.

# North Bucks Monorail City
## Gillian Darley

In 1960 Ernest Marples, the UK's Minister of Transport, commissioned the report *Traffic in Towns* from a team led by Professor Colin Buchanan, and when it appeared in 1963, the minister pointed out that, since cars had come to threaten the quality of urban life, 'planning of traffic and planning of land use must go together.' Ironically, given Marples's own personal investment in automotive transport (he was founding director of the road construction company Marples Ridgway), it fell to him to push the boundaries, telling members of the House of Commons that he was not ruling out 'new technological developments such as travelators and mono-rails'. He may have had in mind the monorail patents of the French Société Anonyme Française d'Étude de Gestion et d'Entreprises (SAFEGE), which was then exploring the possibilities of their invention as far afield as Japan and California, and had recently entered into discussions for a link between Heathrow airport and central London.

Meanwhile, a far more ambitious monorail project was already in gestation, beneath the radar of central government. In the summer of 1962 the ebullient chief architect-planner of Buckinghamshire County Council, Fred Pooley, asked his assistant Bill Berrett to consider 'the implications of building a new city for 250,000 people'. Trusting Berrett to provide interesting answers to that question, Pooley – whose own experience had been drawn from the replanning of Coventry – then went on holiday to Scotland.

Berrett was assigned a locked room on council premises in Aylesbury, where he worked, alone, for a couple of weeks. So sensitive was the project that he had to write everything by hand. On returning to work, the refreshed Pooley was so delighted with the result that he showed it to key figures in the council, who agreed to take it further. Berrett's pencil notes were typed up and his centre of operations moved over the road into the basement of the Buckinghamshire County Council headquarters. (The secrecy was to provide a firewall against the plans for a major expansion around Bletchley being drawn up by the outside consultant Bernard Engle.) One certainty from the regional plans was that a North Bucks New Town would emerge – somehow, somewhere. The race was on.

In late 1964 the doors to the subterranean office in Aylesbury were flung open to reveal a detailed plan for 'monorail city', reported in near-hyperbolic terms by the influential critic Ian Nairn in his architecture column in *The Observer*: 'Here for once is a city of the future which isn't just an abstract

Monorail for North Bucks New Town, drawing by Bill Berrett, 1964.

diagram or intellectual firework.'[1] Nairn told his readers that the county council was aiming to buy 9,300 ha (23,000 ac) between Bletchley and Wolverton. Subject to the views of the various district councils and a number of 'feasibility studies', all seemed set fair.

Pooleyville, as it became known after the chief architect – behind whose 'bottomless blue eyes . . . many surprising ideas are born', as Nairn put it – began to seem an entirely feasible dream. Berrett produced handsome renderings and copious logistical detail, in part informed and inspired by the monorail systems the pair sought out overseas, at airports and World Exhibitions such as the Lausanne Expo of 1964. But grandfather of them all was the Schwebebahn in Wuppertal, western Germany, an electric suspension railway that had run since 1901 (and still runs today), its carriages bustling along beneath the superstructure like mobile fruit bats on their branches.

Back in Buckinghamshire, the monorail was key to the project, underlining its legitimacy as a visionary kind of urbanism. It would free the new city from congestion, bringing to Middle England a glamorous version of urbanity in which motor traffic was put out of sight, consigned either to a lower level or to an outer region, a vision that still belonged very much in the realm of theoretical speculation among the international planning avant-garde. On plan, the North Bucks Monorail City resembled butterfly wings, with four lobes linking a series of 5,000-home 'pods' – all to be designed by different hands. A maximum seven-minute walk took residents to the monorail stop, from which they headed to the workplace, school, hospital, shops or leisure centre. That free journey into the new city centre would take no more than ten minutes. In Pooleyville, moving vehicles remained on the periphery, while the raised monorail was the curvilinear spine, the *raison d'être* of the plan, and thus of the city. The townships were 'strung like beads on a chain', interspersed with open countryside. This was the polar opposite to American grid-plan cities, let alone their suburbs, expanding on lines entirely dictated and dominated by the automobile. Already, though, from the early 1960s, both models had fallen under the abrasive critique of the American journalist and activist Jane Jacobs, whose acid condemnation of 'the fresh-minted decadence of the new unurban urbanization' had caused her and others to turn their attention to the regeneration of older city centres. Planners, she wrote, 'are at a loss to make cities and automobiles compatible with one another', and her hostility to all attempts to reconcile the two gained many followers.[2] While not the direct cause of Pooleyville's extinction, a growing climate of opinion against modernist urban experiments was not in its favour.

Only six months after its public inauguration, it was reported that the county council had abandoned the 'fully worked-out plan' to build their bold and imaginative monorail city. Farmers were unhappy with the compensation offered, and the government was reluctant to pay for public housing without

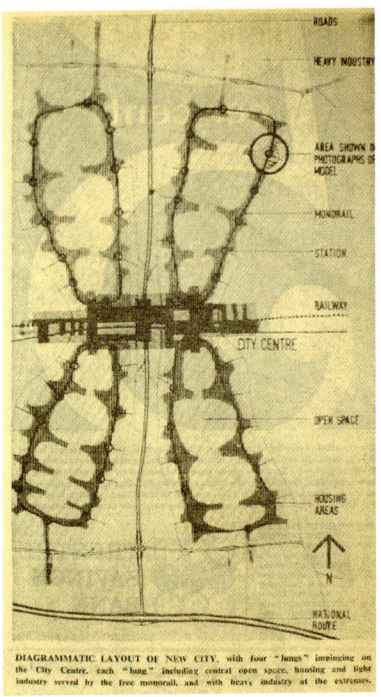

DIAGRAMMATIC LAYOUT OF NEW CITY, with four "lungs" impinging on the City Centre, each "lung" including central open space, housing and light industry served by the free monorail, and with heavy industry at the extremes.

Diagrammatic layout of North Bucks New Town, 1964.

being accorded a stake in the city centre. Central government, particularly the Minister of Housing and Local Government, Richard Crossman, and his 'terrier', the formidable civil servant Evelyn Sharp, had savaged Pooleyville. Now a smaller city would fall under their aegis, and within their budget, conforming to the terms and conditions of the New Towns Act and answering the current political imperatives. But Pooley and his team worked on, and lobbied hard, hoping that at least some of the innovative thinking would survive. Models had been built and feasibility studies completed. Pooley, not mincing his words or diluting his prejudices, argued that the challenge of the car and of the new leisure provided by a thirty-hour week pointed to 'a walloping great city centre that's a winner, not a nasty little neighbourhood effort like Harlow or Stevenage', the best-known of the government's 'first generation' New Towns of the early 1950s, in which the main emphasis had been on the small housing neighbourhoods grouped around essential local facilities, small shops and primary schools.[3]

Even as the North Bucks Monorail City faced extinction, elsewhere it was viewed as a great British achievement. The historian Guy Ortolano in his important new account of Milton Keynes describes it as 'a British project with global coordinates'.[4] The monorail city was not just another New Town but an integrated social vision, which 'promised to take fathers out of traffic, mothers out of the home, and children off the streets'. The international publicity that Pooley generated for what he proudly called a 'city for the '90s in the '70s' had ensured a new perception of mid-1960s Britain as a forward-facing society, its own 'Brasília' signalling ease with modernity and an appetite for innovation.

North Buckinghamshire's eventual New Town, Milton Keynes, was designated in 1967, a grid-planned, low-density paean to the automobile and the assertive industries that encircled it. But Pooley did not admit defeat until October 1968, and continued with admirable tenacity to brave government's lack of interest in a local authority-driven enterprise dependent on a relatively untested technology. In contrast, Berrett had accepted reality and by then was working for Milton Keynes Development Corporation as chief architect-planner of the infant New Town. Meanwhile, confined to 46 box files in Aylesbury, its allure and fame almost forgotten, sits the stillborn monorail city.

DEREINST·IN·UNGEHEMMTEM·SCHAFFEN·WERDE
NEU·MAGDEBURG·ZUR·SCHÖNSTEN·STADT·DER·ERDE

# *Notgeld*
## Tom Wilkinson

In the aftermath of the First World War, the inflation that had dogged Germany since the outbreak of hostilities became increasingly severe. At the same time, coins grew scarce – hoarded by citizens and withdrawn by the government – so, in order to give change, businesses resorted to printing tokens marked with small denominations. These were some of the earliest examples of inflationary *Notgeld*, or emergency money. Later, municipalities also began to print *Notgeld* in an attempt to keep their local economies going. At first the central government approved this strategy, and by the time it became clear that it was contributing to inflationary pressure, the production of *Notgeld* had got out of hand. According to one estimate, by 1923 there were around 4,000 points of issue.[1]

The result was a huge amount of money that looked startlingly unlike conventional banknotes. Whereas these had hitherto been encrusted with national symbols and classical ornament, *Notgeld* often featured local landmarks and scenery, as well as episodes from local history and folklore. Such vernacular iconography reflected the circumscription of this money, which could be used only in its place of issue. Other notes were altogether more macabre, with depictions of witches and devils reflecting the perception that occult forces had possessed the nation's fate; some bore more direct reference to the crisis in the form of piles of burning money, the dance around the golden calf, or graphs showing rising infant mortality.

As well as being geographically limited, *Notgeld* was usually also marked with an expiry date, partly with the intention of preventing hoarding – a phenomenon that, one might have thought, would have been vanishingly rare when the value of money plummeted daily, but which was nevertheless evidently sufficiently widespread to warrant official countermeasures. In contrast to money's usual function as a store of value, therefore, these notes were intentionally obsolescent, being orientated to a future in which they were no longer necessary. They were born to die.

In addition to being hoarded for its fleeting nominal value, *Notgeld* was also withdrawn from circulation for other reasons. It soon became the quarry of collectors, who met at regular fairs across the nation and established at least three *Notgeld* magazines. In the final years of the inflation, municipalities and other organizations responded to this growing market by issuing collectible notes. These were yet more striking than before, with bright colours and attention-grabbing imagery, sometimes of the most grotesque

*Notgeld*, one of a series designed by Kurt Tuch, Germany, 1921; note the centrepiece of 'New Magdeburg', a crystalline building evoking the city's chief architect Bruno Taut's Glass Pavilion.

character (antisemitic notes were not uncommon). They were frequently issued in series, some of which narrated simple stories across several notes. One particularly distinctive feature of such *Serienscheine*, serial notes, was that they often bore an expiry date that preceded their actual issue. Were these true money, or were they, as numismatists called them at the time, 'sham-' and 'swindle-money'?[2]

The dubious pecuniary status of collectible *Notgeld* notwithstanding, some examples are of an impressive aesthetic quality. Local artists of some stature were engaged to assist with their design. In Weimar, for instance, Herbert Bayer, who was then a student at the Bauhaus, designed an elegantly spare collection of *Notgeld*, while in Magdeburg, Kurt Tuch – a professor at the city's School of Applied Arts – produced a series of four colourful notes with a distinctly Expressionist flavour. The reverse of Tuch's notes shows the Holy Roman Emperor Otto I, who is buried in Magdeburg Cathedral, and their obverse sides variously celebrate the historic fabric of the city, an eighteenth-century doctor named Eisenbarth, and the famous 'Magdeburg hemispheres' experiment, overseen in 1656 by mayor Otto von Guericke, who thereby demonstrated the power of atmospheric pressure.

The fourth note, pictured here, departs from the antiquarian concerns of the preceding examples, while retaining their Expressionistic mode of representation. Continuity is thereby implied between the city's past and what the inscription informs us is a vision of the city's future: 'One day in uninhibited creativity/ New Magdeburg will be the world's most beautiful city.' The anticipated transformation is summarized beneath a sky filled with fantastic flying machines. On the left a cathedral rises to the edge of the note, its twin towers not dissimilar to those of Magdeburg's famous church. To the right, builders work on a new and equally tall structure with a steel frame; however, the pointed arches of the completed portion tell us that New Magdeburg, although constructed using modern methods, will harmonize with medieval tradition.

In the centre, a crystalline structure is set on a spikily crenellated podium, rays of light bouncing off the facets of its dome. The latter bears a striking resemblance to Bruno Taut's Glass Pavilion, a pioneering Expressionist structure erected at the Werkbund exhibition in Cologne in 1914. This is no coincidence: Taut was made city architect of Magdeburg in 1921, the same year this *Notgeld* was issued, and he immediately proposed a new city plan replete with colourful and crystalline buildings. He also oversaw the competition to design these notes.

All money attempts a temporal mediation, smoothing the uncertain transition between today and tomorrow by promising its bearers that its value (and by extension the value of their labour) may be redeemed at any time. Recently, the architectural historian Claire Zimmerman has likened architectural drawings to such promissory notes, since these images implicitly assure their recipients that they will one day receive a finished building resembling the designs they bear.[3] Magdeburg's *Notgeld* inserts itself into the field occupied by both money and architectural drawing, mediating between the past, present and future of the city at a moment of crisis. Money had lost its function as a guarantee of futurity, and architecture, while aspiring to create a new world, had seemingly lost all hope of being built. In the hyperinflationary period, banknotes became valueless almost as soon as they were printed, reversing the promise of economic growth with which capital and its political representatives were legitimated, and rendering architecture, capital's most precarious dependent among the arts, almost entirely desolate. Commissions were scarce in the post-war years, ensuring that designers were constrained to 'paper architecture' – or, in this case, paper-money architecture.

To collectors, however, such objects had a different value, which, unlike that of banknotes, lay in their aesthetic distinction. Furthermore, they were willing to part with 'real' money in order to acquire its aestheticized relation. The result was that town halls made substantial sums from the sale of *Notgeld*, since collectors never sought to redeem it (and, in the case of notes that were issued already expired, could not do so in any case). This money was therefore, despite being stillborn, endowed with a busy afterlife, circulating via a network of dealers and fairs. It still moves in similar circuits today, albeit largely via online vendors.

The other promise embodied by this particular example of *Notgeld* was not fulfilled, however. Although portions of Taut's plan for Magdeburg were constructed, it never attained the kind of dramatic culmination depicted here. The financial difficulties of the early 1920s that motivated the production of this note certainly contributed to that disappointment. The future architecture promised by these colourful designs was just as extinct as the money on which it appeared – and yet, like the money, it continues to have an afterlife, recurring (if only formally) in buildings such as London's 30 St Mary Axe by Foster + Partners, the expressionistically domed insurance company headquarters more commonly known as 'The Gherkin'.

# Oil from Coal
## David Edgerton

The making of oil from coal is one of those processes that, once invented, are abandoned, only to be revived, then abandoned again. The cycle is repeated over and over again, on each occasion in different circumstances. 'Extinction' is never absolute or complete, although at times it may seem so.

Volatile fuels, without which the history of the twentieth century would have been very different, have come principally from refining crude oil – petroleum – but petroleum is by no means their only source. Ethanol from vegetable matter, gas from charcoal and oil from coal have all provided alternative fuels for internal combustion engines. In each case, the manufacturing process is complex and costly, and the choice of which fuel to develop has often been as much a matter of national political expediency as of rational scientific or economic decisions at a global level.

The story of oil from coal begins in the years between 1910 and 1930, when it was thought demand for petrol would outstrip the supply of petroleum. In the search for alternatives, it was discovered that petrol could be made by chemical operations on coal and coal-like substances. In principle, what is involved is straightforward – the addition of hydrogen to coal (which is largely carbon) to form what were called hydrocarbons (petrol or gasoline is a mixture of hydrocarbons of five to twelve carbon atoms each; for example, octane has eight). In practice, however, this operation is extremely difficult. Different processes were thus devised, all involving heavy and complicated plant.

The Bergius process involved directly hydrogenating coal with hydrogen, under immense pressure, and using catalysts. This was hard enough, but so was the making of hydrogen on a large scale. Hydrogen production was a key part of the process of making synthetic ammonia through the Haber–Bosch process, and for the making of margarine, which is hydrogenated plant and animal oil. The key processes were developed in Germany, a land without petroleum, but with experience in high-pressure hydrogen and coal chemistry. In the 1920s the new Fischer–Tropsch process involved the hydrogenation of carbon monoxide, $CO$ (itself derived from coal). Both processes were brought into operation in the 1920s, at great expense. They were not restricted to Germany. In the 1920s Standard Oil and the British chemical company ICI were developing the hydrogenation process, with licences from I. G. Farben. Patent-sharing arrangements were made between I. G. Farben, Standard Oil, Shell and ICI.

Braunkohle Benzin's oil-from-coal plant, Magdeburg-Rothensee, Germany, c. 1939.

The cost of developing these projects was staggering. They involved years of work by the biggest corporations in the world, as well as state-backed initiatives. In their scale, they can be compared to the civil nuclear projects of the post-war years. Whether these techniques were to be developed on a large scale was a matter of political and economic strategy. In the UK, for instance, the nationalist British Union of Fascists wanted oil from coal, but liberals were aghast at the economic and moral cost of economic nationalism. Faced with the choice of using tankers to bring petrol from refineries in distant oilfields or building plants to make petrol from local coal, the UK chose the cheaper option of tankers. However, it did build one (highly subsidized) hydrogenation plant in the early 1930s, in Billingham, County Durham, and one in the late 1930s, in Heysham, Lancashire; the latter came into operation during the war, although it ended up hydrogenating heavy oil, not coal derivatives.

Oil from coal, however, won out in the country that had driven most of its significant innovations, Germany, by then in the grip of radical autarchic economic policies. If it was to go to war, Germany, lacking access to petroleum, had no choice but to liquefy coal on a large scale. The Nazi re-armament plan created many plants of both the Bergius and Fischer–Tropsch types. Bergius hydrogenation plants (the largest was the original plant at Leuna, near Leipzig) produced nearly all the aviation spirit as well as much motor spirit, diesel and so on consumed by the Third Reich; the Fischer–Tropsch plants produced smaller quantities of motor spirit and heavier oils. Without synthetic fuel Germany could not have fought the Second World War, and even so, it was still radically short of oil by comparison to the Allies.

The end of the war, and the global expansion of petroleum output – especially in the Middle East, which now emerged as the great petroleum centre of the world – meant that oil from coal was even more uneconomic than it had been in the 1930s. It was developed only in places where petro-leum was not available locally and could not be bought, because of sanctions, or for ideological reasons, as in Franco's Spain and the Soviet bloc, although production was short-lived in both. Oil-from-coal processing then emerged in South Africa, first on a small scale in the 1950s, and then on a huge scale in the 1980s, because of sanctions. Apartheid South Africa used the Fischer–Tropsch process to make petrol from its coal, on a scale even greater than Germany during the Second World War, and production continues today.

Oil from coal might have made a more general comeback after the oil crisis of the 1970s. Notably in the USA, which by then had become an importer of petroleum, there was a revival of interest, and there at least one large 'synthetic fuel' plant was built in anticipation of a future without petroleum. It was not to be, because the oil price fell once more. Yet oil-from-coal processes keep reappearing. In the twenty-first century, more than eighty years after its first use in Germany, China launched a number of oil-from-coal plants using the Fischer–Tropsch process in an attempt to meet growing demand for petrol, and compensate for inadequate domestic supplies of petroleum. This is perhaps not surprising; China is the greatest producer of coal in the world, and its production far outstrips the peak production of the USA and of the UK. By contrast, the USA and Brazil have now turned to the production of ethanol by fermenting agricultural products (itself a recreation of the power alcohol programmes of the 1930s).

Things are created, live and die, in particular contexts. They are imitated over time as well as space, copied, transferred and refined. The case of oil-from-coal processes is an exquisite example. Overall it is a story of creation, near-extinction and reappearance in very particular niches. It is also a powerful reminder that viable alternatives exist for many processes and raw materials, and demonstrates how the nationalist political economy has been a powerful driver of technical choice.

# Optical Telegraph
David Trotter

The optical (or semaphore) telegraph was a creation of Revolutionary France. In April 1793 the National Convention agreed to fund the development of the new telecommunications system that Citizen Claude Chappe and his brothers had devised on the basis of several years of experimentation. The Chappe *télégraphe* consisted of a series of masts erected in prominent positions in plain view of each other. Attached to the mast was a wooden beam with adjustable arms at either end. Manipulation of this apparatus generated a sequence of discrete signals constituting a message. An observer at the next station took a careful record of each signal, before reproducing the entire sequence by means of an identical apparatus, and so the message passed down the line until it reached its destination. The key to the system's success lay in a double condensation: of distance and of language. In the right weather conditions, the telescopically enhanced human eye comfortably out-distances a messenger on foot or on horseback over the hundreds of miles to which some lines stretched, while natural language travels better when broken down into a finite set of abstract elements that can be bundled into packages for transmission (using Chappe's apparatus, it took thirty seconds to compose each separate signal). The optical telegraph was the first purpose-built telecommunications system fully to exploit the potential of digital code: that is to say, any process that involves the conversion of a continuous data flow into a sequence of discrete functions, each of which can take on only one of a finite number of values.

In July 1793 the Convention approved the establishment of a French state telegraph, and in August funds were allocated for the construction of a first line of fifteen stations between Paris and Lille, then on the border of the Austrian Netherlands (now Belgium). Chappe was given the title *Ingénieur Télégraphe*, as well as permission to cut down trees and to place telegraphic apparatus on church towers or other prominent structures of any kind. The first official message sent down the line, on 15 August 1794, concerned the recapture of the town of Le Quesnoy from the Austrians. Napoleon Bonaparte, who came to power in 1799, understood the military and diplomatic uses of the telegraph perfectly. His armies may have marched on their stomachs, but they did so in the direction indicated by telegraphic intelligence. By 1805 the system covered much of France, with extensions into Belgium, Germany and Italy. A network had come into being.

Claude Chappe demonstrating his telegraph system in 1793, from *Le Petit Journal. Supplément illustré*, 1 December 1901.

Speed, evidently, was the network's unique selling point. But it has only ever been possible to guarantee speed by dint of massive investment in material infrastructure. A mail service requires some men with horses. A digital telecommunications system has an extensive, costly footprint: signal stations for the Chappe *télégraphe*; submarine cables and wireless transmitters for the electric telegraph; and satellites and server farms for the Internet today. But its very public infrastructure meant that Chappe's system operated in full view of anyone who knew how, when and where to seek out its traffic. The wooden beams his engineers had stuck on the top of town halls and church towers the length and breadth of France could be seen to do a strange little dance when activated: 'wooden arms with elbow joints are jerking and fugling in the air,' Thomas Carlyle wrote, 'in the most rapid mysterious manner.'[1] To fugle is to signal or motion as if signalling. While the fugling happened in the open, the code by which messages were sent was known only to senior officials at either end of the line. But all such systems can be hacked. Hence the priority they give to encryption. The optical telegraph remains exemplary because it was at once ostentatious and hermetic.

European governments responded to the arrival of a network on their military and diplomatic doorsteps by commissioning rival systems. In Britain, the Admiralty built a comparable system connecting its London headquarters to bases at Chatham, Sheerness, Deal and Portsmouth. In 1805 a branch of the Portsmouth line reached Plymouth. But after the end of the Napoleonic wars, in 1815, the Admiralty lost interest in the optical telegraph. Although one or two local military and commercial lines remained in sporadic use well into the 1850s, it was soon to be rendered largely obsolete. The electric telegraph began slowly, in the late 1830s, but by 1851 Britain was linked to France, and by 1866, after a first, unsuccessful attempt, to North America. Morse code converted language into a series of signals far more effectively than the manipulation of a wooden beam. Those signals travelled further, faster. In 1847 *Punch* was touched to discover the Admiralty's old Portsmouth line in desultory action as a kind of al fresco asylum for redundant telegraphists, who continued to amuse themselves by sending nonsensical messages back and forth. Obsolescence, however, created a second life for telegraphy, in language rather than on the ground.

That second life arises out of the optical telegraph's combination of secretiveness with shameless self-display. According to the dictionaries, to 'telegraph' is to convey information by means of a gesture, expression or other sign or signal, in circumstances requiring a degree of exaggeration. The florid gesture inadvertently advertises an intention. A football player who 'telegraphs' a pass has just told the opposition where the ball is about to go. Furthermore, we might on occasion suspect that the performance of messaging has begun to matter rather more than the content of the message.

Telegraphy is a paradoxical behaviour that blurs the distinction between public and private. It seems to want either to publicize privacy, or to remain private while in public.

Writers, who like paradox, cottoned on rapidly. By the 1830s telegraphing had become, if the satirists are to be believed, an essential element of the discourse of lovers at all levels of society. Flirtation was now to be conducted not merely in code, as had long been the custom, but by sly hyperbole, by a performance of encryption.[2] Charles Dickens found much to enjoy in telegraphy's combination of exposure and concealment. In *The Pickwick Papers* (1837), Sam Weller and his father at one point exchange a 'complete code of telegraphic nods and gestures' ('after which, the elder Mr Weller sat himself down on a stone step, and laughed till he was purple').[3] The optical telegraph's legacy lies less in its technological innovations than in its uncanny dramatization of the failure of modern telecommunications systems to separate public from private expression. Who hasn't witnessed the elder Mr Weller in action, mobile phone to one ear, regaling a packed train carriage with information that is of no conceivable meaning or interest to anyone but himself?

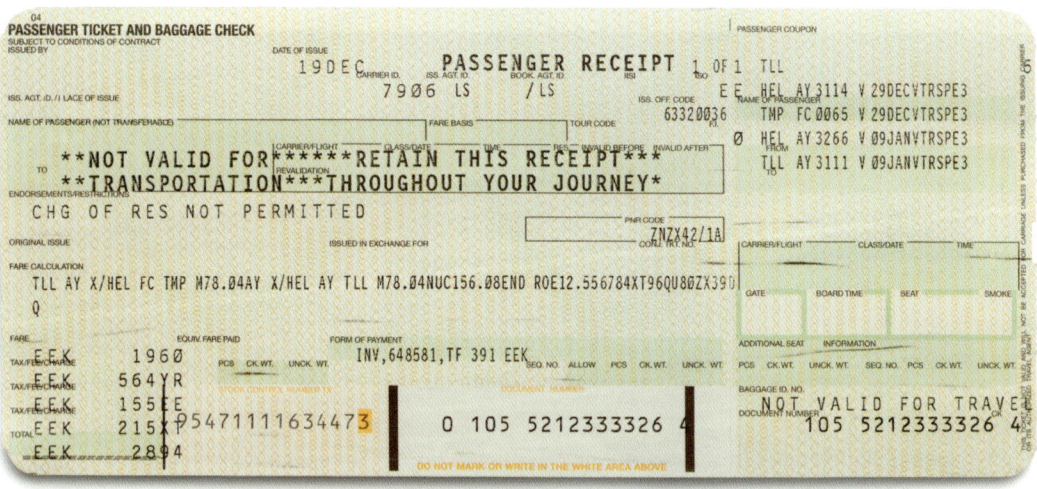

04
**PASSENGER TICKET AND BAGGAGE CHECK**
SUBJECT TO CONDITIONS OF CONTRACT
ISSUED BY

PASSENGER COUPON

DATE OF ISSUE   19DEC   CARRIER ID.   ISS. AGT. ID.   BOOK. AGT. ID.   **PASSENGER RECEIPT** 1 OF 1   TLL

7906 LS   /LS   ISS. OFF CODE   EE   HEL AY 3114 V 29DEC♥TRSPE3
63320036   TMP FC 0065 V 29DEC♥TRSPE3

NAME OF PASSENGER (NOT TRANSFERABLE)   FARE BASIS   TOUR CODE   NAME OF PASSENGER
Ø   HEL AY 3266 V 09JAN♥TRSPE3

CARRIER/FLIGHT   CLASS/DATE   TIME NOT VALID BEFORE   INVALID AFTER   FROM
TO   **NOT VALID FOR*****RETAIN THIS RECEIPT***   TLL AY 3111 V 09JAN♥TRSPE3
**TRANSPORTATION***THROUGHOUT YOUR JOURNEY*   REVALIDATION

ENDORSEMENTS/RESTRICTIONS
CHG OF RES NOT PERMITTED

ORIGINAL ISSUE   ISSUED IN EXCHANGE FOR   PNR CODE
ZNZX42/1A   CARRIER/FLIGHT   CLASS/DATE   TIME

FARE CALCULATION   GATE   BOARD TIME   SEAT   SMOKE
TLL AY X/HEL FC TMP M78.04AY X/HEL AY TLL M78.04NUC156.08END ROE12.556784XT96QU80ZX39D
Q   ADDITIONAL SEAT   INFORMATION

FARE   EEK   1960   EQUIV. FARE PAID   FORM OF PAYMENT   PCS   CK.WT.   UNCK.WT.   ALLOW   PCS   CK.WT.   UNCK.WT.   PCS   CK.WT.   UNCK.WT.   SEQ. NO.   PCS   CK.WT.   UNCK.WT.
TAX/FEE/CHARGE   EEK   564YR   INV,648581,TF 391 EEK   SEQ. NO.   ALLOW
TAX/FEE/CHARGE   EEK   155EE   BAGGAGE ID. NO.   NOT VALID FOR TRAVEL
TAX/FEE/CHARGE   EEK   215XT   STOCK CONTROL NUMBER TXT   DOCUMENT NUMBER   DOCUMENT NUMBER
TOTAL   EEK   2894   P5471111634473   0 105 5212333326 4   105 5212333326 4

DO NOT MARK OR WRITE IN THE WHITE AREA ABOVE

# Paper Aeroplane Ticket
## Gökçe Günel

In the summer of 1995 I took a one-week trip to Italy with my family, short-circuited by the unwelcome realization that we had lost our plane tickets back home to Istanbul. Instead of soldiering through the must-sees of Rome and Florence, while recounting trivia and eating gelato, we took the train to Ancona, a destination that seemed grey and rebarbative, to find the only travel agency that would reissue our tickets and ensure our timely return. Sure enough, we located the missing tickets deep inside one of our bags soon after we left the agency.

We can no longer lose tickets the way we did in the 1990s. Instead, most are sheltered in in-boxes and circulate digitally on screens, sometimes printed to serve as evidence of our itineraries to overzealous immigration officers, everyday gatekeepers of our contemporary border regime. Airlines started promoting electronic tickets in the late 1990s, offering extra benefits to passengers who would opt for computerized receipts. Starting in 2004, airlines researched the logistics of eliminating paper tickets altogether, mainly to reduce costs at a time of rising fuel prices. Given the increasing prevalence of digital communications around the world, it was also promised that electronic tickets would open the doors to an epoch of passenger convenience.

In 2008, following its Annual General Meeting in Istanbul, the International Air Transport Association (IATA) officially announced the industry's conversion to full-scale electronic ticketing. Its bold move was illustrated by a ceremonial photograph of seven male industry executives, and two unnamed female flight attendants in Turkish Airlines attire, all holding a giant cardboard ticket cut-out. The red stamp on the oversized, now defunct ticket read: RETIRED. The paper ticket had fulfilled its duties, and was ready to move to its cottage in the country. 'Today we say goodbye to an industry icon,' said Giovanni Bisignani, IATA's then Director General and CEO, and one of the men in the photograph. 'The paper ticket has served us well, but its time is over. After four years of hard work by airlines around the world, tomorrow marks the beginning of a new, more convenient and more efficient era for air travel.' He concluded, happily, 'If you have a paper ticket, it's time to donate it to a museum.'[1]

Not many objects enjoy such perspicuous rituals to mark their expiration. During the twentieth century tickets of all kinds were most likely printed on paper. When physical airline tickets, standardized by the Warsaw Convention, were used for the first time in 1933, they showed that

Aeroplane passenger ticket, early 2000s.

Ceremony photograph of the International Air Transport Association (IATA) bidding farewell to the paper ticket, 2008. Pictured are Wolfgang Mayrhuber (Lufthansa), Fernando Pinto (TAP Portugal), Giovanni Bisignan (IATA), Samer Majali (Royal Jordanian), David Bronczek (FedEx Express), Temel Kotil (Turkish Airlines) and Douglas Steenland (Northwest Airlines).

a passenger had paid the fare and could board the flight named on the ticket. At a time when air travel was not as common as it is now, paper tickets functioned as indicators of a possible wanderlust, icons of a desire to journey through faraway places; they were issued from airline ticket offices, now mostly erased from urban landscapes. Paper airline tickets conveyed privacy, too; the only copy of the ticket, prepared by the travel agent and (it is to be hoped) photocopied by the ticket's careful and cautious owner, would be kept in a purse or bag, alongside other such important documents as travellers' cheques, visas or passports. As printed objects, such tickets were stable and sturdy, but they could also be tattered, drenched or erroneously laundered in a washing machine. Paper tickets were frequently torn to flag past use. Leftover fragments survived inside paperbacks, waiting to be recovered by new generations of readers in second-hand bookshops, sometimes decades after the travel date. By announcing the extinction of the paper airline ticket, the IATA officers also signalled the end of these affective, institutional and technological connections.

As digital objects, e-tickets build new connections and accrue new qualities. The means by which they appear in our in-boxes has changed. Instead of visiting airline ticket offices, we now explore itineraries via online travel agencies, intermediaries striding across intricate global distribution systems, and always promising the best fares. Once we make up our minds and book flights, the tickets arrive by email. Confirmation numbers, the keywords that appear in those emails, are jumbled-up digits and letters that grant entry to customer-services hotlines and online check-in platforms. The disappearance of paper airline tickets has not eradicated the practice of ticketing altogether, but rather significantly altered the logistical networks that enable the production, use and validation of a ticket.

In retaining their fundamental function as identity markers for travellers, e-tickets still rely on existing infrastructure, such as electric power and computational capacity in passengers' homes and across airport terminals. But they are disembodied and can easily be distributed across space and time. One reviewer on TripAdvisor noted, 'an eTicket is but 1's and o's, and as stated, exists only in the computer world.' Another commentator on the same platform suggested that 'since the eticket lives inside a computer system, you can't lose it, damage it, whatever.' Yet, unlike paper fragments, they are erased in their entirety on the completion of flights.

The word 'RETIRED' on the photograph above signifies a certain finality for paper tickets, and paints them as extinct objects with nowhere to go except museums. However, paper tickets persist in settings that are deemed marginal by the airline industry, and excluded from the IATA's regulatory zones. They continue to commodify travel in regional airports that suffer from frequent power cuts and computer failures, and they are used in authorizing trips with more than sixteen legs. And they occasionally reappear: when a man from Tennessee found a long-forgotten paper ticket under his bed, the airline decided to honour his request for credit. The holder of an e-ticket today would be lucky to receive such generous treatment.

# Paper Dress
## Olivia Horsfall Turner

Between 1966 and 1970 the USA and the UK enjoyed a 'paper caper' that appeared to be the future of fashion. Companies specializing in paper products such as toilet tissue and disposable plates started selling dresses made from high-performance papers. What began as a marketing gimmick became a hit in popular fashion and high-end couture alike, and presaged the fast-fashion phenomenon of the early twenty-first century.

Paper clothing came to embody modernity, but it had been used historically in a variety of contexts. In addition to the long-standing use of paper for accessories such as hats and fans, it featured in fancy-dress costumes from the 1890s onwards, when crêpe paper first became available commercially. Far removed from such whimsy, wool shortages during the First World War prompted further experimentation with paper. Its suitability for occasional use resulted in a patent in 1917 in the United States for an emergency paper raincoat, and in the 1920s paper collars and cuffs were produced so that men could avoid doing laundry while travelling. One of the positive associations of disposable paper products, in addition to portability and convenience, was hygiene; the 1920s brought the development of items such as tampons and facial tissues, followed in the 1950s by single-use nappies. This was also reflected in the development of the largest application of paper to clothing production: disposable hospital gowns and scrubs, which were both sanitary and labour-saving.

The paper dress craze in everyday wear took off in 1966, when in the United States the Scott Paper Company, manufacturer of paper napkins and plates, introduced a marketing campaign to promote Dura-Weve, its new material comprising 93 per cent paper and 7 per cent rayon, originally developed in 1958 for single-use protective clothing. Customers paid $1 plus 25 cents for postage and sent off a coupon; in return they received a Dura-Weve dress in one of two designs: red paisley or black-and-white op art. The fact that 500,000 mail-order paper dresses sold in just eight months took the company by surprise. Other businesses took note and began offering their own versions. Hallmark, for example, expanded its party kits, which usually featured cups, plates, placemats and invitations, to include a matching dress for the hostess. By the end of 1966, over $3.5 million had been spent on paper clothes; this was a fraction of the $30 billion spent on women's clothes as a whole, but still significant for a cheap, novelty product.[1]

Paper dress, manufactured by the Scott Paper Company, USA, 1966, printed bonded fibre paper.

Strictly speaking, paper clothing was not pure paper; wood pulp or a cellulose substitute was mixed with synthetic fibre to make it more durable and resistant to tearing. It could also be treated to make it water- and fire-resistant. The resulting material could be easily printed, refreshed by ironing on a cool setting, and in some cases even washed (although usually at the sacrifice of any protective coating). Kaycel, for example, marketed $12 fire-resistant men's suits of paper on the basis that they could be laundered up to twenty times. For those who delighted in the frisson of wearing such an improbable item, there were even paper bikinis; they would last for three uses.

By spring 1967 paper clothing was sufficiently in vogue that *Time* magazine profiled the available merchandise, and at Expo '67 in Montreal, paper dresses were included in the pulp and paper pavilion. The appeal of paper clothing swiftly expanded from promotional and commercial circles to the fashion world. The designer Elisa Daggs conjured four themed paper outfits to be worn by TWA flight attendants, and in the UK Ossie Clark and Celia Birtwell devised throwaway shift dresses printed on paper manufactured by Johnson & Johnson. Demand for paper clothing that year was so high that in the States there was a paper shortage. This contributed to the development of paper alternatives, from organic solutions, such as Fibron (made of pressed rayon fibres with perforations), to plastics, such as Tyvek (made of high-density spunbound polyethylene). The disposable status of non-biodegradable, plastic-based products was in line with the expectations of the time.

Paper clothing appealed on numerous counts. Novelty was key, since paper products decorated with inventive designs and bright colours could be produced quickly, making them ideal for ephemeral moments from the launches of beauty products and food to presidential campaigns; both the dress and the context became conversation points. In an era that embraced mass production and promoted consumerism, wearing a dress emblazoned with the graphics of a chocolate wrapper or overhauling your wardrobe several times a season was in tune with the fast pace of popular culture. The epitome of the resulting self-referentiality was the Campbell's 'Souper' dress, printed with vegetable soup can labels, inspired by Andy Warhol's *Campbell's Soup Cans* of 1962.

Affordability and value were also advantages. The magazine *Mademoiselle* advised, 'In terms of how much pow you get for your pennies, the paper dress is the ultimate smart-money fashion.' Not that paper clothing was necessarily cheap; newsreel footage records that some ensembles cost up to $60 each. Indeed, the journalist Marylin Bender reassured her readers that the paper 'democratic dress' did not herald the demise of fashion's aristocracy: 'The woman of wealth and social contacts can commission an artist to create a special paper dress for a special event, then donate it to a museum, provided the garment hasn't deteriorated on the dance floor.'[2] This comment was prompted by the Paper Dress Ball of October 1966 at the Wadsworth Atheneum, Connecticut, where guests wore designer paper dresses.

The most powerful lure, though, was the conscious modernity of paper clothes. The aesthetic of a sleek and streamlined future was embodied in the very style of paper clothing, which was typically simple and boxy, with no darts, and fastenings of Velcro or snaps rather than buttons. Paper does not drape elegantly as conventional fabrics do, but it does form dynamic lines and clean silhouettes. An unspoken but potent appeal was surely also the risqué nature of paper clothing: paper was precarious and therefore provocative. Being easily altered with only scissors and glue or adhesive tape, paper garments challenged gendered expectations that women should necessarily have sewing skills. Disposability also liberated wearers from the tyranny of having to clean and care for their garments, suggesting that they would be able to lead more exciting lives. As the textile designer Julian Tomchin commented, 'It's right for our age. After all, who is going to do laundry in space?'[3] Appealing to youthful reluctance to be responsible, the original Scott Paper Company campaign boasted, 'Won't last forever . . . who cares? Wear it for kicks – then give it the air.' The idea that convenience enhanced enjoyment led to the proposal that people would eventually purchase entire disposable holiday wardrobes after arriving at their destination.

The paper dress was ultimately, however, destined for the wastepaper basket. Anecdotes relating the danger of lit cigarettes, and scenes at cocktail parties where men intentionally spilled drinks on women's dresses, suggest sufficiently good reasons for paper clothing to be eschewed as anything other than a brief experiment. The material qualities that made it appeal to individuals seeking a modern, space age aesthetic also rendered it less suitable as taste shifted in the early 1970s to a more eclectic look. Finally, the rising influence of the green movement and hippie culture meant that disposability was no longer regarded as a virtue, and throwaway culture was rejected in favour of ecological responsibility.

In the 1990s there was a brief resurgence of paper clothing from high-fashion designers including Helmut Lang, Hussein Chalayan and Sarah Caplan, perhaps the product of *fin de siècle* futurism. Likewise, since the early twenty-first century, globalization and the resulting low cost of clothing production has led to 'fast fashion', one aspect of which is that consumers throw away cheaply purchased items rather than washing them. Against the backdrop of the climate crisis and acute awareness of the scarcity of resources, however, sustainability is again winning against disposability. Tomchin's vision that paper clothes would be sold from tear-off rolls has not come to pass, and paper garments remain the preserve of the hospital ward.

Les Escargots non sympathiques.

# Pasilalinic-sympathetic Compass
## Richard Taws

Snails are on the front line of extinction these days. The first animal to become extinct in 2019 was George, a fourteen-year-old Hawaiian tree snail and final representative of *Achatinella apexfulva*, who breathed his last on New Year's Day. George's plight is not unique, as a combination of climate change and the introduction of invasive species throws invertebrate numbers worldwide into rapid free fall. Dubbed 'the loneliest snail in the world', George spent more than a decade as the last of his kind, his caretakers seeking in vain to find him a mate. His pathetic isolation is all the more striking because there was once a time when enthusiastically mating snails promised a future of free communication, not just with one another, but between people, too. This dream died swiftly, even before the snails themselves, but it illuminates how snails have operated, not only as repositories for anxiety about obsolescence, but also as conduits for strange kinds of future thinking.

In late 1850 Parisian newspapers had a field day with the announcement of a singular invention. The 'pasilalinic-sympathetic compass' – sometimes referred to as a 'snail telegraph' – was brought to the public's attention via an account by one Jules Allix, published over two issues of *La Presse* in October that year. A former law student, Allix followed an eccentric path through life. Not long before the article was published he had presented himself for candidacy of the Constituent Assembly following the Revolution of 1848, and he later endured periods of imprisonment and banishment for his political activities. He subsequently played a role in the Paris Commune of 1871, during which time he proposed, among other initiatives, replacing the crucifixes in girls' schools with an image of the fearsome ancient Greek wrestler Milo of Croton. Recognizable for his long hair, his outlandish dress and his habit of scrutinizing colleagues through a *lorgnon* while talking non-stop, Allix was sent to the Charenton asylum on 21 May 1871, sparing him involvement in the *semaine sanglante* that began that day, in which thousands of Communards were murdered.[1]

In his report in *La Presse*, Allix recounted an experiment with the snail telegraph that had taken place in the company of the sham occultist Jacques Toussaint Benoît, who, working with an 'American' referred to only as Monsieur Biat-Chrétien, sought to garner sponsorship from the investor Hippolyte Triat. The experiment took place in Benoît's apartment, with Allix and Triat as witnesses. Biat-Chrétien served a more distant role, participating – allegedly – from the other side of the Atlantic.

Honoré Daumier, *Les Escargots non sympathiques*, lithograph on newsprint paper, second of two states, from 'News of the day' (*Actualités*), *Le Charivari* (25 September 1869).

The pasilalinic-sympathetic compass was based on the claim that mating snails create a lifelong sympathetic bond, forming a connection that could be maintained across great distances and over long periods. Most terrestrial gastropods are hermaphrodites, prompting speculation about their unique ability to act as both transmitters and receivers. Annexed to a simple code, the possibility that this nebulous snail love could be appropriated for human communication soon presented itself. In the long history of telegraphy there were several precedents for forms of sympathetic communication. Around the same time, the Irish physician and scientist William Brooke O'Shaughnessy, knighted by Queen Victoria for his work on the telegraph in India and a pioneer in the field of medical cannabis, experimented with skin-to-skin contact as a technique for long-distance information exchange. Even the highly successful optical telegraph developed by Claude Chappe and his brothers, in operation across France from the mid-1790s to the 1850s, hewed closely to the body in its dependence on anthropomorphic sema-phore. However, Benoît's investment in non-human participants marked his invention out as distinct from these systems. It might be considered part of a genealogy of information networks in which natural phenomena take priority. As the academic Justin E. H. Smith has argued compellingly, such a genealogy would take seriously the deep history of placing living beings at the centre of the science of nature, licensing the comprehension of more recent systems – the Internet, most notably – as yet another chapter in natural history. A recent emanation of the symbolic operations of the natural world, rather than a wholesale breach with them, the Internet is replete with animal drives and dependent on natural resources.

The snail telegraph's mechanism was very simple, even if its operation was shrouded in mystery. A large scaffold of wooden beams supported 24 zinc bowls, lined with cloth soaked in copper sulphate solution and held in place by a copper ring. In each bowl resided a snail, glued down to prevent a (slow) escape. Each snail was associated with a letter of the alphabet. At the other end of the line, an identical device contained 24 snails paired with those at the transmitting apparatus. The touch of the operator, via an electri-cal stimulus, activated a snail, which would cause a reaction – an 'escargotic commotion' – in its partner. Copper is unpleasant for snails (it is thought to produce a sharp electro-neural response), thereby intensifying their agi-tation. By this method, messages could be sent back and forth between the amorous gastropods. The experiment was only a very partial success, to Triat's irritation, since most words arrived with significant errors and the scientific method was openly violated in the set-up. Still, Allix was impressed enough by the results to publish his hyperbolic report, which invoked Luigi Galvani, Franz Mesmer and Alessandro Volta as historical authorities legitimizing the novel discovery. Scale, he admitted, remained a problem. One of Allix's more delightful prognoses, in the second installation of his feuilleton report, is that in time the pasilalinic-sympathetic compass would be made ever smaller – there exist, after all, snails no bigger than a pinhead – and that miniaturized versions of the device would in due course be incorporated into furniture, into 'interesting' jewellery and even 'the waist-chains [chatelaines] of ladies'.

Although Allix requested that other newspapers republish his account in full (in the interest of scientific disinterestedness, naturally), the experiment was not received with quite the seriousness he desired. On 3 November 1850 the satirical journal *Le Tintamarre* reported the hilarity that followed its publication and issued a series of absurd possible uses of its own for the 'escargotphiles' new invention. *Le Tintamarre*'s correspondent predicted the demise of the established press, to be replaced by a giant snail, and prophesied the replacement of numerous human functions, institutions and structures by gastropod overlords.

While Allix himself dipped into and out of the news, coming into focus again following the events of 1871, for many years afterwards the story of the thwarted invention continued to have currency. As *L'Aurore* reported on 28 January 1911, 'escargot sympathique' had entered the French language as an idiomatic expression so pervasive that its precise origins were often no longer remembered, although mention of the system's protagonists always recalled the bizarre failed experiment. In 1869 Honoré Daumier contributed the defining image of the pasilalinic-sympathetic compass in a print satirizing the lethargic progress of social reform under the Second Empire. While Benoît and Biat-Chrétien's device no longer existed in physical form, and the snails were long gone too, it was left to an earlier communicative technology, lithography – itself supposedly superseded at this point by more recent innovations in print media and photography – to place the system fully in its social world. Two decades earlier, in a similar vein, a vaudeville performed for the first time at the Théâtre de Montansier, Versailles, on 17 November 1850 sent up the invention by locating it in a longer history of technological succession and improvement, in which human achievements were bound to extinction by psychic information machines, resplendent in their spiral shells:

> Here is progress . . . Well! Everything is like that . . . You didn't have the telegraph, and the Chappe brothers invented it . . . after the ordinary telegraph came the electric telegraph, after the electric telegraph came the underwater telegraph, after the underwater telegraph here come the snails . . . after the snails . . . Ah! My word, after the snails, walk away . . . it's over, there's nothing more to do, the world is perfect.[2]

George the Hawaiian tree snail would no doubt have approved.

# POPULAR SCIENCE

March 1949

UN 3

25c

Monthly

HEAT BIN

RAY PIOCH

## Sun Furnace in Your Attic

# Phase-change Chemical Heat-storage Barrel
Daniel Barber

In the architectural and engineering discussions that took place immediately after the Second World War, there was widespread interest in the role of solar energy for space heating. Félix Trombe and Jacques Michel were experimenting with high-heat solar condensers in France; there were experiments in India and elsewhere with solar cooking systems; in the USA, a group at the Massachusetts Institute of Technology (MIT) explored solar house heating. As Mária Telkes, an engineer and one of the main proponents of solar heating at MIT, recognized in 1947, the 'problem of solar heating was actually a problem of heat storage'.[1] It was easy to design a house so that solar radiation was absorbed by external panels or even entered the interior, and it was relatively easy to insulate those panels, or that house, to keep some of the heat in. However, for a solar heating system to be viable – that is, to stand up against fuel heating systems of various types – it was necessary to store the radiation absorbed when the Sun was out for use when it was not (at night, or under cloud cover). Much of the challenge of solar energy, from ancient times to the present, has revolved around heat storage.

The early post-war period brought numerous and varied elaborations of heat-storage methods. At stake was the refinement of passive solar systems, where solar energy was used to heat a storage medium (air or water were the defaults) that would in turn heat a space. An early experiment that set the standard, carried out at MIT just before the war, involved a sealed and insulated solar panel with water circulating in copper pipes; that heated water was then stored in a heavily insulated underground tank. When heat was needed, air was blown over the hot water in the tank and then into the room. It worked, but it was expensive – the insulation for the tank cost more than the building it was heating. Other experiments used air, or capitalized on the thermal properties of masonry and other materials, to allow some heat storage. One engineer in Colorado, George Löf, used river pebbles; warmed by solar-heated forced air and stacked inside a column on the interior of the house, the pebbles would radiate heat into the interior as the air around them cooled.

Telkes was a proponent of using chemical salts as a storage medium. She was appointed as a researcher to MIT's department of chemical engineering in 1939, just as the institute's first solar house experiment briefly described above was being completed. She had emigrated to Cleveland from Budapest in 1925; in the 1930s she worked at Westinghouse and experimented with

Illustration of the Dover Sun House system, cover of *Popular Science* (March 1949).

different alloys for direct conversion of solar energy through thermocouples. When she learned of the MIT experimental programme, she applied for a job. For MIT's second experimental house, she developed a system that promised to eclipse all the others. She proposed the use of Glauber's salt, a solution of sodium sulphate decahydrate. This chemical solution contained crystals into which water was bound. When heated to 32°C (90°F), the crystals melted, absorbing the heat. This heat was stored by the compound in its liquid form. When the temperature dropped, the compound recrystallized and the heat was released. Under ideal conditions, this phase-change chemical heat-storage system could retain solar radiation for a few days. Unfortunately, there were numerous problems with MIT's second house, including leaks, poor insulation and poorly trained graduate students, which compromised the experiment.

In 1945 Telkes found another chance: she began a collaboration with the architect Eleanor Raymond to design a house based on the phase-change system, known as the Dover Sun House, just outside Boston. By this time Telkes had been pushed out of MIT. The house was funded by Amelia Peabody, a Boston artist and socialite and frequent client of Raymond, and celebrated in the press as an all-women house. After experimenting with a number of locations for the panel and storage elements, Telkes and Raymond decided on a design that involved a huge solar panel on what looked like the second floor. This panel drew heat into enclosed interior 'heat bins' that contained the chemical solution, stored in metal barrels. These heat bins were distributed according to thermal need: a large one in the bedroom; smaller ones in the kitchen and living room. The storage barrels were hidden, and standard vents on the walls distributed the radiant heat to the interior.

The house didn't work very well. The large collector panel included a backing layer of carbon-black-painted metal, to maximize solar gain. This material became so hot that the caulk used to seal the panel system cracked and leaked, so that water got in and heat got out. The whole panel had to be rebuilt every summer of the house's operation. The heat barrels themselves leaked under the pressure of the salt and the expansion and contraction of the phase-change process; it didn't help that the barrels were repurposed from the earlier MIT house and many were already compromised. Other contingencies extended the house's liveability – the tenants were Telkes's cousins, and sympathetic; they would often turn on the stove and sleep near the kitchen on cold winter nights – but it didn't last long. By 1952 the heating system was abandoned and the house converted to a forced-air system.

The phase-change concept, however, had a longer life. Indeed the Dover House experiment was affected by external factors that had little to do with any technical facility; notably, doubts cast on Telkes's abilities as an engineer had more to do with sexism and misplaced anti-Communist sentiment than any failing of the technology. Had it been given increased financing and

Phase Change Storage Heat Bins, design by Eleanor Raymond and Mária Telkes, at Dover Sun House, near Boston, Massachusetts, USA, 1947–52.

attention to maintenance, the house might have been more effective. Larger geopolitical concerns also played their part; the initial interest in solar heating right after the war soon diminished as oil began to flow freely into the USA.

Phase-change chemical heat storage has since been the subject of substantive research and development. It was made to work, once some of the aforementioned contingencies were mitigated. Telkes herself promoted the process at the United Nations, New York University and the Curtiss-Wright Corporation, and when she took a permanent position at the University of Delaware. Phase-change heat-storage systems slowly but persistently matured; however, not with the use of metal storage barrels. It is now possible to buy sheetrock (drywall) and other standard building materials with phase-change materials (PCMs) embedded, increasing the capacity of the wall to store and radiate heat. In most cases, however, PCMs are seen to accompany, rather than replace, a mechanical heating and cooling system – a less comprehensive and less radical vision of solar heating than the one Telkes advanced.

Historians are trained to avoid counterfactuals, yet it is difficult, in the face of these distorted technological trajectories, not to imagine a world in which the corporate investment, tax subsidy and policy decisions that have supported the global proliferation of oil were directed instead at solar technology research and experimentation. Would we now all live in solar houses, with phase-change chemicals in the seats on solar-powered buses to warm us through a frigid morning commute? The story of Dover Sun House and the phase-change storage barrel helps us to recognize the narrowness and contingency of the conceptions we have collectively designated as 'progress', especially at a moment when alternatives, in energy and in lifestyle more generally, are so desperately needed.

# Player Piano
## Hal Foster

The player piano, which operated by way of scores registered on paper or metal rolls that were driven mechanically, is an extinct object that comes to me at second hand, by way of the great American novelist William Gaddis (1922–1998). For more than half a century Gaddis worked intermittently on a manuscript about the short life of the player piano in the United States. Why toil so long on an outdated form of entertainment? '*Agapē Agape* is a satirical celebration of the conquest of technology and of the place of art and the artist in a technological democracy,' Gaddis wrote in a proposal for the book in the early 1960s. 'As "The Secret History of the Player Piano" it pursues America's growth in terms of the evolution of the programming and organizational aspects of mechanization in industry and science, education, crime, sociology and leisure and the arts, between 1876 and 1929.'[1] This is a vast project, and it got away from Gaddis. Yet 'The Secret History of the Player Piano' did not disappear, and in fact Gaddis often borrowed from its themes; a figure furiously at work on an unwieldy treatise is a staple of his fiction (in *JR* of 1975, a character named Jack Gibbs struggles over this very manuscript), and over the years he wrote several essays on related topics, which were collected posthumously in *The Rush for Second Place* (2002). In early 1997 Gaddis was diagnosed with terminal cancer, which prompted him to distil his mass of notes, cuttings, outlines and drafts into a fiction of 84 manuscript pages. This is the version of *Agapē Agape* (2002) that was left when he died a year later.

Like his other novels, *Agapē Agape* is nearly all spoken, the soliloquy of a dying man in bed, who is and is not Gaddis. In a frantic state of distraction he struggles to patch together his book; he has a deadline without extension:

> No but you see I've got to explain all this because I don't, we don't know how much time there is left and I have to work on the, to finish this work of mine while I, why I've brought in this whole pile of books notes pages clippings and God knows what, get it all sorted and organized . . . that's what my work is about, the collapse of everything, of meaning, of language, of values, of art, disorder and dislocation wherever you look, entropy drowning everything in sight, entertainment and technology and every four year old with a computer, everybody his own artist where the whole thing came

Player piano, c. 1910.

from, the binary system and the computer where technology came from in the first place, you see?[2]

For the dying man, this struggle between order and entropy is personal, but it is also the philosophical crux of his book and the practical question of his age.

Eccentric as a displaced artefact, the player piano is central to the secret history that is the pretext of the novel, the open sesame to 'the patterned structure of modern technology and the successful democratization of the arts in America'. The outmoded objects that intrigued the Surrealists possessed this rich ambiguity, too, and Walter Benjamin also looked to such things for the 'profane illumination' of historical dreams that they encrypted (in his essay on Surrealism in 1929 he listed 'the first factory buildings, the earliest photos, objects that have begun to be extinct, grand pianos'). In this light the *idée fixe* of the dying man, which Gaddis telegraphs as 'my whole thesis entertainment the parent of technology', might well have interested Benjamin, who is given a cameo in the book. Yet the notion that mechanization is internal to art, even initiated there, seems to reverse the Benjaminian view that it revolutionized art from the outside, from the world of industry. In diabolical fashion, the dying man laments the effects of mechanization even as he implicates art in this technological degradation.

For this Gaddis double, the relevant history of mechanization begins with the Enlightenment automata of Jacques de Vaucanson. The inventor of flautist and shitting duck automata in the 1730s, Vaucanson became Inspector of Silk Manufactures in France, and in 1756 he transformed the industry with mechanical looms controlled by pierced cylinders. It was literally from the pieces of this apparatus, restored at the Conservatoire des Arts et Métiers, that Joseph-Marie Jacquard devised his punch-card loom in 1804, which, thirty years later, inspired Charles Babbage in his Analytical Engine, a signal predecessor of the contemporary computer. For the dying man the player piano is a lost term between these last two machines, the vanishing mediator between industrial and digital ages: 'the beginning of key-sort and punched cards and IBM and NCR and the whole driven world we've inherited from some rinky-dink piano roll'. It becomes the magical cipher in his own account of automatization, which is why 'getting the whole chronology in order 1876 to 1929' is so important to him.

Why these dates? A chronology in *The Rush for Second Place* tells us that the Pianista player was first exhibited in 1876 at the Philadelphia Exposition, along with an electric organ; 1876 is also the birthdate of the telephone. The demise of the player came in 1929 with the spread of gramophones and radios (the latter grew from 5,000 in the USA in 1920 to 2.5 million in 1924), the advent of sound film and the first public demonstration of television.

This history is intentionally potted. Under 1876, for example, Gaddis also lists the Christian Science Association, whose charismatic founder, Mary Baker Eddy, is taken at her word: 'I affix for all time the word *Science* to *Christianity*; and *error* to *personal sense*; and call the world to battle on this issue.'[3] For Gaddis this 'elimination of failure through analysis, measurement, and prediction' is the great power of technoscience, but it is one that can overwhelm all matters of the spirit, in religion and in art, 'where truth and error are interdependent possibilities in the search for unpredetermined perfection'. In addition, Gaddis argues in his proposal that

> the career of the player paralleled the zeal for order and patriotic proclivities for standardization and programming contributed by McCormick (patents), Rockefeller (industry), Woolworth (merchandising), Eastman (photography), Morgan (credit), Ford (assembly line, plant police), Pullman (model town), Mary Baker Eddy (applied ontology), Taylor (time studies), Watson (behaviorism), Sanger (sex) etc., etc.

Clearly Gaddis is concerned less with mechanization per se than with the 'more pervasive principle of organization' that continues to govern 'automation and cybernetics, mathematics and physics, sociology, game theory, and, finally, genetics'. Clearly, too, this project pushed him into a semi-paranoid 'zeal for order' of his own, even if at the same time his life was perpetually on the edge of tipping over into chaos. For Gaddis the player piano was not an ending, it was a beginning, the beginning of a world in which, thanks to automation and the computer, chance, unpredictability and failure would all be wiped out. In such a world the arts had no role and no future; the extinction foretold by the history of the player piano – 'You push the button – We do the rest' – was that of the artist. Yet somehow enough disorder remained in the world for the player piano to become defunct in its turn, and for artists, Gaddis among them, to outwit the systems.

# Pneumatic Postal System

Jacob Paskins

Long before email and SMS, the pneumatic post was the quickest way to send a written message of condolence, a declaration of love or even a demand for ransom. Systems for delivering messages through tubes by the force of compressed air originate in experiments undertaken by George Medhurst and William Murdoch at the turn of the nineteenth century. Josiah Latimer Clark created the first pneumatic mail system on an urban scale in London in 1853 to transmit shipping information to the Stock Exchange. Hamburg (1864), Berlin (1865), Vienna (1875) and New York (1898) followed with extensive networks, while other cities including Algiers, Birmingham and Marseilles operated more modest services. The Paris system, dating from 1866, was however notable for its density, its longevity and its accumulation of cultural resonance.

The Paris pneumatic post originated to deliver telegrams rapidly between the stock exchange and the central telegraph office. The system was always a branch of the telegraph (and later telecommunications) division of the postal service (PTT). Opened to the public in 1879, the pneumatic post expanded fast thanks to the installation of subterranean delivery tubes in the city's existing sewer network. By the 1930s some 427 km (265 mi.) of pneumatic pipes connected 130 post offices in the city. The service was available to all addresses within the Seine department, with couriers connecting suburban towns to the inner-city pneumatic network. Initially only small cards could be sent on special stationery (a *petit bleu*), but from 1898 a sender could post any letter weighing less than 30 g (just over 1 oz) within Paris or the suburbs. Fictional *pneumatiques* sent between lovers featured in stories by Maupassant and in Proust, while a notorious historical *petit bleu* was a torn-up, unsent letter that became a crucial piece of evidence in the Dreyfus affair.

François Truffaut's film *Baisers volés* (Stolen Kisses, 1968) provides a glimpse into how the pneumatic post worked. Once a letter was posted into a *pneumatique* letterbox, an operative would cancel the stamps and roll up to five letters into a cylindrical metallic capsule. A machinist would insert the capsule into a despatch machine that projected it into the vast network of subterranean iron tubes. (There are several shots in Truffaut's film of the one-hundred-year-old rusty delivery pipes inside the sewers.) The capsule was propelled through the tubes by compressed and depressed air produced by steam, and later electric, pumps. Capsules would be rerouted manually at post-office junctions until automatic sorting, introduced in the 1930s,

Pneumatic post, Paris, 1956.

allowed them to travel more directly to their destination. Once in its capsule, a letter could cross Paris in about 25 minutes. For the final leg of the journey a telegraph worker would deliver the letter, commonly referred to from the 1920s as *un pneu*, by bike or motorbike. Standard delivery time was less than two hours after posting, with an average of just 45 minutes. Some women supposedly exchanged *pneus* simply to have the pleasure of greeting the young delivery boys.

The pneumatic post remained popular long after the introduction of the telephone, because the French telecommunications network was relatively slow to expand. In the 1960s new subscribers often had to wait several months to be connected to the overstretched telephone system, and sending a *pneu* was an appealing alternative to standing in a draughty telephone cabin. The venerable *pneu* long remained a familiar aspect of urban culture even after the emergence of computers and satellites, not least because of the guaranteed delivery, fixed cost and ability to send original documents. The afternoon newspaper *Le Monde* would sometimes receive *pneus* from readers responding to an article published that day, *before* the evening's second edition had gone on sale. Although a *pneu* was as rapid as a telegram, it was more economical, since a whole letter could be sent for the price of a pneumatic stamp rather than charged per word transmitted. And, crucially, a *pneu* was the original document written by the sender, unlike a telegram, which was the transcript of an electronically coded message. Signed by hand, a *pneu* enjoyed good administrative and legal authority; it was possible to authorize bank transfers by *pneu*. A separate pneumatic network connected government ministries, providing secure communication of confidential state documents.

Although the pneumatic post long coexisted with the telephone, its popularity began to dwindle as investment in telecoms increased. After a record 30 million *pneus* sent in 1945, deliveries dropped to 3.9 million items in 1964, the lowest since the 1890s. In the 1970s, as the infrastructure aged and labour costs rose, the system went into decline and started to lose money. *Pneus* became increasingly uncompetitive with other forms of communication; from three times the price of a standard letter in the early twentieth century, by the 1950s they were five times more expensive. In 1976 the cost of a single *pneu* rose to 8.40 francs, while a monthly telephone subscription cost 39 francs plus call connection charges of 0.39 francs. Delivery times became longer and less reliable as technical failures increased, such as the loss of air pressure in the corroding pipes. The existing postal infrastructure was increasingly neglected in favour of telecommunications; the PTT preferred to invest in telex, fax and Minitel than in a mechanical system that many considered to be technologically obsolete. Only 648,000 *pneus* were sent in 1982, and quieter parts of the system had closed before the service ended definitively at 5 p.m. on 30 March 1984.

Although the service would eventually be superseded by ever more instantaneous means of communication, the *pneu* did anticipate the perils of careless despatch and rapid delivery familiar to today's users of electronic mail and messaging systems. If *Baisers volés* demonstrates the speed of sending a *pneu*, Jean-Luc Godard's film segment 'Montparnasse et Levallois' (part of *Paris vu par . . .*, 1965) warns of the hazards of sending letters in haste. Monika sends two *pneus* from Montparnasse station to two lovers, Ivan and Roger, but moments after posting them she realizes she has mixed up the letters and unsuccessfully tries to retrieve the envelopes from the postbox. She runs to Ivan's nearby sculpture workshop to explain the mix-up before his *pneu* arrives, and confesses she has another lover. When Ivan kicks her out, she hurries to Roger's garage in Levallois-Perret, a suburb linked to Montparnasse by the no. 94 bus. Monika's journey is slower than the post; Roger has already received his *pneu*. She similarly spurts out the truth, but believes Roger has forgiven her. However, Roger reveals that she had not mixed up the envelopes after all; she had needlessly revealed her infidelity twice, pressurized by the speed of the *pneu*.

Antiquated, yet also ahead of its time, the pneumatic post was a sustainable concept that congested cities today can only dream of: an energy-efficient system that reduced road traffic. The closure of the Paris network did not mean the final death of the *pneu*; a service in Prague continued until 2002, and the French governmental pneumatic post outlasted the public system by twenty years. While city-wide pneumatic postal services can now be declared extinct, pneumatic tube delivery systems still thrive on a smaller scale in hospitals, banks, airports and libraries to deliver messages or small items. Renewed interest in the transport potential of long-distance pneumatic tubes is evident in the Hyperloop project being developed in the United States. The aim of this system is to transport passengers in pods that are propelled through a pressurized tube at 1,200 kph (745 mph). If it is ever constructed, this impressive idea could provide a rapid alternative to road and air travel, but time will tell if Hyperloop meets the same fate as 'atmospheric' railways and failed projects to link Paris to London by pneumatic postal tube in the nineteenth century.

# Polaroid SX-70
## Deyan Sudjic

Almost forty years after the last Polaroid sx-70 came off the production line at a specially built factory on the edge of Boston, Massachusetts, Edwin Land's instant camera is still a beautiful and entirely contemporary-looking object. With a brushed-metal finish and inset leather panels, it has the material quality of a cigar case. Pull open the intricate folding mechanism and the viewfinder clicks into position for the mirrors inside to show you the image that the neat square of Polaroid film, lying flat on the base of the camera, will capture. Launched in 1972, the sx-70 was the culmination of Land's determination to liberate photography from the darkroom, a determination that went back to 1948, when he manufactured his first camera, the clunky-looking Polaroid 95. The 95 was tricky to use and involved loading two different rolls, one of photosensitive negative film, the other of positive paper. You took a picture, then you had to wind the mechanism to force the chemically coated paper into contact with the negative. As the image developed it went through a cutter to trim the paper. Finally, as the resulting sandwich of paper, emulsion and chemicals emerged, the user had to peel away the film to see the finished result.

In 1947, when Land turned his company away from war work, photography was still a painfully long-drawn-out process. It began with the opening of the cardboard box in which a roll of celluloid film came packaged in a foil wrapper to keep it safe from direct sunlight. The roll was then threaded on to the spool inside the camera. Only once the whole roll was used up – and with room for up to 36 exposures, that could take days or even, for a frugal user, weeks – the film was extracted and processed in a darkroom, where it had to be developed, fixed, washed and dried before a print could be made, and that in turn had to be developed, fixed, washed and dried. Even the fastest darkroom could not complete the process in much less than a couple of hours for black-and-white processing, longer for colour. In comparison, the Polaroid 95 produced an image in less than a minute.

Over the following decades Land went through a number of iterations to simplify the basic concept, before coming up with the sx-70, which did away with all the complications. It put everything – paper, negative and chemicals – into a single ten-shot cartridge that automated the whole process of making colour prints. When the button on the front of the machine was pushed, the shutter opened to deliver a carefully measured dose of light, then closed again. Seconds later the film was excreted through the mechanism

sx-70 camera and case, designed by Henry Dreyfuss and manufactured by Polaroid Corporation, c. 1972.

propelled by a burst of electricity from a tiny battery. As it passed through the camera, the chemicals contained in a thin pouch that ran across the bottom of each Polaroid cartridge, that would develop and fix the image in less than a minute, were smeared across the surface of the film. A wafer-thin paper rectangle was ejected into the open air, accompanied by a satisfying whoosh. All this was made possible by the 200 transistors and a complex of moving mirrors, light sensors, gears and solenoids packed into the case, and the chemicals that Polaroid insiders called 'the goo'.

The American design critic Phil Patton once described pushing the button on an sx-70 as 'igniting a surprising mechanical rumble, like the sudden lowering of airliner landing gear, that with the camera pressed to your eye travels through your cheekbone. And when you are finished, the device snaps shut with a sound as satisfying as the closing door of a fine sports car.' He was even more poetic about the way the surface of the plasticized paper would go from milky opacity to pin-sharp, full-colour clarity: 'Adriatic milky green slowly acquired the Caribbean blues and rose reds that characterized the sx-70 palette.'[1]

Manufactured from 1972 to 1981, the sx-70 was perhaps as close as analogue technology ever came to achieving the dematerialized immediacy of digital image-making that was already being developed in the laboratories of Japanese mobile-phone companies. Polaroid's response to the threat of digital photography was to try to create less expensive, more popular versions of instant photography, while clinging to the belief that its lucrative business capturing instant images for identity cards and driving licences would insulate it from the need for more far-reaching changes. It was a strategy that led to the company's first bankruptcy, in 2001. The sx-70 is a plausible ancestor for the iPhone. Land saw his camera as something as ubiquitous as a smartphone, always on hand to capture an idea or image. In *The Long Walk*, the film Polaroid commissioned in 1970, he talked about making a tool as ubiquitous as the pencil.

The Polaroid camera was more than a magic trick. It was also highly addictive. Pressing the button, hearing the whoosh and seeing the image take shape produced an overwhelming urge to see and hear the magic repeated. Yielding to the urge, you had to use more and more of the costly instant film. Land was just as adept a marketeer as Steve Jobs. In order to ensure that his product was taken seriously as a piece of photographic equipment, he paid the photographer Ansel Adams a retainer. He commissioned Charles and Ray Eames to make a film to launch the camera. He gave artists, from Robert Rauschenberg to Andy Warhol, access to copious quantities of his products. Some, such as the satirist Ralph Steadman, manipulated the image as it developed, applying heat and pressure. David Hockney, who used several images to create a kind of photographic cubism, was perhaps the most celebrated user. The Italian architect Carlo Mollino was also an avid Polaroid photographer; his archive of erotic images was a foretaste of the possibilities of a medium that no longer had to go through the scrutiny and censorship of a processing laboratory.

Polaroid's associations with high culture, and the sx-70's insatiable appetite for film, helped to shift vast numbers of the company's main consumable product. At its height, at the end of the 1980s, Polaroid had a turnover of $3 billion, and 21,000 employees. Ten years after Land's death in 1991, however, the company was forced into bankruptcy. What killed Polaroid was much what killed Kodak – the inability to reinvent itself quickly and convincingly enough and to find an alternative to the huge profit margins it had enjoyed from the sales of its film. After two successive Chapter 11 filings, first in 2001 and then again in 2008, Polaroid stopped making film, rendering the instrument itself useless. Recently, with what began as a crowdfunding project and has subsequently turned into a business, Polaroid film is once again being manufactured to satisfy a hunger for things that people can touch and feel, rather than the slippery elusiveness of pixels. At the same time, manufacturers have re-entered the instant camera market with models that conspicuously lack the conviction of the original sx-70, sold as a nostalgic novelty for the selfie generation. But nostalgia was not what Land had in mind, and we are left with an object stripped of its original, forward-looking purpose.

# Public Standards of Length
David Rooney

When the Palace of Westminster burned down in 1834, Britain's official standards of weight and measure – a set of physical weights and lengths held there for safe keeping – perished with it. Four years later the Astronomer Royal, George Airy, was appointed chair of a commission to restore the standards and protect them from further loss. Crucial to Airy's approach was the dissemination, far and wide, of public standards, partly to hedge against the possibility of total loss, as had happened in 1834, but more significantly to embed accurate measurement visibly into public life. A recurring theme in discussions of public measurement at that time was the accuracy of surveyors' chains, the 100-foot and 66-foot devices used to measure land, which were prone to stretching with repeated use. A legal case heard in 1844 led the judge to pronounce that every surveyor in the country should be able to prove the accuracy of his measuring chain. Manufacturers, merchants, builders and craftspeople needed access to smaller measures, too – yards, feet and inches. But how?

The answer, according to Airy's Standards Commission, lay in publicly accessible standards, verified by government officials, mounted in walls and pavements in every town and city, where people could check their chains and yard rules. Standard metres in public places had been displayed across Paris since the 1790s, and Airy's report of 1841 commented that Prussian law demanded 'the fixing of mural standards of length in public places'. This was the model proposed for the UK.[1]

Airy, in his official capacity, was the first to act, setting up public length measures outside the Royal Observatory, Greenwich, on 30 January 1859. His measures took the form of a metal tablet showing certified measurements of 1 yd, 2 ft, 1 ft, 6 in. and 3 in. Visitors could place their rule on the appropriate set of pegs between end-stops. Rules of the correct length would fit exactly. Soon afterwards, similar installations could be found elsewhere in London, at the National Gallery in Trafalgar Square and the premises of De La Rue printers in Bunhill Row. In the 1870s public standards were installed in Trafalgar Square itself and at the City of London's Guildhall. But length was an industrial matter, and the industrial powerhouses were in the north. Manchester, the 'Cottonopolis', set the trend by installing length standards in front of its town hall in the late 1860s, and in 1876 the government wrote to every British local authority encouraging them to follow suit.

Public standards of length tablet mounted outside the main gates of the Royal Observatory, Greenwich, UK, in January 1859. Still in situ.

Standardized units of length had been around at least since Roman times, and by the nineteenth century accurate measurement was regarded as a public good. The science historian Simon Schaffer has commented that in Victorian Britain 'exact measurement was advertised as a vital accompaniment of commercial, military, and thus imperial triumph. Its role was most vivid in the context of metrology, the process through which the British state constructed and disseminated standards.'[2] The government's letter of 1876 came at a time when local authorities across the UK, particularly in the northern industrial centres, were constructing bold new civic buildings that transformed the urban landscape.

If accurate measurement represented probity, then wall-mounted standards of length embodied morality, and a set of length measures became part of the commonplace architecture of these industrial people's palaces. Many were set up in the 1880s, and by 1914 public standards of length could be found on civic buildings in Birmingham, Bradford, Dublin, Edinburgh, Glasgow, Gloucester, London, Manchester, Sheffield and Reading, as well as fifteen industrial towns in Lancashire under the control of the County Palatine of Lancaster, including Ormskirk, Rochdale and Wigan. When Manchester set up new standards outside the Assize Courts in 1877, county officials proudly advertised the service in the *Manchester Courier and Lancashire General Advertiser* (26 October) to 'builders, surveyors, artizans, and the public generally . . . for the purpose of affording the public a ready means, without expense, for testing the correctness of all measuring chains or tapes, yard wands, foot rules, &c'. For towns that made their money from cotton, cloth, steel or wool, length mattered.

But this pride was short-lived. The outbreak of the First World War marked the end of the age of the public standards tablet, and no more were installed after 1914. There is, in fact, little evidence that they were much used even when new. For most of the tablets, official reverification petered out soon after installation, leaving their accuracy open to question. It is hardly surprising. In 1887, just eleven years after they were installed, the Trafalgar Square standards on their hefty granite base had become, in the rueful words of the Board of Trade official carrying out reverification that year, 'a lounge for the outcast whose manners and customs are unhappily not conducive to the comfort of those who have to use our standards'. In 1916 officials found the Edinburgh standards 'hidden and rendered practically useless by the near proximity of a YMCA hut and . . . covered with packing cases'.[3]

There are technical reasons for the extinction of the public length standards in Britain's towns and cities in the decades following the First World War. By the turn of the twentieth century, surveyor's chains were rapidly being replaced by steel tapes, which were easier to use and far less prone to stretching. Length measures in engineering and manufacture had also

Public standards of length at the Royal Observatory, Greenwich, from a postcard published *c.* 1930.

become far easier to verify with the invention in 1896 of slip gauges, precision metal blocks that could be stacked together to make standard lengths, which were both accurate and portable. By 1963 the leading metrologist Kenneth Hume was able to claim that 'slip gauges are used as standards of measurement in practically every precision engineering works in the world.'[4] No longer must provincial workshop managers travel to distant town centres to check their yardsticks. Standard lengths had been disseminated to every toolroom and workshop – the ultimate realization of Airy's desire to spread standardization far and wide.

The writing was on the wall for physical standards at the scientific level, too, in this period. Experiments in the 1890s showed that length could be defined in relation to the wavelengths of light, which is a fundamental property of matter, although practical standards had to wait until work during the Second World War on atomic reactions led to chemical isotopes that were suitable for defining length. In 1957 an international committee decided that the light from a krypton-86 hot-cathode lamp was the answer to this technical problem, and three years later the world's standards body approved its adoption as the standard. Physical length standards were not only toppled from the apex of the measurement hierarchy, but consigned to history.

But the extinction of the public standards came from wider shifts in industry, too. In the Victorian age, standards of length on public buildings had promised a future in which standard measurement, as a public good, was highly visible, but this cut across the ambitions of the surveying and engineering professions that wanted to protect their measurement culture behind high disciplinary walls. In the twentieth century a global infrastructure of standards testing and verification grew up that far outstripped modest tablets on the walls of Victorian town halls. Today, most of the public standards of length remain on show in town centres across the country, just where they were installed almost a century and a half ago, but their usefulness was rendered extinct long ago.

# Pyrophone
Tim Boon

Owing to the environmental imperatives of climate change, it seems the gas era is drawing to a close. In Britain it has been nearly two centuries since piped town gas became normal in cities and towns, and virtually a century since the universal availability of standardized electricity via the grid threatened to eclipse it. Gas has kept its place, however, not only as an energy source, but as a signature part of our technological culture; today not just domestic boilers and hobs run on gas, but also many of the great steam turbines of our electricity-generating stations. In the 1930s the Gas, Light and Coke Company employed the modish modernist architect Maxwell Fry to design the Kensal House flats in west London. The striking feature of the flats was the deliberate absence of that modernist energy source, electricity, which was rapidly making inroads into the market previously dominated by the older fuel. At Kensal House all lighting, space and water heating was achieved by gas, or that other product of the gasworks, coke. It sounds like a counterfactual modernity, but it did happen, and the company made a film in 1937 to promote its merits. If only they had known about the pyrophone, they could have boasted about music made by gas, too. A short trip south from the flats in Ladbroke Grove to the acoustics gallery of London's Science Museum in South Kensington in the 1930s would have revealed to them one of the rare surviving examples of the this device.

Pyrophones – fire organs – make use of a physical phenomenon known as 'the singing flame'; the natural philosopher Bryan Higgins observed in 1777 that, under the right circumstances, and correctly placed, a gas flame in a tube will produce a musical note. Several decades later the physicist Ernst Chladni established that the longer the tube, the lower the pitch (just as with an organ pipe), and it was further shown that two or more flames in a tube make a sound when separated but not when united. This was the phenomenon applied by the minor French physicist Frédéric Kastner when he invented the instrument he named the pyrophone, patented in 1873; each of the thirteen tuned tubes in the Science Museum's single-octave instrument has a cluster of pivoted miniature burners that are mechanically separated by the pressing of the corresponding note on the keyboard.

In an unlikely historical detail that would jar if it were used as a plot device in a second-rate novel, Kastner's wealthy mother secured the services of Henri Dunant, founder of the Red Cross – who was then down on his luck – to promote her son's invention. As a result, the instrument was demonstrated

Pyrophone, patented and made by Frédéric Kastner, France, 1876.

at the Royal Institution in London in January 1875, accompanied by a lecture given by the celebrated physicist John Tyndall. A second performance occurred at the Society of Arts in the following month. Dunant, in his lecture to the Society of Arts, was eloquent about the instrument's sonic qualities:

> The sound of the pyrophone may truly be said to resemble the
> sound of a human voice, and the sound of the Aeolian harp;
> at the same time sweet, powerful, full of taste, and brilliant;
> with much roundness, accuracy, and fullness; like a human and
> impassioned whisper, as an echo of the inward vibrations of the
> soul, something mysterious and indefinable; besides, in general,
> possessing a character of melancholy, which seems characteristic
> of all natural harmonies.[1]

After the instrument developed a mechanical fault the following year, Dunant deposited it at the Science Department of the South Kensington Museum (later the Science Museum). It was quickly included in the acoustics room of the Special Loan Collection of Scientific Apparatus in 1876, the largest temporary exhibition the museum has ever mounted, featuring some 20,000 objects. In common with many instruments in this 'loan' collection, it never left the museum's collections after 1876, and, indeed, remained on display for decades. In many ways it is emblematic of the Science Museum's attitude to collecting musical instruments; it was collected as a curiosity, an object encapsulating the ostensibly musical side effect of a strange natural phenomenon. There, it joined Charles Wheatstone's 'Magic Lyre' (a demonstration of sonic resonance) and an Aeolian harp, among other devices.

We may ask what made the pyrophone become extinct. It is true that the history of music is replete with 'lost' instruments such as the medieval serpent, a bass reed instrument sometimes used to accompany church services. Such instruments may occasionally come off the 'endangered species' list by virtue of a revival of the repertoire to which they are attached; that has happened with both serpents and harpsichords. But where there is no established repertoire, such instruments tend to become museum pieces. For example, Jean-Baptiste Vuillaume's 'octobass', a 3.48-metre (11½ ft) bowed bass string instrument considerably larger than a double bass, suggests an amusing musical path not taken; it is shown, like the pyrophone, in a museum – in its case, the Cité de la Musique in Paris – but both are curiosities. It is said that both Hector Berlioz and César Franck played the pyrophone at Kastner's house in Paris, and the museum's documentation includes sheet music for 'Petit Prélude', a composition especially written for the instrument by Théodore Lack. But this scarcely amounts to a significant repertoire demanding the instrument's revival.

Sounding museum objects are likely to retain their 'extinct' status even more than their non-sounding fellows in the museum stores, because they must be played to reveal their characters and escape obsolescence. It is fascinating, therefore, to note that in the early 1950s the Keeper of Physics, David Follett (later director of the Science Museum), became intrigued by the pyrophone and had it renovated to working order; there are photographs of him playing it during an event to mark the 124th anniversary of Dunant's birth, at the venue – the Royal Society of Arts – where it had been demonstrated by Dunant 77 years earlier. It was even played on a novelty music radio programme, *Lend Me your Ears*, on 6 November 1951. A witness to a performance of the reconditioned instrument was somewhat less fulsome about its sound qualities than Dunant had been back in 1875; it was described as 'melancholy . . . but more reminiscent of the mournful mooing of a constipated cow, or, in the deeper notes, of the *Queen Mary* in the distance in a fog'.[2] This may well be the true reason for its failure to catch on in the 1870s.

Perhaps the 1930s counterfactual of an alternative modernity in which gas reigns triumphant is the wrong one to choose for the pyrophone. It is far easier to imagine pyrophones providing a soundtrack to the speculative, steampunk world of William Gibson and Bruce Sterling's *The Difference Engine* (1990), in which Charles Babbage's calculating machines have ushered in a Victorian information age more than a century ahead of time. Or pyrophones might equally evoke the age before modernity. It was rumoured that the French composer Charles Gounod planned a pyrophone part for his opera *Jeanne d'Arc* (1887). The notion of a fire organ as the soundtrack for a medieval auto-da-fé seems gruesomely apt.

# Realistic Wax Mannequin

## Maude Bass-Krueger

What kind of mannequin should be used to display fashion? As the burgeoning new department stores and couture houses of the mid-nineteenth century began to contemplate this question, a compelling argument was made for using realistic bodies with lifelike heads and limbs moulded from wax. They were considered the most modern, artistic and luxurious mode of display by Parisian fashion designers and manufacturers, and were a far cry from the inexpressive figures that now populate retail windows.

Animal and vegetal wax had been modelled into masks for funeral effigies by the Egyptians, Greeks and Romans, but it wasn't until the late nineteenth century that wax mannequins were used for fashion display. Ceroplastics, or the art of wax modelling, was popularized in the late eighteenth and early nineteenth century by Philippe Curtius and his housekeeper's daughter, Marie Tussaud. Their wax museums attracted large crowds in London and Paris, and visitors were eager to marvel at the lifelike wax replicas of historical figures and infamous criminals in the Chamber of Horrors. From the eighteenth century onwards, tailors, textile manufacturers and designers had used small porcelain dolls to commercialize their wares. The new department stores of the 1850s, eager to capitalize on the public fascination with wax museums, began to use wax mannequins to sell ready-made fashions. Couturiers also turned to high-quality wax mannequins to display their wares at Universal Exhibitions.

These wax mannequins, which were uncannily realistic, were used to display not only contemporary fashions around the turn of the century, but historic fashions. Old garments and accessories began to be studied, collected and exhibited as objects of historical interest in France, Germany and England from the 1850s onwards. Historians, collectors and curators fought over the proper way to display historic garments in exhibitions of fashion history, which were organized during Universal Exhibitions, at department stores and in museums. Some argued that presenting the garments flat in glass cases or on headless mannequins made from wood and buckram was more scientific. Others were convinced that dressing the fashions on lifelike wax mannequins posed in tableaux conveyed historical experience more authentically. By the time of the 1900 Universal Exhibition in Paris, the two sides were deeply entrenched in their opposition. The organizers of the 'Musée rétrospectif des classes 85 & 86: Le costume et ses accessoires', an exhibition of historic dress on the ground floor of the official Palais du

Wax mannequin bust, designed by Pierre Imans, France, c. 1910, painted wax, residual hair, silk ribbon, cotton net and resin.

Vêtement, used free-standing glass cases and headless mannequins to present eighteenth-century garments. Across the esplanade, at the Palais du Costume, a privately funded exhibition organized by the couture house Maison Félix, visitors walked past 34 tableaux vivants dioramas where recreated versions of historic dress were mounted on realistic wax mannequins.

The proponents of wax mannequins for the display of both historic and contemporary fashions won over their more conservative colleagues, and the public clearly favoured the spectacle of looking at the arrestingly lifelike wax figures. Over the course of the twentieth century, as wax mannequins became increasingly refined, their makers thought of themselves as artists and sculptors, rather than industrial manufacturers or modellers. In 1896 Pierre Imans, a Dutch-born ceroplastician, had begun making a kind of wax mannequin that he considered to be far superior to those on display at the popular wax museums. Imans's 'waxes' (he refused to call them mannequins) featured face paint, resin eyes, and eyelashes and hair made from human hair. The heads were made by pouring hot wax into a form; the eyes, eyelashes and hair were fixed into the form while the wax was still warm. Imans initially placed wax busts on rigid mannequin forms, but by 1911 he had created articulated bodies as well.

Jeanne Paquin's lifelike displays for the Turin International Exhibition in 1911 represented the apex of the fashion for realistic wax mannequins. 'So real one believed them to be alive', a reporter for *Les Modes* wrote about the über-realistic wax mannequins Imans made for Paquin. Yet, as taste shifted from realism towards modernism after the First World War, lifelike mannequins became, in the words of the critic Guillaume Janneau, 'frightful wax cadavers, disturbing facsimiles'. The abstracted bodies favoured by the post-war Cubist and Futurist artists were seen as more modern than the mimetic bodies of pre-war realism, which recalled a bygone era. More crucially, the advent of electric lighting forced manufacturers of wax mannequins to find new materials that – unlike wax, which melted when hot – would hold up under the new lights. In 1922 Imans began manufacturing his mannequins from carnesine, a mixture of plaster and gelatin.

At the 1925 Pavillon de L'Élégance – the couture pavilion at the Paris Exposition Internationale des Arts Décoratifs et Industriels – the mannequin manufacturer Victor-Napoléon Siégel designed papier-mâché mannequins with stylized features that reflected the new modernity of the 1920s woman. Like Imans, Siégel thought of himself as an artist, but he differed from his mentor in his belief that mannequins reached maximum artistry by distancing themselves from nature, not by seeking to reproduce it. He abstracted the features and bodies of his mannequins and painted them in unnatural colours such as red, gold, silver and green. Shop windows no longer recreated lifelike scenes through narrative *tableaux vivants*, but instead captured the

Jeanne Paquin display
at Turin International
Exhibition, photograph
from *Les Modes*,
August 1911.

consumer by taking their attention off the mannequin and its setting in order
to focus their eye on the clothing. By the time of the Universal Exhibition of
1937, realistic wax mannequins were no longer sought out by designers and
department stores for fashion display. Terracotta giants made of roughly finished plaster and oakum were commissioned from the young sculptor Robert
Couturier, who described them as 'intentionally devoid of any pleasantness,
of the gentleness that usually goes with elegance'.

Although some fashion designers and department stores kept using
mannequins with lifelike features, they were no longer made of wax, but
rather plaster and then fibreglass. Lighter, less fragile and better adapted
to mass production, these new materials are used to create the increasingly
abstract and elongated mannequins that are preferred by fashion designers,
department stores and museums today.

# Rotring, Letratone, MiniCAD
Tony Fretton

Steady north light fell through the windows of Arup Associates' drawing studio in Fitzroy Mews, London, each day – and on one Monday in 1971 through the roof, when part of the Telecom Tower fell in, dislodged by an IRA bomb (detonated in those more considerate days on Sunday, with prior warning). Ignoring this, we set to work, patiently fixing 112 g tracing paper to our drawing boards with Scotch tape, smooth side up, and laying out our drawings with a parallel motion, set square, compasses and scale rule, first with a clutch pencil honed in a rotary sharpener, then with Rotring pens, starting at the top and working from left to right so as not to smudge, and finally stencilling on lettering and applying Letratone when needed. Revisions were made by scratching out part of the drawing with a razor blade, burnishing the tracing paper with the metal cap of a clutch pencil and then drawing on the new parts. The personality of the draughtsperson could be seen both in the quality of scratching out and the style of the drawing. Some were admirably clear and to the point; others had more going on in the background than necessary, an outcome of the priority given to free ex- pression over competence in UK schools of architecture. At the Architectural Association, from which I had temporarily escaped, the chairman Alvin Boyarsky famously said, 'We fight for Architecture every day with drawings on the wall.' Brilliant though he was, I imagine he never had to construct a building or defend claims with such drawings. In contrast, American architects passing through the office drew with sterling clarity, entirely in pencil and lettered by hand in a standard style that could be seen from SOM to Frank Gehry.

Drawings for issue to contractors and consultants were sent out to a print bureau, and the returned prints were sorted into duplicate sets, put into envelopes and taken to the office post room or the post office. A set of reference prints were kept in the drawing studio, clamped together in a rack, and the original tracings were edge-bound and hung in a closed cabinet during the life of the project, before being archived somewhere else forever. Office cultures grew around these objects and procedures. It was etiquette not to dust your board off on to your neighbour's, invade their layout space or borrow their equipment without permission. And there were styles of dress: suits and ties for management, Levis and desert boots for board jock- eys, or black roll necks and tapered trousers if you were at Norman Foster's. There were practical jokes, office love affairs, surreptitious reading of job

Isometric drawing of Lisson Gallery, London, architect Tony Fretton, 1984, ink on tracing paper, 59.5 × 84 cm.

advertisements in the architectural magazines, Olympic levels of smoking, and prodigious alcohol consumption on Fridays after work.

All the ways of producing and disseminating drawings that I have described are now obsolete. Even the words 'Rotring' and 'Letratone' attract a spellcheck query. But obsolescence is not confined to the past; it is an ongoing process in which some things quietly disappear, others remain and are reinterpreted, and new things acquire meaning, and all are incorporated into the ongoing narratives and habits of our lives and working practices.

In my own office a decade or so later, in the narrow living room of my apartment, we still produced and disseminated drawings in the same way, but in a very compressed manner and with an office culture consisting mainly of drinking powerful Bloody Marys on Friday evenings. At a certain moment when projects of the right scale converged with affordable computing, we decided to change to digital. Given the scale of the room, we had to get rid of the drawing boards to make way for the computers. Suddenly we found ourselves looking at monitors and keyboards without knowing what to do, until a recent graduate patiently taught us.

Over time and in a series of different studios, a new digital world opened up. Its connectivity and speed were exhilarating. You could view a wide range of drawings and documents concerning different aspects of a job on screen at the same time, and see them while travelling or out of the office. Design discussions could take place in front of the screen, using three-dimensional models of the design that could be moved and viewed from any angle. Meetings could take place over Skype. The studio could be smaller, because everything was stored in a server that was the size of a small suitcase, then a briefcase and now a pencil case. Digital drawing could be expressive, ironic and amusing, as I found when misusing a drawing program to make freehand sketches, and later when drawing with a finger on my iPad. Working drawings still showed the personality of those who made them, because the same judgements had to be made as in hand drawing about line weight and colour, establishing foreground and background, using the expressive power of unfilled space and composing the drawing within the boundaries of the page.

Digitalization was greatly extending the practices we knew and adding breadth, richness and encounters with the unexpected. But a darker unexpectedness has opened up. As operating systems, drawing programs and computers are being updated with increasing frequency, drawings from even five years ago cannot be opened except by laborious conversion between programs, and perhaps in the future not at all. Their inaccessibility puts them at risk of being abandoned, and then forgotten. If similar things are occurring across the digital world, significant data is becoming inaccessible and hence lost without our even knowing, affecting our access to the past

Screenshot: 'There is no
application set to open
the document . . ., 2020.

and our sense of the present. For us as architects, the symbol of hope in
the digital world has been extreme compactness and great reach, while the
symbol of its inbuilt obsolescence is an array of black exec files overlaid
with the error message 'There is no application to open these documents.'

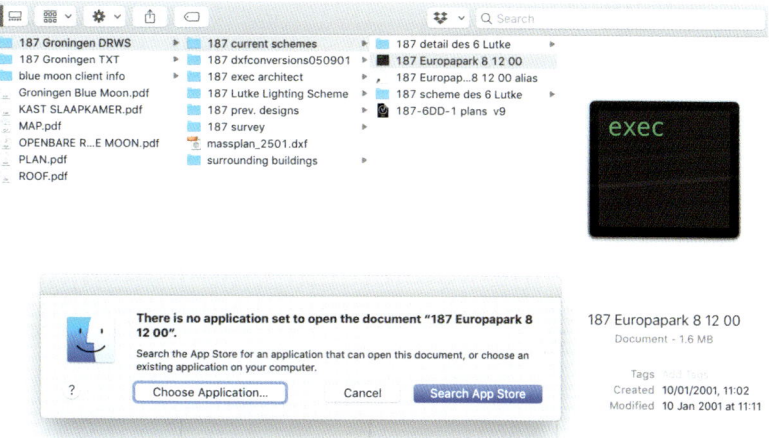

# Scaphander: 'Man-boat'
## Steven Connor

The Abbé Jean-Baptiste de La Chapelle (1710–1792) was a mathematician who made a reputation for himself writing books on the teaching of mathematics, arguing optimistically that, of all forms of knowledge, mathematics was most naturally adapted to children. He was admitted in 1747 as a Fellow of the Royal Society in London, and also became a fellow of the academies of Lyon and Rouen, as well as a Royal censor. He contributed 270 articles to Denis Diderot and Jean le Rond d'Alembert's *Encyclopédie* during its development in the 1750s and 1760s. During the 1760s, La Chapelle started to extend his interest into other areas. One was the history of ventriloquism, which led to the publication in 1772 of a compendious account of ventriloquial phenomena and performances through history, *Le Ventriloque, ou l'engastrimythe*. The other was the development of a device that he called the *scaphandre*, or man-boat, from the Greek σκάφος (boat) and ἀνήρ (man).

La Chapelle's treatise of 1775 on the construction and use of the scaphander ran to 328 pages, but the device it concerns is described succinctly enough in its prospectus, as 'a corset, by means of which men and women may, while fully clothed, much better than without clothing, swim without ever having learned to, while remaining completely upright, floating immersed only up to the level of the chest'.[1] In fact, the corset was more like a bodysuit, combining waterproof trousers with a cork-lined linen tunic. La Chapelle's research was principally motivated by the conviction that it is unnatural for a human, as a creature that walks upright, to adopt the horizontal posture of a quadruped while swimming, especially given the difficulty of breathing when facing down into the water. He observed the advantage accorded to the elephant over other quadrupeds, in the form of a trunk, which may be used as a breathing tube. In fact, one of the uses La Chapelle predicted for the scaphander was that it would make learning to swim easier, although it is not easy to see how.

In fact, La Chapelle seems to have been unsure whether his invention was to be thought of as a means of propulsion or an aqueous habitat. He demonstrated it to the Académie Royale des Sciences in August 1765, by entering the Seine and proceeding to perform various actions that would be impossible while swimming, including eating, drinking, writing, taking snuff and, somewhat surprisingly, firing a pistol. La Chapelle included in his design a hat in which writing materials and comestibles could be kept dry. The firing of the pistol may have been designed to show the utility of the

Scaphander suit, engraving by J. Robert, from Jean-Baptiste de La Chapelle, *Traité de la construction théorique et pratique du scaphandre, ou du bateau de l'homme* (1775).

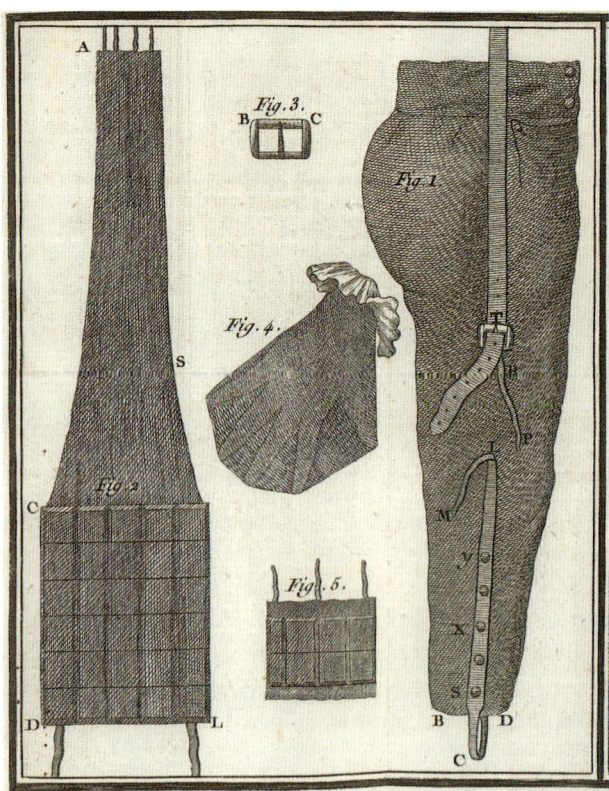

Bottom half of the scaphander suit, engraving by J. Robert from Jean-Baptiste de La Chapelle, *Traité de la construction théorique et pratique du scaphandre, ou du bateau de l'homme* (1775).

scaphander for hunters of water birds (a startled swan perches on the head of one of the figures used to illustrate the device, not providing very much in the way of camouflage, if that is its purpose), or to suggest its military applications, for La Chapelle envisaged it being used by invading armies to cross rivers (as well as for retreating armies to make their escape by water).

These military possibilities may have been to the fore when, in 1768, La Chapelle sought to repeat his success in front of Louis xv in the Seine in the Forest of Sénart, south of Paris. On this occasion, the drawback that the scaphander could be propelled only by a kind of pedalo action of treading water was made painfully clear when the strong current swirled La Chapelle away from view so swiftly that the king was unable to make out anything of his display. This put paid to any hopes the inventor might have had for celebrity and reward as a result of his device.

It would be pleasing to find some connection, other than that of their shared oddity, between La Chapelle's parallel enthusiasms for voice-throwing and water-walking. He certainly seems to have associated them closely. He inserted an announcement of his forthcoming treatise on the scaphander into his book on ventriloquism, and then, in the opening pages of the treatise, devoted a long footnote, extending over some seven pages, to answering some of the objections to the earlier work (including the objection that its long footnotes get in the way of the reading). Since it was La Chapelle's aim in *Le Ventriloque* to show how the apparently supernatural exercise of ventriloquism may satisfactorily be explained by physiological mechanisms, the two projects are perhaps held together by an Enlightenment faith in the power of practical reason. Despite claiming that anyone could easily learn how to construct a scaphander, La Chapelle uses the occasion of his book to see off the claims of a rival, a tailor named Bailli, who, having been employed in the development of the scaphander, then tried to pass himself off as its originator, despite having 'not the slightest notion of mathematics, geometry, physics, mechanics, hydrostatics, anatomy, etc'.[2] There is rather an extensive discussion of ventriloquism in the *Encyclopédie*, although it is not by La Chapelle, and the scaphander is not among the many pieces of technical apparatus treated in it. This would suggest that, despite La Chapelle's enthusiastic promotion of his device, it never came near to being put into production, nor did it prompt any gadgeteers to produce their own versions.

Nevertheless, if Chapelle's invention never established itself in the waterways of Europe, the name he devised for it has had more staying power. In 1810 the malacologist and expert on giant octopuses Pierre Denys de Montfort introduced 'scaphandre' as a name for a genus of sea snails that includes *Bulla lignaria*, the woody canoe-bubble, giving rise to the family known as Scaphandridae.[3] The word 'scaphander' began to be used in English in the early nineteenth century to refer to a cork flotation band, and in 1895 it was applied to an asbestos suit developed for the Berlin Fire Brigade. *Scaphandre* is now the usual French term for a diving suit, adapted as *scaphandre autonome* for the aqualung and, as *scaphandre spatial*, extended to the space suit. Something of La Chapelle's absurd but utopian conception of a costume that is magically also a space of habitation and means of locomotion survives in the idea of powered exoskeletons, as embodied in the Iron Man Marvel superhero and the automated leg-gear in the Aardman Animations film *The Wrong Trousers*. Perhaps this is the dream the scaphander briefly exemplified – of a kind of technologically perfected second body that could at once shelter and project, allowing its user-occupant to calmly abide while boldly going.

# Scarificator
## Thomas Kador

The object is made of polished brass with an angular shape; a cuboid, with an octagonal profile when viewed from above. Fitting neatly into the palm of an adult hand and cool to the touch, it sports an almost irresistible trigger on the top, as well as a circular dial in the centre and a smaller round button on one of the sides. It is clearly a mechanical device of some sort.

Careful manipulation of the trigger, dial and button reveals more about its function. When we pull back the trigger, overcoming considerable resistance, twelve blades briefly emerge out of six linear grooves at the bottom of the object, before disappearing again as the trigger clicks into place. When we press the button on the side, the instrument makes a clicking sound, as if a spring has snapped, and the trigger quickly pops back into its original position. At the same time, and almost imperceptibly, the twelve sharp blades re-emerge from their grooves and snap back to their initial position. Anything held close to the bottom of the device would be cut by those blades, receiving twelve small but deep incisions. The dial on top allows the depth of the blades to be adjusted.

People encountering this object for the first time are endlessly intrigued and offer wild guesses regarding its function, ranging from the processing of various foods – carrots, potatoes, eggs – to cigar cutting, to paper or fabric incising. These suggestions tend to become less and less assured as observers practically think through how the instrument would be operated. Having exhausted other possibilities, they usually at last surmise – correctly – that the device is designed for cutting human or animal skin in a quick, clinical fashion. It operates in a similar way to the finger-stick test, used routinely for extracting a single drop of blood in order to check haemoglobin or insulin levels as well as to test for HIV, but obviously this device – the scarificator – works on a much larger area.

Over the second half of the nineteenth century, scientists such as Louis Pasteur in France and Friedrich Henle and Robert Koch in Germany put forward theories that most human and animal diseases were caused by micro-organisms or germs, which we now largely know as viruses and bacteria. Up to that point, the most popular theory was that infectious diseases such as cholera, tuberculosis and the bubonic plague were brought about by miasma (bad air). Other illnesses, which we would nowadays characterize as metabolic, neurological or psychological, were largely attributed to the four basic humours that every human was thought to possess: phlegm, black bile,

Scarificator (probably of Coxeter's Roto-Lever type), UK, mid-19th century.

yellow bile and blood. This theory dates back to the ancient Greek forefather of Western medicine, Hippocrates, who further suggested that each of these humours resided in a specific organ or region of the body, and that people's personalities – phlegmatic, melancholic, choleric or sanguine – were determined by how the humours were balanced. It was believed that most illnesses came about if the balance between the humours was substantially out of kilter. Consequently, the most appropriate treatment would be to restore that balance by reducing the amount of any excessive humour.

Largely thanks to the writings of the late Roman medical scholar Galen of Pergamon, blood was considered the dominant of the four humours, and thus bloodletting became a standard treatment for a range of illnesses and conditions. The most effective way of bleeding a person without – at least immediately – killing them was through venesection, that is to drain one of their main veins. Several historical texts and depictions demonstrate that, as in a blood-donor clinic today, the vein of choice appears to have been the median cubital on the inside of the elbow. For this purpose, a simple cut with a scalpel, lancet or fleam would suffice. However, as an alternative with a more localized focus, capillary bleeding could be effected by cutting the skin on virtually any part of the body, frequently assisted by a cup to create suction and encourage greater, more rapid bleeding. The process of administering localized capillary cuts is known as scarification. For a quick, precise, consistent and clinical application of these cuts, a device such as the one discussed here would have been the tool of choice, especially in the nineteenth century.

The oldest known mechanical scarificators seem to date back to the second half of the seventeenth century in central and southern Europe. Their practical application is discussed in a collection of treatises on 'bloodletting, cupping and scarifying' published by Nicholas Culpeper et al. in London in 1663. One beautifully adorned example from 1669, produced by Giovanni Battista Boeller's Italian workshop, can be seen at the Science Museum in London. However, the example depicted here is most probably one of J. Coxeter's Roto-Lever Scarificators, patented in London in 1845. The key innovation of this particular model is that the rotation lever allowed the operator to adjust the depth to which the blades would penetrate the skin. This helped to ensure that the cuts were deep enough to cause the desired amount of bleeding, depending on which body part the device was applied to, but also that they did not go too deep, since excessive blood loss was a major risk of scarification. Subsequently, further developments of similar instruments were patented in both Europe and the USA into the first decade of the twentieth century.

The effectiveness of bloodletting as a general treatment began to be disputed from the mid-nineteenth century onwards, as physicians in both Europe and North America reported on the practice's harmful outcomes. Apart from the discovery of pathogens as the true cause of many infectious diseases, the advent of medical trials with large patient cohort studies in the late nineteenth century helped to demonstrate that blood-letting frequently resulted in higher mortality rates than alternative treatments. Yet the medical establishment held on to its belief in the efficacy of venesection well into the twentieth century. Several key medical textbooks, including Robert Hutchinson's *Index of Treatment* (1936) and William Osler's *Principles and Practice of Medicine* (last revised in 1942) recommend it for a range of conditions.

Today, there is little dispute that blood loss, such as that induced by bloodletting, can significantly weaken patients and even cause their death, and therefore draining a person's blood is no longer a major part of the repertoire of contemporary medicine. There are a few exceptions, however, including conditions, such as hemochromatosis, for which bloodletting brings real benefits. And, in a limited number of procedures, such as reconstructive surgery, blood-sucking leeches are still used. At the margins of medical practice, scarifying also gets an occasional mention. For example, in her book *The Art of Cupping* – currently in its fourth edition (from 2007) – the popular complementary medicine practitioner Hedwig Piotrowski-Manz includes a brief chapter on wet cupping and explicitly discusses scarificators, although even she suggests that disposable surgical needles might be preferable, to reduce the risk of infection. While the practice of bloodletting may not be entirely extinct, scarificators will not be encountered in the hands of medical practitioners again any time soon.

# Serving Hatch
Tim Ainsworth Anstey

A radical innovation from the 1920s, the serving hatch became widespread in British homes after the Second World War. It was embraced during the 1960s by a middle class whose houses lacked servants but included technological innovations such as wall-to-wall carpet or an eye-level grill. Its popularity declined in the 1970s as social ideas about domestic space changed and as another new invention, the recirculating cooker hood, spread. With the unconscious eyes of a child, I noted the effect of these developments on the domestic arrangements of my near relatives.

My Aunty Joyce lived in a 1950s architect-designed house in The Keep, Blackheath, London. She had a shelving unit that acted as a screen between her kitchen and dining area, but this was permeable to airs and vapours: a design feature, yes, but not a serving hatch per se. The serving hatch proper characterized a more provincial form of spec-building. My Uncle John and Aunty Joan lived in a house called 'Kennet' in Church Crookham, Hants, which conformed to the type. Bought new in 1965, it had two adjacent doors in the back wall of its entrance hall, one leading to the kitchen, the other into the dining room. These were separated both functionally and aesthetically, the kitchen in powder-blue Formica, the dining room with an Axminster carpet and heavy curtains. On the wall between them was a little opening. In the dining room it floated, framed on the green- and gold-patterned wallpaper like a too-low-hung picture. In the kitchen it was wedged between worktop and overhead cupboards. It was painted in white gloss enamel, had small shutter doors and was repeated in thousands of homes across the country.

What was the appeal of this ubiquitous shuttered opening? There were pre-war English precedents: during the 1920s and '30s posh homes with serving hatches appeared in society magazines such as *Country Life* and *Punch*.[1] And there were international and modernist ones. The kitchens of the Weissenhofsiedlung housing in Stuttgart by the Dutch architect J.J.P. Oud had serving hatches, images of which circulated widely after its completion in 1927. Yet the connection between such examples and the later popularity of the hatch seems somewhat tenuous, partly in terms of understanding (it's a long way from Stuttgart to Church Crookham, and I doubt John or Joan had heard of, or could even pronounce, *Weissenhofsiedlung*), partly in terms of taste: Oud's kitchens were hardly what the average British householder might have called homely.

Serving hatch, no. 1 Shorefield Way, Milton-on-Sea, Hampshire, UK, *c*. 1969.

But serving hatches had another identity in post-war Britain, at once more prosaic and more nostalgic, which does perhaps speak to their appeal. Most people who bought new homes in the 1960s had served in wartime. And war made ubiquitous the experience of another kind of serving hatch, previously associated with factory sirens and the humdrum lives of blue-collar workers: that of the canteen. Aunty Joan had been a land girl; Uncle John was imprisoned for five years in Germany and walked home to England. These dramatic experiences in their lives – young love, separation, reunion – played out against a backdrop of dinner queues in Army messes and gossip at serving hatches in blacked-out village halls. Even 25 years afterwards, the trauma of wartime hung like a gentle mist over my early childhood. In terms of substitution, dinners at 'Kennet', served through that faithful opening in the dining-room wall, re-enacted a ritual that war had made dear.

Whatever its exact origins and connotations, a sure explanation for the sudden popularity of the serving hatch lies in the way contradictory technologies were fitted into 1960s homes. John and Joan had a gas stove with an eye-level grill, an appliance that fed smells, vapours and occasionally smoke directly into the upper atmosphere of their kitchen. The result of much cutting-edge experimentation, the eye-level grill – or 'igh-level grill; colloquial pronunciation made its etymology hard to pin down – quickly became the subject of class-tinted ridicule.[2] 'I am delirious about our new cooker fitment with the eye-level grill,' joked the comedian Michael Flanders in the hit musical review *At the Drop of a Hat* in 1959. 'This means that without my having to bend down – the hot fat can squirt straight in my eye.' For Flanders and his co-performer Donald Swann, this was one of a number of innovations that, in combination, produced absurd tensions in post-war 'Designs for living':

> (*both*) We've planned an uninhibited interior décor!
> (*Swann*) Curtains made of straw!
> (*Flanders*) We've wallpapered the floor . . .

The 'wallpaper' they had in mind was probably wall-to-wall carpet, first used in UK homes in the 1950s. The product of new industrial processes, fitted carpet created a whole new sensation of domestic comfort. But in the vicinity of an eye-level grill it would have been a disaster. Like John and Joan's Axminster, it was allergic to fat. Unlike the Axminster, it was fixed to the floor, so couldn't be removed for cleaning. The solution? A serving hatch between dining room and kitchen, located at a height that made passing plates easy, but which restricted air-exchange. Open only to the lower, less fat-laden regions behind it, the serving hatch eased the task of combining

'At Home: the Epicure',
Graham Laidler ('Pont'),
*The British at Home* (1939).

deep-pile dining with hi-tech cooking, keeping the atmosphere of one from interfering with the tactile pleasures of the other.

Serving hatches became less ubiquitous in the 1970s and '80s. The sociological reasons for this are many. Women disliked the separated kitchen as a type; housework began to be identified by its difference from, rather than its similarity to, any other kind of work performed by free people in a free society, and hiding it behind a hatch was too similar to other kinds of cover-up. But the technical reasons for the decline can be pinpointed with some precision. In 1972 my parents bought a house, new to us, but dating from the 1930s. Modernizing now meant knocking down the division between cooking and eating. The kitchen they installed was dominated by Formica, just as John and Joan's had been in the mid-1960s. But above the new electric stove, instead of an eye-level grill, they mounted a recirculating two-speed cooker hood with a charcoal filter.[3] In our kitchen airborne fat was, if not banished, then largely captured before it could contaminate other surfaces. This allowed soft furnishings to enter the room. The dinette was upholstered; the windows had curtains matching its surprising pattern. The requirement for a near-hermetic seal between cooking and dining had dissolved.

I cannot say that serving hatches were an exclusively British phenomenon (there were Belgian and German variants in the 1960s; serving hatch is *doorgeefluik* in Dutch, so Oud at least had a noun to describe his odd invention). But I would say that, in order to bloom, the serving hatch required a delicate mixture of environmental conditions more evident in Britain than elsewhere. In Scandinavia, where domestic wall-to-wall carpeting never caught on, neither did serving hatches; Swedish kitchens had *matplatser* (integrated dining areas) from the 1930s. To be sure, in those cultures too the advent of the cooker hood softened kitchen furnishings. A Swedish advertisement from 1980 shows a dining table laid for two, dominated by upholstery and fabrics, in front of a smoke-red Electrolux 'Poppy' kitchen, complete with hood. But while culturally enlightened Swedes may have appreciated a new extractor over the stove, the fundamental spatial relation between cooking and eating remained unaltered when they upgraded their kitchens. In the UK, on the other hand, the cooker hood participated in a specific extinction. Serving hatches can still be found in 1960s building stock, where they appear as fossils, reminders of another form of living, another familiarity with fat. The arrival of the cooker hood, as much as what Aunty Joan used to call 'women's lib', deprived them of their prime regulatory function.

# Sinclair C5
## Simon Sadler

As a veritable icon of failure, it would be hard to find an industrial design more extinct than this. The Sinclair C5 was a recumbent, open-cockpit, single-seat, polypropylene-bodied, three-wheeled, pedal-assisted battery electric vehicle, steered by handlebars below the user's knees, which the British tech entrepreneur Sir Clive Sinclair envisaged as the first in a suite of electric vehicles that would revolutionize personal mobility. After several years in development by its 44-year-old inventor, it suffered a high-profile and ill-starred launch in January 1985 at Alexandra Palace in London.

Until that point Sinclair, who had been knighted two years earlier, might have been described as a sort of British Steve Jobs. He launched a series of elegant hit electronic products in the 1970s and '80s, including the Sinclair Executive of 1972 (the first slimline electronic pocket calculator, shown at the New York Museum of Modern Art and feted by the UK Design Council). He also manufactured affordable home computers, starting with the Sinclair ZX80 (1980), which helped the UK to be the biggest consumer of personal computers by household in the world at the time. Sinclair's ability to defy the odds meant that scepticism over the development of the C5 was suspended until that unhappy midwinter day in London.

Although he was aware of the inefficiency of electric vehicles, Sinclair had a lifelong interest in them. Their viability increased with the abolition of tax on electric vehicles in 1980 and the introduction by the Department of Transport of a new legal category of vehicle, the 'electrically assisted pedal cycle', in 1983. The C5 would be the first such vehicle to market, operable by a user of any age, without insurance, licence or helmet. Sinclair avoided the expensive and complex route to efficiency – to innovate with batteries, a fraught field as he knew all too well – in the hope that the success of his machine would force other companies to innovate for him. He focused instead on aerodynamics, an aspect that is often overlooked in small, low-speed vehicles. The chassis was designed by the esteemed sports-car company Lotus, and assembly was subcontracted to the domestic appliance giant Hoover in a deal sweetened by government. Sinclair's goal was to produce 100,000 C5s annually at a base price of £399 (a little over £1,000 now), before accessories. But with negligible demand, production was slashed by 90 per cent within three months of the launch, and ceased entirely in August 1985; 9,000 of the 14,000 manufactured C5s remained unsold as Sinclair Vehicles went into receivership.

Sinclair C5 electric vehicle at its launch, UK, 10 January 1985.

The C5 was a memorable farce of Thatcher-era entrepreneurship, although, as an affordable electric cycle, its fate also speaks to our own moment of personal mobility innovation. 'On a one-to-one basis Sinclair could be chillingly convincing about the project's potential,' reported two tech writers, but journalists covering the Alexandra Palace launch struggled to get uphill in those demonstration vehicles that were not already discharged or inoperative.[1] The maximum legal speed of this new category of vehicle was just 24 kph (15 mph), and the C5's claimed range of 32 km (20 mi.) was doubted by testers. The face of the rider was at the level of the billowing emissions of buses and trucks, and many argued that the 'High-vis Mast', offered only as an option in an effort to reduce price, was essential for safety.

Sinclair issued a publicity photograph showing the vehicle's reluctant designer, Gus Desbarats (recruited via a Sinclair scholarship at the Royal College of Art in London), in a C5 alongside a cardboard cut-out of the Austin Mini car to illustrate the C5 pilot's superior height relative to that of the car driver. But, unlike thronging compact post-war cars such as the Mini, or the mopeds to which the C5 offered competition, or, for that matter, cyclists (who were reasserting their right to the road through organizations such as the London Cycling Campaign, founded in 1978), the C5 was unable to colonize its environment. In a do-or-die bid to increase public receptivity for the C5, Sinclair hired teams of unemployed teenagers (a growing resource in Thatcher's Britain) to ride them around, to no avail.

Truly cult innovations shape their own environments. The Ford Model T famously remade early twentieth-century America around its needs, and a century later Apple drew the entertainment industry into its hardware and software. But to rely on such exceptional moments of acceptance for new technology is magical thinking, nourished by righteous appeals to empowerment and the remaking of the world as a better place. In a pitch reminiscent of Apple's celebrated '1984' advertising campaign against IBM the previous year, Sinclair grandly yet obscurely characterized his vehicle as a weapon, 'cutting giants down to size, turning impersonal tyrants into personal servants . . . with the C5, Sinclair Vehicles puts personal, private transport back where it belongs – in the hands of the individual.'[2] Like Richard Buckminster Fuller before him and Elon Musk after him, Sinclair was blessed – and cursed – with the belief that the world existed to accommodate his design. He operated in a 'bubble', according to a despairing Desbarats, and 'failed to understand the difference between a new market, computing, and a mature one, transport, where there were more benchmarks to compare against'.[3] Sinclair commissioned no market research, and, as the director of his advertising agency explained at the launch, the project proceeded 'purely on the convictions of Sir Clive'.[4]

Magical thinking needs magical technology. A push on the pedals of one of the electrically assisted bikes and scooters available for rent in many urban centres today converts even meagre effort into laugh-out-loud speed, whereas the C5 called to the mind of a writer for the *Sunday Times* a 'Formula One bath-chair'.[5] The experience of the C5 was debilitating, when, to succeed, it had to feel superhuman, or social, or of a critical mass; the bicycle, for instance, has in its long history possessed all these qualities – nimble, convivial, swarming. In fairness, new-generation electric bikes and scooters are beneficiaries of the lithium battery technology that has rapidly improved since the late 1990s; as Sinclair later conceded, the C5 'was early for what it was'.[6] New-generation mobility devices are also the post-Fordist beneficiaries of networked information technology that permits easy rental without capital outlay, a financial model beyond Sinclair's Fordist strategy of relentless cost-cutting. In the scramble to design personal mobility after the automobile, it is more productive to understand the reasons for the C5's early extinction than to recall it as a maligned herald.

# Skirt Grip
Amy de la Haye

To contemporary eyes the function of this object is not immediately obvious. Yet, from the mid-nineteenth to early twentieth centuries most fashionable women owned at least one, often several, items of this type. It was known variously as a skirt grip, fastener, holder or lifter; a hem holder; a page; and a *porte-jupe* or *porte-robe*. Here it is referred to by its most common name, skirt grip, although the term 'skirt lifter' is more descriptive.

These gadgets were designed to enable a woman to raise the hem of a long skirt, to prevent it from trailing in dirt or getting wet when walking outdoors. Poor working women necessarily wore their skirts shorter. Since textiles were, to all sections of society, highly valued commodities and laundering laborious (a day – 'washing day' – was required to complete the task), the skirt grip proved most useful. It also helped to facilitate movement – when entering a carriage, using stairs, dancing, horse riding or cycling – at a time when fashion was generally impractical, and to reject or deviate from the prevailing trend was to risk social ridicule and/or exclusion. A product of the Industrial Revolution, skirt grips were introduced in the mid-1840s and – not surprisingly – became immediately popular.

Throughout this period foundation garments constructed the modish silhouette, with a corset sculpting the upper body. Cumbersome, horse-hair-padded petticoats were worn over crinoline forms until the advent in 1856 of the lightweight, steel-framed cage crinoline, which was succeeded by various bustle forms from 1870 to 1890. Etiquette demanded that a fashionable woman change ensemble up to seven times a day. For walking and visiting during the day she donned a tailored coat and matching skirt, known as a costume; for the afternoon, dinner, the theatre or a ball she would wear a dress of two pieces, bodice and skirt. Only the most progressive women wore bifurcated attire for cycling and other sporting activities.

Worn singly (usually on the right-hand side, since most people are right-handed) or as a pair, skirt grips were clipped or pinned to the waistband. These fastenings were attached to cord (often silk-covered jute) or metal chain, which supported the body of the object that was suspended just above the hem. The grip opened to pinch a section of fabric. Padding between the metal discs protected finely woven silks, linens and cottons from snagging, while robust woollen cloth was clenched with serrated clasps. The wearer pulled the cord, which sometimes operated on a pulley

Skirt grip, c. 1866,
brass, length 10 cm.

mechanism, to raise the grip/s and the skirt. On lightweight spring or summer dresses the action created a fetching polonaise (gathered) effect.

Skirt grips were manufactured in France, Germany and America, but Britain dominated the market, and there production was centred on Sheffield. They were made from polished steel, silver or silver plate, brass, nickel or bronze, and were occasionally – for the ultra-rich – crafted in gold. The dominant construction forms resemble either tongs or scissors. However, from the 1860s onwards entrepreneurs lodged patents for new and novelty designs, some of which they gave memorable names, including 'Eureka', 'Bicycle' and 'Invincible'. To entice fashion-hungry consumers further, the design of the skirt grip became more decorative, in form and/or with its applied ornamentation. The natural world was a fertile source of inspiration for designers, and many grips are decorated with designs that resemble flowers, insects (often butterflies), peacocks and shells. Love hearts and items that were considered lucky, such as horseshoes, were also popular. Another option was a grip made in the form of a human hand, masculine or feminine, with historical pleated, modern tailored or dainty scalloped cuff detailing.

Victorian society was fascinated by the hand as an indicator of socio-economic status, of labour and leisure, of character and for its sensuality. Symbolically it was associated with loyalty, strength, romance and fidelity. Not surprisingly, it was a popular design applied to a variety of consumer goods, notably tableware, letter-openers, jewellery and door-knockers. The skirt grip pictured here, missing its chain and pin, is made from brass. When new, it was a brighter yellow and the surface would have been shiny from having been dipped in acid, burnished and lacquered. A very similar design now in the collection of York Castle Museum in the UK, with extant black cord and clip, is stamped 'Regd. Novr. 14 1876'. How we read a simulacrum of a man's hand lifting a woman's skirt does of course shift over time and space, and with individual interpretation. Likewise, perceptions of gendered identities have become more fluid in the twenty-first century. But in the mid-to-late nineteenth century, society was constructed rigidly around the binary distinctions.

If a family could afford for a woman not to work, that was the convention, and patterns of consumption and taste, particularly women's, were subject to close scrutiny by a society preoccupied with social status. Etiquette books offered guidance, especially to those with 'new money' obtained from manufacturing and commerce, on how to negotiate the quagmire of nuanced social codes. The author of *Manners in Modern Society* (1877) cautioned, 'There is no easier method by which to detect the real lady from the sham one than by noticing her style of dress. Vulgarity is readily distinguished, however costly and fashionable the habiliments may be, by the breach of certain rules of harmony and fitness.'[1] Choosing which skirt grips to purchase and when and how to wear them, and using them with elegance – ensuring no more than a glimpse of ankle was shown – was a mark of such discernment.

Immense quantities of skirt grips were manufactured from durable metals, and it is perhaps surprising that these objects are not more commonplace and familiar today. Many museums with dress collections have one or more examples, but I do not recall ever seeing one displayed in a public gallery. In today's market contexts, the skirt grip often survives as an incomplete object, missing the pin, cord and/or pads, which renders the original use opaque. Dress history texts are peppered with occasional references, and there are two short and heavily illustrated, subject-specific books compiled by private collectors. Websites such as Pinterest are also useful for comparative analysis. Contemporaneous evidence can be drawn from patents lodged in Public Record Offices in the UK and from editorials in ladies' papers (precursors of the fashion magazine), while close inspection of contemporaneous portrait paintings, fashion plates and photographs occasionally reveals a skirt grip being worn.

Why did the skirt grip become extinct? As women started to live more active and independent lives, partly as a result of their patriotic activities during the First World War, the skirts they wore became shorter and more practical – no longer overly full nor so narrow as to be described as 'hobbled' (a style worn between about 1908 and 1914) – making the skirt grip redundant.

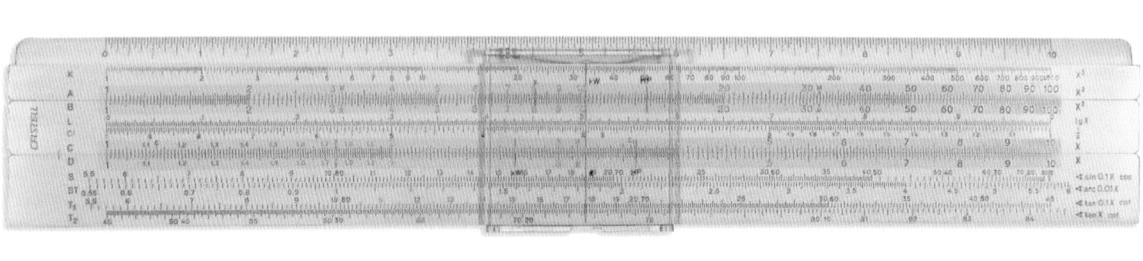

# Slide Rule
Adrian Forty

For more than a century, until their sudden extinction in the mid-1970s, slide rules – or slipsticks, as they were sometimes known in the USA – were the principal means of calculating most mathematical problems, apart from addition and subtraction. The slide rule was the attribute of the engineer: the sign of professional expertise, as the stethoscope was to the physician. Invented by the mathematician William Oughtred around 1630 on the back of John Napier's invention of logarithms in 1614, and the translation of logarithms to a linear scale by Edmund Gunter, slide rules became common in England in the latter part of the seventeenth century. Samuel Pepys, always one for novelties, had a slide rule made for him, and there were specialized versions for measuring building quantities (Coggeshall's rule) and for gauging – calculating volumes of barrels and the like by brewers and excise men (Everard's rule).

Slide rules seem to have been less common in continental Europe – probably because the English were accustomed earlier to decimal fractions, a prerequisite of the slide rule – until the introduction of the metric system, after the French Revolution. By the 1820s, applicants to the French *grandes écoles*, the schools of public administration, were required to know how to use a slide rule. A standardized one with logarithmic and trigonometric scales and a sliding cursor, developed by the French military engineer Amédée Mannheim in 1850, was adopted by the French army for use by engineers and artillery officers, and was manufactured in quantity by the French instrument-maker Tavernier-Gravet. The Mannheim rule became the pattern for mass-produced slide rules in Germany, England and the United States in the latter part of the nineteenth century.

In 1886 there appeared the first rules with the scale of numbers printed on white celluloid, instead of boxwood or metal, making the markings and numbers much easier to read. From then on, celluloid scales became more or less universal. By the mid-twentieth century, slide rules were sufficiently cheap for every engineer and every university and high-school student of science to own at least one, and often several. High-end, precision rules were still made of boxwood or aluminium, but standard models were plastic. Engineers would carry their slide rules in special leather holsters hung from their belts. Smaller, pocket slide rules were personal accessories carried around in jackets, along with comb and penknife; I remember that the architectural historian Reyner Banham, whose assistant I became in

Faber-Castell slide rule 57/88, late 1960s.

1973, kept one in his briefcase for calculating the word length of his articles. It is estimated that over its final century, around 40 million examples of the slide rule were manufactured globally.

The advantages of the slide rule over printed logarithm tables were the speed with which calculations could be made, and the lower chance of error. Log tables called for pencil and paper on which to add or subtract the logarithms, while the process of extracting the logarithm from the table and then re-entering the result in the log table to find out the value meant that altogether there were three stages at which mistakes might occur. With the slide rule, on the other hand, the entire working process was visible. Its disadvantages were its inability to do addition and subtraction, the limited accuracy of all but a few specialized rules to three significant figures, and the fact that slide rules did not indicate where to put the decimal point – but the last feature was held to be an advantage as well, since the user had to compute an approximate answer mentally, and this acted as a check on the calculation. Slide rules, it was said, put people closer to numbers. But the limitation to three decimal places meant that engineers, knowing their calculations were imperfect, would design conservatively, building in redundancy so as to be on the safe side; the slide rule could be said to have been responsible for much wastage of materials and energy.

Slide rules were sufficiently ubiquitous for the 1969 Apollo Lunar Mission to have had one on board. Yet within a decade they were to be

Employee with slide rule, Hercules Powder Co. Experiment Station, Wilmington, Delaware, photograph by Robert Yarnall Richie, 1947.

completely obsolete, superseded by electronic pocket calculators – whose very development had been made possible by calculations made on slide rules. Ironically, the slide rule was an agent of its own extinction. The first electronic calculator, the HP-35, brought out by Hewlett-Packard in 1972, was described in its instruction manual as 'a high-precision portable electronic slide rule'. The comparison was unjust, though, for calculators not only gave far higher accuracy than slide rules, they could also do addition and subtraction, making them attractive to the general user, and not just the techies, who had been the slide rule's main market. By 1975 pocket calculators had become cheaper than most slide rules, and by 1980 slide rules had almost entirely gone out of production. A curious relic of the short change-over period was the hybrid Faber-Castell TR1/TR3 calculator that included, on the back, a slide rule: a belt-and-braces device for those who mistrusted the calculator, or feared sudden battery failure. Specialized slide rules, dedicated to a particular function, survived for longer: among them the circular E6B slide rule, developed for air navigation in the late 1930s and still in use; the 'pregnancy wheel' for determining ovulation dates; and a slide rule for scaling organ pipes.

The extinction of the slide rule was regretted by some, who maintained that the instrument gave a comprehension of numbers and calculation that was lost with the invisible workings of the electronic calculator. Curiously, much the same objection had been given three and a half centuries earlier by the slide rule's originator, William Oughtred, to explain his reluctance to publicize his invention. The slide rule, he wrote, belonged to the 'superficiall scumme and froth of Instrumentall tricks and practices', that would only undermine sound theoretical knowledge of mathematics.[1]

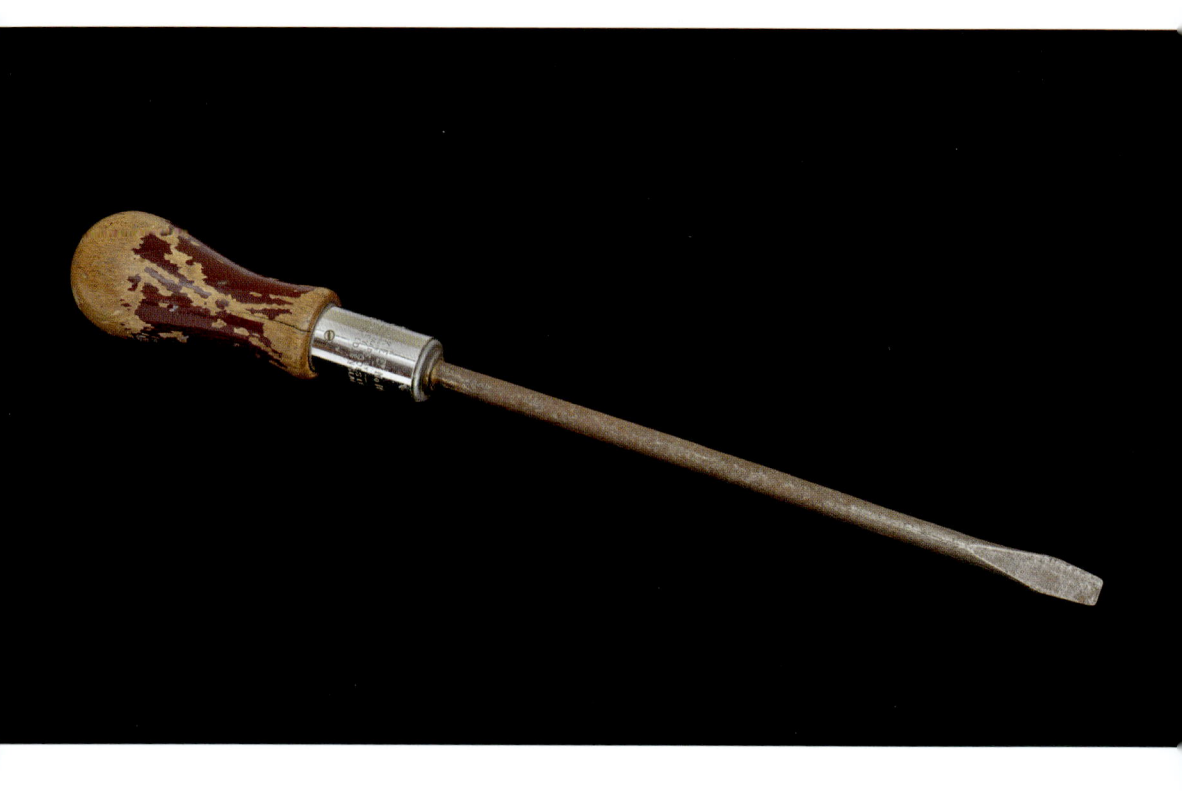

# Slotted Screwdriver
## Richard Wentworth

In my studio there's a drawerful of flat-tipped screwdrivers, and two drawers of slotted screws. 'Why not get rid of them?' my son asked. 'We thought it amazing you still had them.' Cross-head – Phillips, Pozidriv – and now star-head – Torx – screws have all but taken over from slotted screws, because, well, try tightening a slotted screw with an electric screwdriver without the bit slipping out. I might have had the screwdriver illustrated here when I was fifteen. It may be the oldest I have. All the paint is coming off the wooden handle. On the collar it says 'Yankee' in inverted commas, 'no. 10A' and then 'Stanley Tools Ltd, Sheffield, England'. There are no postcodes. There's no customer-service number. There's a sort of pride, a confidence in that inscription – it's funny how much you get from just a little bit of language. The first 'Yankee' screwdrivers, with a reversible ratchet mechanism, appeared in 1899. The American firm that manufactured them originally was bought up in 1946 by Stanley, which went on making 'Yankees' until 2007, when they were discontinued. I probably never oiled this one, but it's perfectly serviceable, and may end its life prising open a tin of paint.

Historically, the screwdriver is a late arrival in the toolbox. The Romans had hammers, saws, drills and planes, but screwdrivers didn't show up until the late fifteenth century, when the first screws appeared, used by gunsmiths. Then, from the mid-sixteenth century, watch- and clockmakers started using them, too. But these screws were clumsy things, and not until the mid-eighteenth century did instrument-makers start producing threaded screws with any precision. These were straight, machine screws, and they needed screwdrivers for them. There are screwdrivers – *tournevis* – in Diderot's *Encyclopédie* of 1765, and in English the direct translation 'turnscrews' was what screwdrivers were called well into the nineteenth century. Tapered wood-screws started to appear in the eighteenth century; they were needed for butt hinges, another eighteenth-century invention, which can't be fixed with nails. Making wood-screws must have been hellishly difficult, each one made by hand, turned on a lathe. It was impossible to make a pointed tip, and they weren't very good. Only in the 1830s did the Americans perfect the pointed screw, with automated production, and from then on, screw production shot up. Meanwhile, standardized straight-threaded machine screws had been perfected by the British engineer Henry Maudslay, whose screw-cutting lathe came in around 1800. All these kinds of screw had slotted heads – easy to make, but troublesome to use. You need to use both hands:

'Yankee' slotted screwdriver no. 10A, Stanley Tools Ltd, UK, 1960s.

one to hold the screw and guide the tip of the screwdriver, the other to turn the screwdriver. And the screwdriver can easily slip out, damaging the screw or the surface into which it is being screwed.

A Canadian, Peter Robertson, was the first to come up with a decent alternative, in 1907. His screw had a recessed square socket head, and worked well with power-driven screwdrivers; Henry Ford was an early adopter. The Canadian man who once ran Clerkenwell Screws – the specialist screw supplier in London – used to go on at me in a patriotic way about Robertson. But it was Henry Phillips's cross-head screw, patented in 1937, that really took over from slotted heads. A better businessman than Robertson, Phillips persuaded General Motors to use his invention, and from then there was no going back, as the Second World War assured their future use for everything mass-produced in metal. But the screwdrivers for these cross-head screws relied on connection to a power supply, either compressed air or electricity. That was fine in a factory but no good on a building site, and slotted wood-screws, driven by human muscle, hung on in the building and woodworking trades until the arrival of cordless, battery-powered variable-torque electric screwdrivers in the 1980s made cross-head screws viable anywhere.

The long, slow extinction of flat-tipped screwdrivers hasn't just been a matter of a new technology displacing an older one; it's about a change in how our bodies relate to stuff. I was probably screwing things together – badly – by the time I was seven or eight. It's not an aesthetic for me, although I don't like seeing it done badly, I don't want it to be visibly horrible. I'm not a church finish or bank finish fusspot. But when it comes to the stage of 'I'm going to screw it up now', the seating and the setting are all cumulatively known to me. The relationship between eye, fingertip and tool is dynamic. There's that moment of the right slot to the right head, to the right depth, to the right pressure, that feeling of completion that is utterly bodily – and it doesn't make a noise. In the act of driving a screw home you could feel it torque right up to your shoulder, even if you weren't using one of those beautiful wooden oval-handled screwdrivers and were instead slumming it with a Yankee.

In my brief folly of working at the Royal College of Art, I disliked the fact that the students had all become rpm victims. Once, in the sculpture studio, I was filing something in the vice. A good file makes a comforting noise, and coarse ones are even called 'bastards'. The act of filing is incredibly sophisticated in a two-handed way, with questions of balance, length of stroke and awareness of all the spatial conditions, and the implicit awareness of another plane, the one that is desired yet invisible. It can't be done idly, and it brings an immense sense of concentrated focus. A passing student, the kind of young man who loves 'kit' and the public performance of 'making' as 'art', said in a much too cocky way, 'You want to get the angle grinder.' I remember flinching and not quite knowing whether to play the grandfather figure or say nothing. It was a typical piece of twenty-first-century rpm-speak. You waste time looking for extension leads, checking whether the disc on the tool is the right one, searching for the appropriate spanner (always missing), looking for ear defenders and protective glasses (a disaster if you already have prescription specs) and going through all the other tedium of getting the tool ready to go. This of course is followed by one of the most unpleasant noises to have been introduced in the last forty years.

What's it like, that moment of recognition that something has become obsolete? My son, who's an engineer, said to me recently, 'Do you realize that Phillips is over?'

'What do you mean?' I said.

'Well, I've been watching the builder working on my house, and he buys these new screws and they're about £30 a box.'

'Fucking hell!' I went.

'But yes,' he said, 'I've watched the time–completion ratio' – typical engineer-speak – 'so I've started buying £30 boxes too. They're specially hardened steel, they don't shear, and they're self-countersinking.' They're probably star-headed. First slotted screws go, and now cross-head Phillips and Pozidriv are on the edge of extinction, too. 'I'm a convert,' my son said. It didn't take long.

Not long ago, a furniture-restorer acquaintance bought up the entire stock of slotted screws from a joinery business in Somerset that was closing down. He knew they were going to become a rarity. So what will happen to my drawerful of slotted screwdrivers? I think I will probably put a note for my granddaughters in the drawer saying, 'Dear Lucy and Rosa, these are a comedy of my life.'

# Space Frame
## Catherine Slessor

As extinction events go, it was suitably apocalyptic. On 20 May 1976 a spark from a welding torch ignited acrylic panels cladding the geodesic dome of the Montreal Biosphere. Within a short time, the huge spherical structure was alight, like a colossal, blazing snow globe. Flames raced across its surface and a crepuscular plume of smoke rose over downtown Montreal. Mercifully, there were no human casualties, but there was a death: the death of an idea and a system of construction that had gripped the imagination of architects in the modern era. Once, the Biosphere was the apotheosis of space-frame technology, but just over a decade later the Montreal fire turned out to be its funeral pyre. Riding high in April, burned down in May, to paraphrase Frank Sinatra.

The brainchild of the visionary engineer Richard Buckminster Fuller, whose famous interrogation of Norman Foster – 'How much does your building *weigh?*' – gave high-tech architects sleepless nights, the Biosphere was the United States' contribution to Expo 67. Conceived as an architectural *coup de théâtre*, it was intended to demonstrate the power and progressiveness of American technological ambition. Two years later American men would set foot on the Moon. With its suggestive intimations of planetary colonization, the Biosphere was a vision of a lightweight, hygienic, utopian future, in which technology wipes away all tears. Bucky put the space age in space frame. So how did the space frame fall to Earth?

As its name implies, a space frame is simply a structural network. Its basic geometric module is the triangle, and through this a highly efficient three-dimensional lattice can be constructed, capable of spanning much greater distances than a conventional post-and-beam system. Endowed with this inherent flexibility, the space frame was capable of adopting an array of guises, from hotel atriums to geodesic domes, and effortlessly transcending scales, from the monumental to the bijou. The space frame's structural modus operandi is collaborative and egalitarian. It literally shares the load. When any single structural member reaches capacity, other members pick up and carry additional load, so that the entire system functions as an integrated and independent network. Unifying space and structure, the space frame could, in theory, extend forever, enveloping the Earth in a planet-sized biosphere.

Depending on how far back you go, the space frame has its roots in the work of many architects. Joseph Paxton, Gustave Eiffel and Andrea

Montreal Biosphere, the former u.s. Pavilion for Expo 67. The plexiglass covering on fire, 20 May 1976.

Palladio all figure in its prehistory. However, its unlikely godfather was the Scottish inventor Alexander Graham Bell, whose experiments with tetrahedral kites between 1903 and 1909 demonstrated the potential of a new kind of lightweight structural system. A delightfully bizarre photograph of the time shows Bell kissing his wife, Mabel, who is wearing one of his kite frames.

From this intimate tableau, the scene shifts to the interwar and post-war era and the more dauntingly impersonal visions of Konrad Wachsmann, in which towering space frames dwarf mere humanity, reminiscent of the nineteenth-century painter John Martin's terrifying biblical apocalypses. In 1941 Wachsmann emigrated to America, ending up at Chicago's Institute of Design, where he became convinced that architecture must adopt the techniques of industry. 'The machine is the tool of our age,' he pronounced. 'It is the cause of those effects through which social order manifests itself.'[1]

Commissioned to research and design hangars for the U.S. Air Force, Wachsmann developed three-dimensional structural systems with highly sophisticated nodal connections. These megastructures set the tone for the space frame: modular, industrial, rational and efficient, yet still visually compelling in the curious delicacy of the repetitive, lattice forms. Wachsmann's brush with the military–industrial complex seemed to open up new horizons for construction, populated by hypothetically extendible exoskeletons that stimulated the imagination of the Japanese Metabolists and assorted European avant-garde architectural movements of the 1950s. In practice, however, the exceptional precision demanded of the joints was costly and hard to achieve, limiting what could be built.

The 1960s ushered in a more hazy and informal era of clip on, plug in and tune out. Brushing aside reservations about buildability, the space frame popped up with cheeky-chappie regularity in the Fun Palace of Cedric Price, the megastructures of Yona Friedman and the urban fantasies of Constant Nieuwenhuys's New Babylon. The English architectural provocateurs Archigram hijacked the space frame as a vehicle for their pseudo-psychedelic streams of consciousness, without the buzz-killing stress of actually having to build one. In this respect, Bucky's Biosphere stands out for making it off the drawing board and into Expo reality in 1967.

At Expo 70 in Osaka the space frame almost matter-of-factly hogged the limelight in Kenzō Tange's Festival Plaza. Beyond the self-regarding carnival of Expo it came to feature in more quotidian contexts, lending its lustre to swimming pools, exhibition halls, transport interchanges and hypermarkets. It was also turning the heads of the incipient generation of High Tech imperators, so that space frames and exposed structures figured prominently in buildings such as Paris's Pompidou Centre (1977) designed by Richard Rogers and Renzo Piano.

Alexander Graham Bell kissing his wife Mabel Gardiner Hubbard Bell, who is standing in a tetrahedral kite, Baddeck, Nova Scotia, Canada, 16 October 1903.

Yet the stray welding spark of May 1976 amplified what was already an existential crisis. The exogenous shock of the 1973 OPEC oil embargo had done its work. Hugely profligate of energy, space frames seemed a luxury no longer either affordable or desirable. There were also problems with maintenance. With hundreds of individual struts, space frames were extraordinarily difficult to keep clean. The rise of Postmodernism also contrived to seal the space frame's fate. Exposed structures suddenly seemed hopelessly fussy and passé, to be swept aside by a tsunami of irony and pastel classicism.

Today the 'dressed' space frame, as opposed to its 'nude' antecedent, still skulks under the carapaces of many superstar buildings, but as an unseen supporting player rather than the heroic protagonist. It could be argued that because the space frame is still employed to create architecture, it is not technically extinct, but nonetheless it has moved some distance from its original lightning-in-a-bottle moment, when a space frame could itself be considered architecture.

As for the fire-ravaged Biosphere, it had a melancholic, lingering afterlife. Although the conflagration stripped it of cladding, the space-frame structure remained more or less intact. It languished in this partially ruined state until 1995, when, having been remodelled, it reopened as an environment museum. While the Biosphere's structure itself is no longer a harbinger of the thermally controlled, space age future, it seems right, and almost inevitable, that Bucky's structure should now be used to explore the environmental concerns that may render our entire planetary existence obsolete.

# Stanley 55 Combination Plane
## Nikos Magouliotis

Hand planes – tools for shaping wood by pushing a cutting blade over its surface – have been around since antiquity, and are perhaps as old as carpentry and woodwork. Until the eighteenth century, all such planes consisted of a main body of hardwood, into which was fixed a metal blade. As the tool was forced along a wooden surface repeatedly, the blade carved its surface. The primary purpose of a basic plane (with a straight blade) is to create a smooth surface. Moulding planes, however, have curved blades that carve a piece of wood into different profiles in order to form cornices and other ornamental finishes on building facades, interiors and furniture.

In the nineteenth century hand planes underwent an innovation boom. Advances in metallurgy allowed manufacturers in Britain and the USA to make planes entirely from cast iron, rendering the tool more durable and adjustable. In the 1890s the Stanley Rule and Level Company, an American designer and manufacturer of craft tools and equipment, launched the Stanley 55 Combination Plane, a tool that marked the peak of hand plane sophistication.

Dubbed 'the Swiss Army Knife of Handplanes', the product consisted of a kit of 55 different cutter blades – of different cutter blades – and a metal body on to which they could be mounted. The body, or rather, frame of the Stanley 55 consisted of different movable parts, which could be adapted to various angles and widths, allowing almost any moulding to be carved with just one tool. The idea wasn't new; custom-made adaptable planes had been made earlier, out of wood. But the cast-iron frame of the Stanley 55 made the tool more durable, more precise and much more adaptable. 'Packed in a neat substantial box', like a 'planing mill within itself', the Stanley 55 kit was presented as a compact, sophisticated tool that promised to replace the numerous single-moulding planes a craftsman had to have at hand to carve different profiles.[1]

Even though it promised to render obsolete a whole set of traditional crafting tools, at the threshold of the twentieth century the Stanley 55 was faced with factors that foreshadowed its own potential obsolescence. Its modern form, with numerous metal compartments, bolts and screws, made it look like a sophisticated machine. But it was, in essence, only a mild modernization of a traditional tool that still depended on human force. The invention of machine-powered routers (at first as flat-bed machines, and later as handheld tools) allowed for the cheap and fast production of wooden

'55' Moulding Plane, Stanley Tools, with box and packaging, USA, early 20th century.

Drawing of the Stanley '55' Plane, from *Stanley Tools Catalogue No. 34* (1915).

### STANLEY "FIFTY=FIVE" PLANE.

mouldings and thus threatened the Stanley 55 and all the previous traditional moulding planes with extinction. Although the Stanley 55 presented itself as a practical and economic solution, it was not affordable to every craftsman. And, accompanied by a 22-page instruction manual, it was seen by many as too complicated and time-consuming to use. The Stanley 55 was, from the beginning, a luxury item for a niche market.

Apart from the conditions that rendered the tool technically outdated and inaccessible, the early twentieth century brought cultural changes that threatened to make the forms it produced aesthetically obsolete. The decorative mouldings it was designed to produce were largely incompatible with the modern aesthetic vision promulgated by avant-garde architects and designers. Its cast-iron adjustable framework was the result of industrial modernization, but the decorative profiles it produced was a thing of the past: they came from eighteenth- and early nineteenth-century builders' and craftsmen's manuals, which translated the European classicism of early modern architectural treatises into an easily reproducible and traditional formal repertoire that craftsmen used to appeal to the taste of their clientele. By the first decades of the twentieth century, modernist architects in Europe and North America were announcing the end of ornament and the beginning of an era of smooth, undecorated surfaces for buildings and everyday objects. A few hundred miles south of Massachusetts and Connecticut, where the

Stanley company manufactured its moulding planes in the 1920s, Frank Lloyd Wright lectured to Princeton University architecture students and announced, in 1930, the 'passing of the cornice' and all sorts of other 'fancy fixings' – among them the mouldings that the Stanley 55 was made to produce. Wright argued that American architecture ought to resist all this curved and sculpted 'enfolding' and instead 'unfold': get rid of mouldings, straighten surfaces and simplify corners in order to create modern facades, interiors and furniture.[2]

However, despite all the conditions that threatened to render it obsolete, the Stanley 55 did not die quickly. The Stanley company dominated the hand-tool market from the late nineteenth century until the end of the twentieth, producing all sorts of advanced and simple planes and other tools. Hand-powered tools could not compete financially with the increasing mechanization of moulding production. The company continued to produce and market the Stanley 55, but reoriented to do-it-yourself enthusiasts and gradually diminished the construction quality and precision of the plane. It wasn't until 1962 that production of the Stanley 55 ceased. This paradoxical survival of a moulding tool well into the era of Mid-century Modern design and smooth lines seems incompatible with established chronologies of modernism. But, if the rejection of mouldings was almost unanimous among the high avant-garde, it was certainly not immediately embraced by the ranks of tradesmen and craftsmen that worked in the construction of buildings, furniture and other everyday objects. The Stanley 55 was a vehicle through which the moulding lived a longer, albeit quieter life, parallel to the modernist urge for formal reduction and largely unnoticed by its historiography.

# Telephone Table
## Edwin Heathcote

The telephone table is a microcosm of defunct domestic products, rituals and ideas. It was usually an asymmetrical object with a seat and, to one side, something a little like a bedside table with a drawer beneath. It probably had spindly turned legs, possibly splayed; it was most likely to be made from teak. The first domestic telephones were mounted directly on the wall, but in the early twentieth century free-standing versions appeared and tall tables arrived to accommodate them. The low-level versions that became ubiquitous had only a brief period in the limelight, from about the early 1950s to the late 1970s, coinciding with the period now defined as Mid-century. The surface was intended for a phone – usually a 711 standard phone with its springy, coiled cord and rotary dial (for most of the period only one phone was available from what was then the General Post Office) – and the drawer for a telephone directory, which was a thick book of names, addresses and phone numbers, and perhaps a *Yellow Pages*, which was a primitive commercial search engine printed, as its name suggested, on yellow paper. This assemblage sat in the hall, and every single component of it is now extinct: the idea of sitting in the hall to make a phone call, the landline, the telephone directory. All are obsolete.

It's difficult to believe now but once, not that long ago, people used to go into the corridor to make a phone call. This was the draughtiest space in the house or the flat, as well as the least comfortable and the most public. When I was a teenager you used to go into the hall to phone friends or girlfriends, often attempting a little intimacy in exactly the space where this was least possible. Parents would pass by, frowning at you in an attempt to shame you into getting off the phone, on which this call was clearly costing them money; older brothers or sisters would hover, wanting to talk to their potential lovers. And once they were on it, no one ever wanted to get off the phone. The wandering by and scowling became a de facto guilt-inducing queuing system. It's a curious thing that we once felt the need to have some privacy in our phone conversations, but were deprived of it by having a phone in a fixed, explicitly public location. It was the domestic analogue of the phone box in the street with a queue outside – only without the privacy of the box.

The telephone table's association with technology put it in a bracket of furniture that included the television cabinet, the radiogram and, later, the computer desk. By having a piece of furniture dedicated to it there was also

Telephone table, manufactured by Ercol Ltd., UK, mid-20th century, teak.

an implicit acknowledgement of the status of the telephone as a thing apart. It was a kind of transportation device, an idea that popular culture seized on: the phone booths in *Superman*, *The Matrix* and *Doctor Who* all facilitate transformation, teleportation and time travel. Like the front door and the letterbox slot in it, near which it was usually placed in the corridor or hall, the phone was a portal to the outside world. As the architectural historian Robin Evans explained in his essay 'Figures, Doors and Passages' (1997), the corridor as an architectural form arose from the desire for privacy in the domestic sphere, for the separation of rooms and the differentiation of the house's public and private functions.[1] So the telephone's architectural location in the corridor makes a kind of sense, placed in an in-between world of public circulation and communication. It also contains, perhaps, a hint of our current situation in which our communications are constantly but often imperceptibly surveilled, a manifestation in furniture and location of the impossibility of privacy in communication. The presence of the phone book in its drawer (akin to the omnipresent Gideon bible in many hotel rooms) also foreshadows the search engine, as does the address book, notebook and pen on its surface.

Emerging in the mid-twentieth century, the telephone table was part of the proliferation of new furniture typologies aimed at filling the increasingly ubiquitous suburban house. The post-war boom and the consumerism need-ed to maintain it created a need for new products, and these duly emerged. There were nesting coffee tables, TV dinner tables, breakfast-in-bed tables and trays, and low-slung sideboards. There were television and cocktail cabinets, mass-manufactured Formica-topped corner bar units and stools. There were fitted kitchens and bathroom units, sewing cabinets and magazine racks, and footstools that doubled as storage. There were hi-fi cabinets and wall-mounted shelving units, executive-style desks, easy chairs, recliners and breakfast-bar stools. The telephone table arrived to fill a hole in the space where there was too little furniture. What else was there for the corridor? An umbrella stand? A coat rack? Too Victorian. Perhaps a console table if there were delusions of domestic grandeur? The telephone table was a piece that arose from the desire to make more stuff, and yet became extremely practical and then, just like that, disappeared. It's difficult to understand why it vanished so completely. Perhaps it was the arrival of the cordless phone in the early 1980s; certainly, the dawn of the mobile phone made it almost laughably obsolete. But perhaps it was just fashion. The kind of Mid-century teak that characterized it might now look fashionable but in the Neo-Victorian, French rustic or yuppie black-and-chrome interior of the 1980s, it looked like your granny's furniture. It was gone.

We never had one. The telephone table was a piece of furniture made for modern houses with central heating; it wouldn't have fitted in to our over-stuffed Victorian terrace. When I saw them as a child, I envied the modernity, the chic unattainability of this purposeful piece that seemed so suited to its function, so deliberate. Everything of ours seemed to be fulfilling a function it was not intended for; ad hoc, poorly adapted. A Victorian side table with a fussy embroidered Florentine doily for that most modern technology: it all seemed to rebel against the intrusion of a creamy plastic phone.

I saw a telephone table the other day at a market, looking purposeful and elegant, low-slung and modern. And I wondered to myself, what could that be used for now? Incomplete without its phone, it looked forlorn yet dignified. Perhaps it too is due a revival, along with its Mid-century contemporaries. Might a repurposed telephone table become something else? I notice in contemporary interiors the glowing presence of the Wi-Fi router, a kind of sacred object, yet so often found sitting on the floor or the bookshelf. Might it not have its own piece of furniture? Could the telephone table become an altar to the ubiquity of Wi-Fi? It would be a fitting tribute from one age of communication to another.

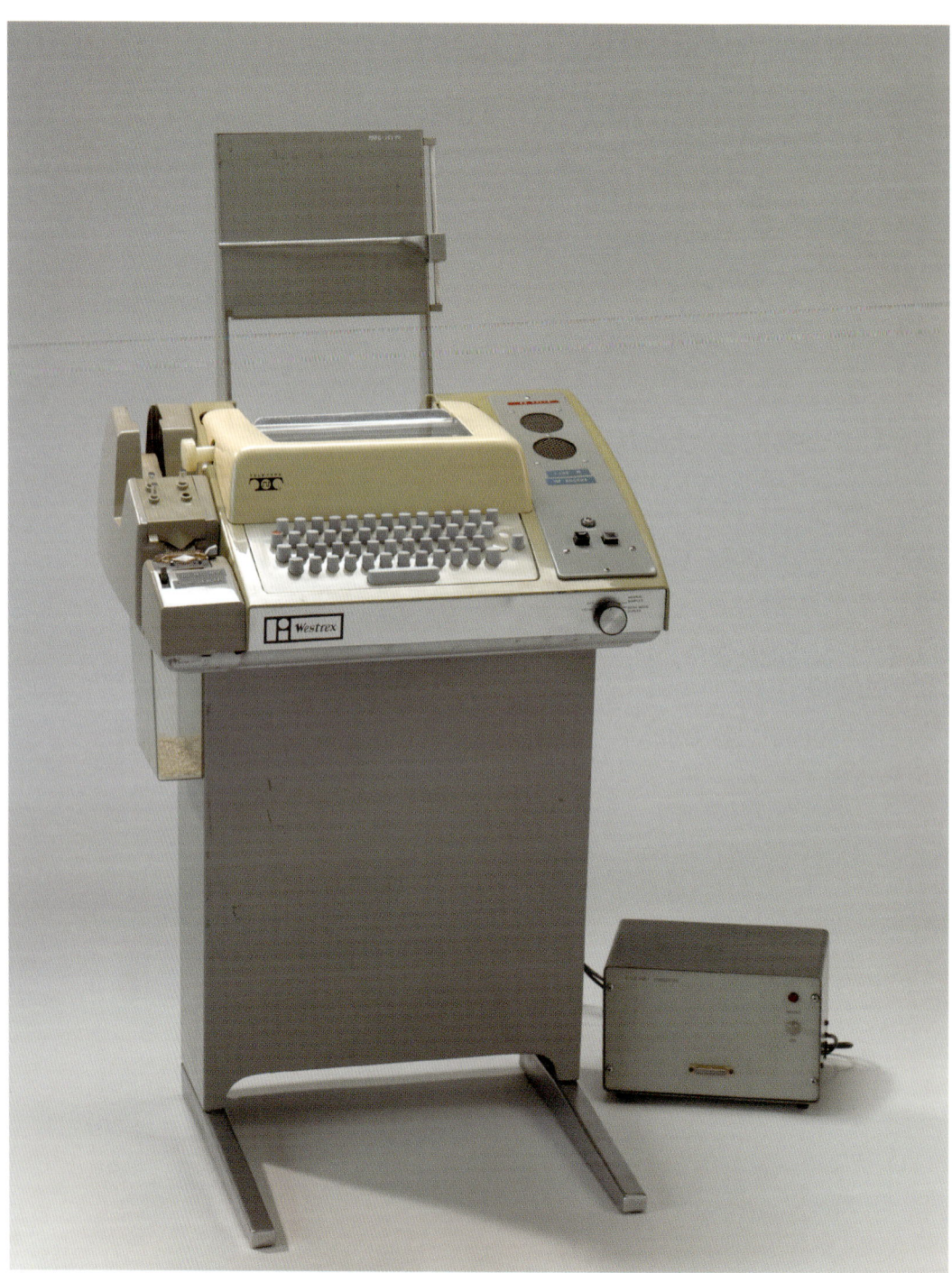

# Teletype
## James Purdon

For most of the twentieth century, teletype carried the vast majority of global written telecommunications traffic, flashing text messages between businesses, government institutions and – since most telegraph traffic after the early 1930s was channelled through the teletype networks – private individuals. The first 'printing telegraph' machines were developed in the nineteenth century by Royal Earl House, David Edward Hughes and Émile Baudot. These complicated devices were controlled using a piano-style keyboard that required extensive training to operate, and sent electrical signals to a receiving 'teleprinter' for inscription in code on a length of ticker tape. By the late 1890s the piano keyboard had been replaced by a QWERTY-style typewriter in various distinct systems developed independently. Despite differences, all these new automatic 'teletypewriters' could be used by any capable typist to send messages through a switched network to a receiving printer fed by a tape – or, later, by a continuous roll of paper – on which they were printed out as plain text.

In 1927 Creed & Co. began supplying the General Post Office, first with the Model 3 – which used an updated code system, designed by Donald Murray, to print on ticker tape at 65.3 words per minute – and then with the revolutionary Model 7, which could print lineated text on a paper roll. By 1933 teletype machines were responsible for more than 70 per cent of domestic British telegraph traffic. Creed also supplied equipment to the Associated Press, Reuters and other news agencies, and teletype rapidly became the primary means by which news and information were syndicated to newspapers, radio stations and television networks around the world. By the mid-twentieth century the teletype machine had come to symbolize the fourth estate itself, as in the screwball comedy *Woman of the Year* (1942), in which Katharine Hepburn and Spencer Tracy's newsroom meet-cute is continually interrupted by the frenetic demands of the newswire. The clacking of the teletype machine became a familiar background sound in offices, factories, ministries and newsrooms, but teletype remained for the most part an institutional medium. Its deployment as a component in the hidden infrastructure of commercial telegraphy meant that while members of the general public sent their telegrams through the medium of teletype, they were not the primary operators of its machinery; messages were instead transcribed by trained teletypists from handwritten slips deposited at the counter of the telegraph or post office.

Westrex teletype terminal, type ASR/KSR33, UK, 1960–80.

Because teletype carried messages between different parts of a closed bureaucratic institution, or from one such institution to another, its appearances in film and fiction frequently mark a boundary between official and unofficial worlds: the point or occasion where the black box opens, offering a rare glimpse of the invisible operations of a self-contained organization. Precisely because of those associations, the conventions and format of teletype acquired the status of a unique aesthetic connoting a certain genre or mood, notwithstanding any content the medium might deliver. In that sense, the medium was the message. Alfred Hitchcock's *The Lodger* (1927), for instance, uses a teletype news report in place of the silent movie intertitle as an efficient way of introducing the audience to the background story of the serial killer stalking London. Where Hitchcock went, others followed, and the use of the teletype readout to deliver information became a standard device in Hollywood noir, as in the tense chase sequence in Joseph H. Lewis's crime thriller *Gun Crazy* (1950). Later the Cold War thriller began to deploy the association of the teletype with official bureaucracy to evoke the kind of cover-up that separated real events from official history. A good example appears in the concluding frames of John Sturges's *Ice Station Zebra* (1968), where the messy machinations of NATO–Soviet espionage at the North Pole are covered up by a teletype press release depicting the two superpowers' covert Arctic showdown as a joint rescue mission.

This double coding of the teletype medium, its association both with conspiratorial bureaucracy and with the public voice of the news, is aptly caught in Andy Warhol's *Flash – November 22, 1963*, a portfolio of eleven screenprints of images connected with the presidency and assassination of John F. Kennedy. *Flash* takes its title both from the camera flash that captures the faces of Kennedy and his assassin on photographic film and from the teletype protocol of the 'FLASH', the highest level of priority marked on outgoing material by the newswires. Each image in the series was enclosed in a set of 'wrappers' printed with extracts mimicking a news teletype readout of the events of the assassination. The linear narrative produced by the teletype record invokes a kind of archival seriousness, a sober, journalistic description of events unfolding in time, in contrast with the muddled collage of double-exposed images that confront the viewer with a barrage of symbols and famous faces. Warhol had selected and edited these teletype 'flashes' from the altogether glitchier record of the actual newswires, appropriating not the exact content but the cultural associations of the medium, in much the same way as he appropriated other media: by extracting, highlighting, revising. The teletype record becomes part of a commentary on the spectacle of the Kennedy assassination specifically as a mass-media event.

What makes this use of teletype so effective is the medium's cultural status as a first draft of history. As the internal monologue of powerful

UPI (United Press International) 'FLASH' teletype dispatch reporting the assassination of JFK, 22 November 1963.

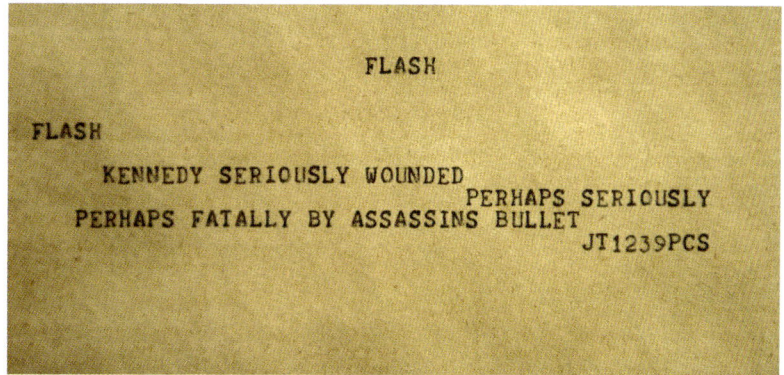

UPI (United Press International) 'FLASH' teletype dispatch reporting the assassination of JFK, 22 November 1963.

institutions, the teletype record suggests a back-channel where the evidence of conspiracies might be secretly encoded; as the raw voice of the newswire, it suggests a public forum where such evidence might enter the historical record. At the very end of Alan Pakula's classic Nixon-era political thriller *All the President's Men* (1976), a teletype montage covers the passage of time between Woodward and Bernstein typing up the story of the Watergate conspiracy (while Nixon's second inaugural address plays live on a foreground television set) and the President's resignation just nineteen months later. Teletype here serves as the medium through which the secret parts of the political unconscious become public knowledge.

As a major medium for the transmission of printed text, teletype survived well into the 1980s, when it began to be superseded by the fax machine (which had the advantage of being able to reproduce images in addition to text) and, eventually, by the Internet. Teletype machines remain in use in a few legacy applications – primarily in shipping and aviation – although these too are gradually being replaced by all-digital systems. Yet, even as its devices and infrastructure have mostly been consigned to museum collections, teletype continues to shape the way we communicate in virtual spaces. The rhythms and conventions of real-time text communication, now ubiquitous in everyday life, evolved from network protocols that it established and which now facilitate twenty-first-century telecommunications. The influence of teletype remains potent technologically, for instance in the ASCII character introduced by the American Standards Association in 1963 to regulate teletype signals, and still used by most computers to assign alphanumeric symbols to keyboard inputs. It remains culturally potent, too – whenever a Hollywood thriller sets the scene with a dateline in a nostalgic teletype font.

# Théâtrophone
## Carlotta Darò

'*Le théâtre chez soi*' (Theatre at Home) was the advertising slogan of the *théâtrophone*, which, for half a century, from 1881 to 1936, transmitted the sounds of performances from the Paris Opera House and other theatres through telephone networks. As an individualized listening device and a citywide infrastructure accessible to audiophiles, the *théâtrophone*, by creating an audience physically dissociated from the place of performance, threw into question the nature of the theatrical experience. At the heart of the *théâtrophone* system was the condition of 'acousmatic listening' – listening to a sound whose source is hidden from view.[1] In essence the system combined the power of long-distance electrical transmission with a more traditional desire for the sociability of physically attending a live performance, resulting in a new tension between source, cause and effect. The *théâtrophone* was an invention of the modern bourgeoisie at the 'apogee of capitalism', to use Walter Benjamin's expression; modern man seeks 'refuge among the masses of the big city', only to isolate himself the very moment he comes into contact with it. The *théâtrophone*, as originally conceived, was therefore a symptom of the quest for individualization within the homogeneous space characteristic of mass culture.

The *théâtrophone*'s inventor, Clément Ader, who had been involved with the installation of Paris's first private telephone network in 1879, presented his first 'telephone auditions' at the International Electricity Exhibition of 1881, allowing live listening to performances from the Opéra and the Comédie française. In two different rooms in the Palais de l'Industrie were some forty listening stations, ancestors of the telephone booth, in which devices with dual earphones manufactured by the Ateliers Breguet isolated each listener in his or her own sound bubble, facing a sound-absorbent wall. The rooms were completely covered with drapery and carpets, with two doors cloaked in thick fabric, low light to avoid distracting the eye and in the centre a table to accommodate a supervisor, whose job was to cut off the sound transmission circuit every five minutes so that a new group could enter and enjoy being immersed in the sonic environment of the great Paris theatres.

The *théâtrophone* belongs to a family of modern inventions (telegraph, pneumatic post, telephone, radio) that rely not solely on the design of a device but also on the establishment of a network. It owed its existence to the invention in 1876 of the telephone. A series of microphone sensors on the theatre stage transmitted the sound signal through underground wires

*Théâtrophone*, domestic receiver, Paris, 1900.

to the listening stations of the Palais de l'Industrie (and also to the Élysée Palace, similarly equipped for President Jules Grévy to inaugurate the system).

In terms of the history of listening practices, the *théâtrophone* is distinctive for introducing stereophony. In order to reproduce the stage space 'realistically', Ader's device allowed binaural listening. Through the wiring, the two headphones were connected to two microphones on the apron at the front of the stage, to the right and left of the prompter's box. The *théâtrophone* was postulated on a synthetic form of the restitution of three-dimensional space, making it akin to nineteenth-century optical devices such as the stereoscope, which reproduced the effects of binocular vision by means of first a process of abstraction and then perceptual reconstitution of the images in the viewer's brain.[2] In the case of the *théâtrophone*, the movement on the stage was perceived through the movement of sound from one earpiece to another. Moreover, as occurs in the real space of a theatre, where from our seat we have a unique visual and auditory perception of the performance, the *théâtrophone* listener similarly enjoyed a unique experience because the system was originally designed so that each microphone was paired exclusively with a corresponding earpiece. While the listener was undeniably elsewhere, several physical conditions of the theatrical experience were maintained, such as the movement of the performers on stage, a form of spatial restitution of the sound (with the preservation of the ambient sounds and acoustic effects of the hall), and the specificity of a distinct listening spot for each spectator. Nevertheless, the *théâtrophone* experience was a reduced version of the more extended theatrical one. The performance was no longer perceived in its entirety, deprived as the listener was of bodily presence, and of the experience through the other senses, of the space, of the ceremony and social codes, from arrival at the theatre, waiting, intermission, finale, applause and the taking of curtain calls, to leaving the hall. Instead, it was a partial restitution of the show that gave a glimpse of the whole event – a sample, as it were, that in some respects anticipated current forms of compulsive listening or channel surfing, typical of more recent media such as television, MP3 players and streaming players.

Ader's device was succeeded by a portable coin-operated machine (fifty centimes for five minutes of listening, one franc for ten minutes), exhibited at the Universal Exhibition of 1889 by Belisaire Marinovitch and Geza Szarvady. The following year the Compagnie du Théâtrophone was set up and began to establish a network of receiving devices in cafés, restaurants, clubs and hotels on the grand boulevards, making it possible to listen remotely to several shows taking place in different venues across the city. From 1891 the company offered a domestic subscription service that allowed the sounds of Paris theatres to be brought into the private home, either by machines operated by the company or through the telephone, although this reduced the service to monophonic listening. From the 1920s onwards, Sunday sermons from the cathedral of Notre-Dame were added to the service's programme. Annual subscriptions were aimed at an affluent audience, and at an optional additional cost subscribers could extend their listening time limit, and increase the number of theatre venues that could be 'pierced'. Access to the performance of one's choice was managed by the *théâtrophoniste*, a kind of musical switchboard operator, who connected the different circuits. Judging by the comments, in 1911, of one of its more illustrious subscribers, Marcel Proust, the sound quality of the service was poor – although, perversely, 'to make up for the acoustic deficiencies' when listening to Wagner's music, he discovered a delight in the way the *théâtrophone* caused him to draw on his previous memories of the work.[3]

In the 1920s technical improvements led, paradoxically, to an impoverishment of the original experience, as the several original on-stage microphones were reduced to a single sensor capable of broadcasting over several networks. The multiple-capture system was thus lost and replaced by a centralized one. Listeners began to tire of the headphones, especially when wearing them for long periods, and adopted loudspeakers instead. The spatial quality of stereo listening was therefore replaced by monophonic listening, which, despite its drawbacks, still managed to attract new subscribers as late as 1930. Ultimately, the spread of new broadcasting and sound reproduction technology – radio and the phonograph – brought about the *théâtrophone*'s extinction. In this sense, the early bourgeois project of 'realistic restitution' (as though the listener *were* at the theatre) was overtaken by the more interactive relationship to sound – and the greater possibility for choice – offered by radio and the phonograph. The free manipulation of frequency and physical sound-storage mediums captivated the modern public more than the passive (although, initially, qualitatively higher) act of listening through the *théâtrophone*. After the Compagnie du Théâtrophone experienced years of financial crises, in a letter dated 6 April 1936, the director of the Paris Opéra, Jacques Rouché, finally requested that the company remove the *théâtrophone* equipment from his establishment.

# Think City Electric Vehicle
Kjetil Fallan

The electric car that *Forbes* magazine called 'a cheerful little slice of
eco-friendly Scandinavian design' in 2010 became extinct only a year later.[1]
This is the story of how a politically willed, technologically precarious,
environmentally sustainable but economically *un*sustainable object became
obsolete at the brink of success.

Norway is pioneering the electrification of automobility, and electric
vehicles (EVs) are now outselling fossil fuel-powered cars there. A prominent
– if paradoxical – part of the oil-producing country's nation-branding as
an international leader in environmental policy, this historic transition has
been facilitated by governmental policy such as tax exemptions, but also
by the fact that Norway has no (conventional) car industry of its own, and
therefore no anti-EV lobby. In the decades immediately before and after the
turn of the millennium, however, there was a sustained – but ultimately
failed – attempt at exploiting this window of opportunity through the
design and development of the Think City electric vehicle, an innovative,
tiny two-seater car intended for urban driving. Despite massive political
goodwill and major investment from international corporations, it proved
a volatile enterprise and finally collapsed in 2011 – exactly on the cusp of
the EV breakthrough in the Norwegian market trailblazed by typologically
more conventional cars such as the Nissan Leaf and the Tesla Model s.
Looking back at this recent piece of automotive design history, the Think
City is an object lesson in *un*planned obsolescence, demonstrating how
deeply design is interwoven with key structural factors such as rapid
technological development, complex industrial infrastructure, precarious
investment schemes and strong political interests.

The Personal Independent Vehicle Company (PIVCO) was founded in
Norway in 1991 for the express purpose of developing a small electric vehicle.
Ten prototypes were ready for demonstration during the 1994 Winter
Olympic Games at Lillehammer, and the first production model, the City
Bee, was introduced in 1995. It was still a minute operation, though, with
only 120 cars made. Of these, 45 were commissioned for the San Francisco
Bay Area Rapid Transit (BART) station car programme, a valuable testbed in
an important potential market for the upstart automaker.[2] Revving up both
the product and the production line, the company finally embarked on serial
production with the first version of the Think City, introduced in 1998, of
which 1,005 units were made over the next three years. Having come a long

Think City electric
vehicle, Norway, 2008.

way in just a few years in terms of design and development, the new model was still a far cry from what most people expected of a car. Ridiculed – not surprisingly – by the self-declared 'petrol heads' of the BBC hit television show *Top Gear*, the performance and handling did leave a lot to be desired.

By this time, though, the company was working hard to design away such derogatory judgement. Following bankruptcy in 1998, the company was acquired by Ford, which renamed it Think Nordic, and invested $150 million in engineering and safety development. Having access to the resources and expertise of a major car manufacturer made the dream of designing a 'proper car' more realistic. However, when in 2003 Detroit succeeded in undermining California's ambitious zero emission vehicle mandate, Ford lost interest in EVs and sold the company to Switzerland-based Kamkorp Microelectronics, which in turn sold it to a group of Norwegian investors in 2006, following another bankruptcy. The continuing financial problems delayed product development, but when the next and final iteration of the Think City was at last launched at the Geneva Motor Show in 2008, it was indeed much improved on all counts – even if the design by then was looking dated, stemming as it did from 2001. The structural and exterior design was led by Stig Olav Skeie, who had been with the company since the early days and had also worked on the previous models, whereas the interior design was managed by Katinka von der Lippe. It was still a very small car, but it was, technically, a proper car. It boasted power steering, central locking and electric windows. More importantly, it met all the international safety requirements, featuring ABS brakes, airbags and so on, and it was the first ever EV to be crash-tested and certified for highway use. With a top speed of 110 kph (68 mph) and a range of 160–200 km (100–125 mi.), it had arrived at specifications acceptable for a micro-car – but its diminutive size and relatively steep retail price (50 per cent more than the similarly sized Smart Fortwo, or the equivalent of a Volkswagen Golf) meant that it could never be anything but a niche product.

But even if a vehicle of this type could never offer all the functionality that most people wanted from a regular car – in terms of passenger and cargo space, for example – the designers wanted it to look and feel like one. Von der Lippe argued that the Think City could not look too different from the mainstream: 'Customers must be reassured that this is a real car, that you can trust it, that it will drive and handle well . . . so we have made it similar to conventional automobiles.' On the other hand, referring to the car's most unconventional design feature – the unpainted, textured, matte plastic body panels – she claimed that 'people want to show they have made a different choice.'[3] These conflicting design intentions proved difficult to resolve, resulting in something of an unhappy compromise that may have impeded the car's marketability. A solid effort at ironing out such issues was in progress, though, with the larger, more advanced four-seater concept model ox designed by Skeie and von der Lippe in collaboration with Porsche Design, first shown in 2008 and intended to enter production in 2012 – but this remained a showpiece, since there were no funds to develop it into a production model.

Arguably the most innovative feature of the Think City was its environmental sustainability, not just in use, but also in production and afterlife. Some 95 per cent of the materials used were recyclable, including the ABS body and the aluminium space frame. This certainly set it apart from other EVs manufactured more in line with conventional cars, which would eventually prevail in the market. The Think City was thus an innovative piece of design. But perhaps it was innovative in the wrong way, demanding too great a change in mentality from the average car driver. A tiny plastic-bodied car, the Think City failed the designer Raymond Loewy's classic dictum of product design, to be the 'Most Advanced Yet Acceptable'. Unlike competing models soon to arrive, it didn't sufficiently emulate the conventions of car culture.

The other main problem was that the enterprise lacked the financial muscle and momentum to maintain a stable operation. In a last hurrah, fresh capital was recruited and production moved to Valmet Automotive's plant in Nystad, Finland, in 2009; another assembly line was set up in Elkhart, Indiana, in 2010. But already the following year, twenty years after the company was founded, it went bust for the last time and production was shut down. Marketed under the over-confident strapline 'Changing the World, One Car at a Time', Think City was, quite literally, a classic case of too little, too soon.

# Trombe Wall
Paul Bouet

A dark masonry wall, set behind a glazed south facade, with small vents at bottom and top. The appeal of the Trombe wall was in large part owing to the simplicity of its appearance and evidence of its function. As in a thin greenhouse, the air was warmed by the winter Sun and conveyed directly to the room at the back, while some of the energy was stored in the wall and released at night. Only basic knowledge and common components were needed to implement this technology, allowing anyone to build their own solar house. Unsurprisingly, the Trombe wall quickly became one of the most popular solar technologies of the 1970s – if not the most celebrated – especially among counter-culture activists across Europe and North America. It represented a perfect incarnation of the solutions advocated by the critics of high technology, from Ernst Friedrich Schumacher to Ivan Illich. It also met the aspirations of the rising environmentalist movement, seeking to establish a softer relationship between humans and nature: a house designed to harness the Sun's rays would not only produce its own energy and avoid pollution, but make its inhabitants more aware of climate and even of their place in the universe.

Charged with all these virtues, the Trombe wall and its principles were widely published. Its canonical section was reproduced and explained in dozens of solar architecture guides inspired by the *Whole Earth Catalog* (1968–72), as well as in major architectural magazines, from *Architectural Design* to *L'Architecture d'Aujourd'hui* and *Casabella*. One could hardly live in the 1970s and be interested in architecture and the environment without having heard about the Trombe wall and its promises. So builders began to implement it in houses, one of the most prominent examples being Douglas Kelbaugh's in Princeton, New Jersey (1974–5), pictured here. A young architectural graduate at the time, Kelbaugh 'fell in love with the Trombe wall', and implemented it on the south facade of his two-storey family house.[1] More than 75 per cent of the energy needed for warming was provided by the Sun, and the house soon became an icon of solar architecture. But Kelbaugh, in common with most enthusiastic partisans of the Trombe wall, rarely questioned the origin of the technology he had fallen in love with. How had it been invented? And to what end?

Ironically, the Trombe wall was a product of late colonization and technoscience. It was invented in the early 1950s by the French scientist Félix Trombe, at a time when France was discovering the Sahara's immense

Kelbaugh House, designed by Douglas Kelbaugh, Princeton, New Jersey, USA, c. 1975.

reserves of oil and trying to exploit them. A pioneer of the research into solar energy, Trombe advocated the use of the wall in remote areas of the desert where industrial complexes were being constructed. There, the Sun could provide direct energy for both extractive processes and domestic needs, without requiring the complex infrastructure necessary to access conventional energy sources. Thus, Trombe and his team experimented with many applications of solar energy between their laboratory in Mont-Louis, southern France, and the military base of Colomb-Béchar in northern Algeria. One of the experiments they conducted, which used radiation exchange between the Sun and the Earth to warm or cool dwellings, gave birth to what would later be named the 'Trombe wall'.

The conclusion of the Algerian War of Independence in 1962, which marked the end of the French colonization of North Africa, could have stopped the trajectory of the Trombe wall. But Trombe and his team continued their research in metropolitan France, where they worked with modernist architects to develop the technology. With Henri Vicariot, the Trombe wall was integrated into the curtain wall of the spectacular Odeillo solar furnace (1962–8), a monumental technological artefact set in the bucolic landscape of the Pyrenees mountains. With Le Corbusier's disciple Jacques Michel, it was adapted to prefabricated houses and implemented in iconic projects to ensure its promotion and encourage its broadest possible dissemination. Direct access to the Sahara had been stopped, but Trombe still believed that solar technology needed to be developed. As fossil energy sources were subject to an accelerated depletion, solar energy represented the whole of humanity's future.

It was the oil crisis of 1973, combined with a sudden fever of environmental anxiety, that boosted the diffusion of the Trombe wall in the media and its success among counter-culture activists, many of them unaware of its controversial origins. But ultimately, neither Trombe's technoscientific project nor the environmentalists' hopes were realized. The dissemination of the Trombe wall remained limited to some hundreds of houses in Europe and North America, and it never inaugurated a new 'solar age'. Indeed, in the first half of the 1980s the Trombe wall was victim of a double backlash. First, passive solar technology, the category to which the Trombe wall belongs, was marginalized by its active competitors, supposed to be more efficient at the cost of greater complexity. Designed to be industrialized, active solar technology, such as solar panels, received significant support from governments and companies, while passive technology was wholly dependent on the initiative of individuals using their own resources.

But beyond this opposition, efforts to develop solar energy collapsed entirely. In the mid-1980s oil prices dropped and the dominance of other conventional energy sources (nuclear, gas, coal), which had been massively developed in response to the oil crisis, removed the need for renewable energy. Solar heating became an idea of the past, and the Trombe wall ceased to be mentioned in publications or implemented in buildings. It was almost forgotten, until in the early twenty-first century solar energy was again envisioned as a solution, this time to face global climate change. But in a sector dominated by photovoltaic panels placed on roofs, producing electricity rather than direct heat, the Trombe wall, with its radical simplicity and its integration into architecture, remains a marginal technology.

# Vi två passar inte ihop!

Den röda triangeln på medicinför-
packningen talar om att läkemedlet
kan vara trafikfarligt. Tar du sådan
medicin finns det all anledning att
vara extra försiktig med alkohol.
Det gäller både sprit, vin och öl! 
Dagen-efter-effekterna blir ännu
starkare, och du fungerar sämre
som förare.

**SOCIALSTYRELSEN
NTF
SYSTEMBOLAGET**

*När du inte längre behöver den här påsen, så släng den inte hur som helst.
Hjälp till att hålla naturen ren! Och tänk på att plastpåsar kan vara livsfarliga
för små barn. Systembolaget.*

AB CELLPLAST, NORRKÖPING
Tänk på att täta påsar kan vara farliga för barn!
Art. 19089

# T-shirt Plastic Bag
## Johanna Agerman Ross

No doubt it would have greatly surprised the Swedish sales manager Sten Gustaf Thulin to know that, fifty years after he filed a patent for a plastic carrier bag, it would be the subject of countless lawsuits, banned from use by entire nations and employed as the ultimate symbol of humanity's throwaway culture.

The 'T-shirt bag', as it is referred to by the plastics industry, was invented by Thulin while he worked at the plastics manufacturer Celloplast in Norrköping, Sweden, in the 1960s. Although the patent credits Thulin alone, the company director Curt Lindquist and the company's production manager Sven-Erik Lövefors were also closely involved in its development.[1] Their invention was to fold, weld and die-cut a flat tube of polyethylene, creating the plastic carrier bag with integrated handles that we are so familiar with today. The resemblance of the patent drawings to a folded or sleeveless T-shirt gave the bag its nickname.

Polyethylene was invented in Germany in the late nineteenth century, but was not industrially produced until the 1930s in Britain by Imperial Chemical Industries, better known as ICI.[2] The type of thin film-like polyethylene suitable for plastic bags came to the fore after the Second World War. The 'T-shirt' bag is therefore part of the move into plastic packaging that began in the 1950s. This was when groceries such as sliced bread were first wrapped in plastic rather than paper, when dry cleaners started using lightweight polyethylene sacks to cover cleaned clothes, and when plastic bags on rolls were introduced in the fresh produce sections of grocery shops. With the Celloplast invention, the plastic bag would, by the 1970s, be the default mode of carrying shopping home in many countries.

We can see why plastic bags were appealing, not least to petrochemical companies seeking to expand their product lines and markets. In the USA, Mobil Oil was then a leading producer of polyethylene film and had a vested interest in pushing the material's many uses. It ran educational programmes on the use of plastic bags for retailers in the 1970s, and indirectly funded advertising campaigns via the Flexible Packaging Association, with slogans such as 'Check Out the Sack. It's Coming on Strong.'[3] Witnessing the popularity of the brown-paper shopping sack without handles in the USA, Mobil Oil had invested time and resources into developing a version in plastic in the 1960s. However, the manufacturing method and use of material made that bag prohibitively expensive.

Plastic bag: 'the two of us do not fit together', Sweden, 1982.

By contrast, the Celloplast design didn't simply try to mimic existing carrier-bag designs in paper, but instead innovated and drove both the form and function forward. It was lightweight and durable, but could expand to hold large quantities of goods. It was capable of carrying more than a thousand times its own weight while costing just a fraction of a paper bag to produce. Thanks largely to its clever design, the T-shirt bag soon became the norm in America, with Celloplast initially holding the monopoly for its manufacture.

In 1977, however, Celloplast's patent was lifted and the market opened up for the manufacture of plastic bags by other producers. At that point Celloplast was making 760 million bags a year. Nowadays it has been estimated by the United Nations that between 1 and 5 trillion bags are produced annually.[4] With so many plastic bags around, how can they be considered extinct? They aren't – at least not yet – but recent measures against the use of the plastic bag have certainly endangered it.

The first move towards plastic-bag extinction happened relatively early in its history, during the oil crisis of the 1970s. Producing plastic bags at this time became much more expensive, and retailers had to pay more for a convenience their customers had grown accustomed to getting for free with their shopping. In Sweden this led to the first grocery stores charging for plastic bags, in 1974. To make the charge palatable, advertisements were published in newspapers encouraging customers to 'Save your plastic bags. A plastic carrier bag is supposed to last 4–5 times,' implying that the more times you use it, the cheaper it gets. Following this introduction of a fee for the grocery-store bag, all grocery stores across Sweden introduced a set price for plastic bags.

The next real push for extinction is far more recent. In the fifty years since the invention of the plastic bag, plastic has become a convenience in many parts of our lives, and single-use plastics now make up 40 per cent of all the plastics manufactured worldwide. This has devastating consequences for the natural world. The plastic bag in particular has a tendency to escape garbage processing and plastic-recycling plants, and often finds its way into natural ecosystems, where it harms wildlife and nature's own processes. In Mumbai, India, plastic bags have been found to get stuck in storm drains and make the monsoon floods much worse. Volunteers on the International Coastal Cleanup in 2018 counted 964,541 plastic grocery-store bags among their beach finds, and 938,929 other plastic bags. As a result, a number of laws have been introduced either to tax the use of plastic bags as a deterrent from use or to ban them altogether. As of July 2018 the United Nations counted 127 nations as having a plastic-bag tax or ban in place. Kenya is a particularly high-profile case; it banned plastic bags completely in August 2017, with fines of £31,000 or up to four years in prison for lawbreakers.

Plastic bag without decoration, Sweden, 1967.

The UK has undertaken a particularly successful drive in reducing the use of plastic bags in food shops by introducing a 5 pence fee for each bag used. Sales of plastic bags in the UK's seven biggest supermarkets dropped by 95 per cent after the charge was introduced in 2015, a decrease from 7.64 billion single-use bags to 226 million.[5] Instead, shoppers have reverted to old carrying solutions with shopping trolleys, baskets and reusable shopping bags replacing their plastic counterparts. However, the development of biodegradable plastic has given rise to a new type of plastic bag with the promise of degrading naturally without harm to the environment, creating an ambiguous message of continued guilt-free shopping. It seems we still hang on to the dream of a disposable society where our consumption leaves no footprint.

Yet the extinction of the plastic carrier bag leaves traces everywhere we go. In grocery stores, the purpose-built furniture for storing and packing plastic bags at checkouts now stand empty. In our homes, the special bag holders invented to deal with the ever-growing quantities of plastic bags are slowly depleting. These newly created voids are a reminder of the short-lived history of one of the twentieth century's most brilliant yet harmful inventions.

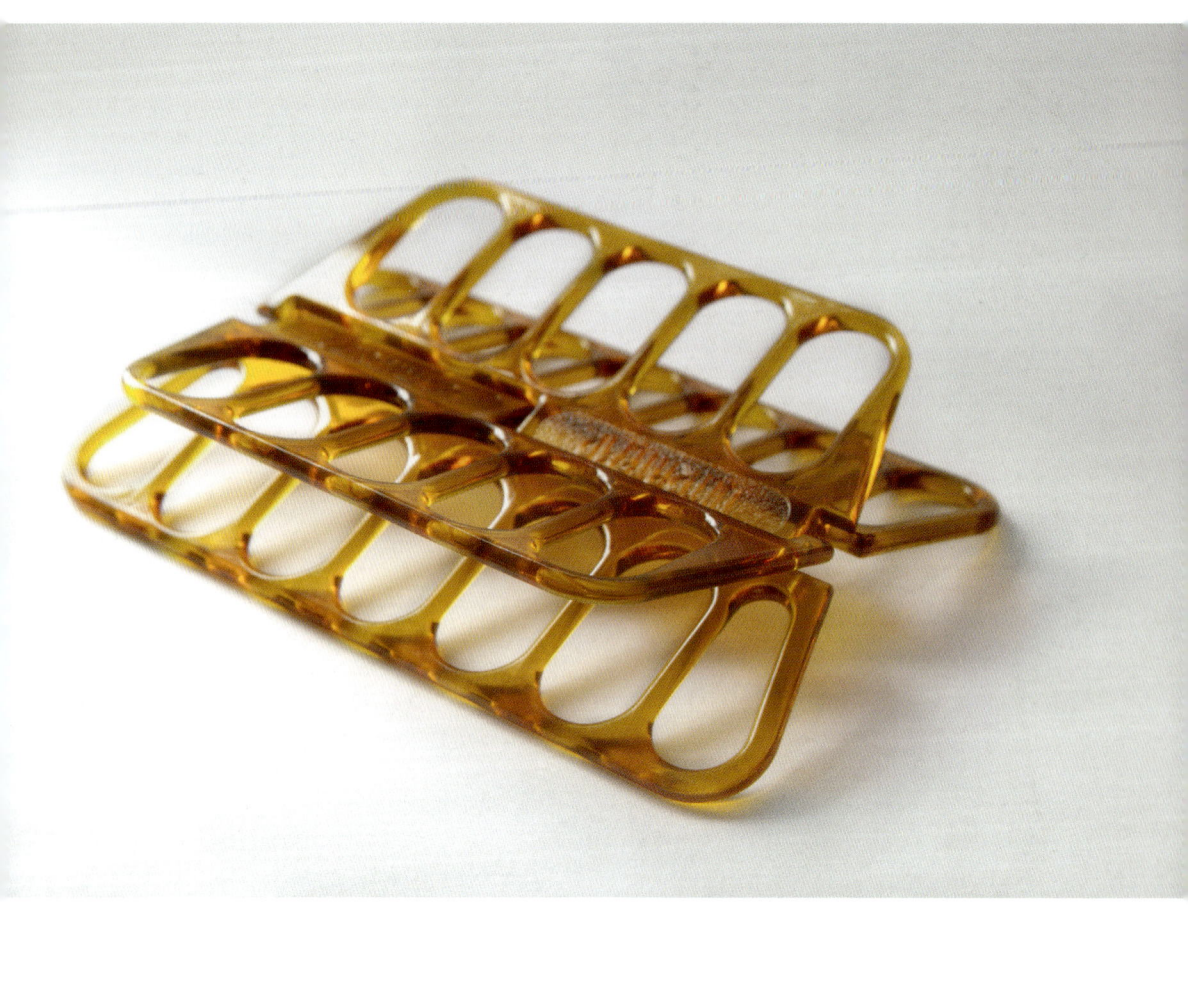

# Ultratemp® Roasting Rack
## Christian Parreno

The Ultratemp roasting rack was a suspended structure to assist cooking. By elevating food off the oven floor, it permitted the heat to be distributed evenly, with the aim of more accurate roasting. It was manufactured in 1990 by the American company Robinson Knife Co., and designed by William A. Prindle, an artist and prolific industrial designer who specialized in kitchen gadgets and transparent surfaces for clients such as Pyrex, Oneida and Sunbeam. According to the patent application, the Ultratemp roasting rack provided a functional alternative to traditional versions, which were made of metallic wire, unsuitable for use at very high temperature and difficult to clean and store, resembling cumbersome trays.[1] The innovative design not only improved a particular utensil, but promised quicker and better cooking, centred on the use of the microwave oven.

The rack appeared as a geometrical composition. Two trapezoidal grids, with elliptical perforations and rounded edges, were interlocked, creating an x-shaped lattice 4 cm (1½ in.) high and 19 cm (7½ in.) long. This precise formation looked like vectors in movement, contributing to its futuristic aura, and permitted flexibility of use. The rack had two positions, high or low, to accommodate roasting as well as defrosting. Furthermore, as an essential feature that informed its trademark name, it was made of polyetherimide with the grade designation Ultem® 1000, an industrial plastic created in the early 1980s and not previously used in consumer goods. The amber-to-transparent material is unaffected by microwave radiation, and capable of tolerating temperatures of 204°C (400°F) while remaining cool to the touch. Because of its non-stickiness and strength, the rack facilitated the manipulation and removal of food and did not absorb odours or rust; it could hold up to 23 kg (50 lb) in weight and was safe for use in the dishwasher.

Devised to perform in conjunction with other implements, the rack required a pan below to collect the oil, grease and fat resulting from roasting. The packaging box included messages recommending that such a receptacle should be either CorningWare, a glass-ceramic cookware introduced in 1958 in the USA, or Visions, a translucent variety offered in 1983. Characterized by thermal shock resistance and modernist design, these products complemented and validated the qualities of Ultratemp, encouraging the need for related appliances in the kitchen. As part of this lineage of products, the rack belonged to a patented collection of nine other instruments, including spatula scrapers, long ladles and forks, slotted spoons, cake slicers and spaghetti

Ultratemp® Roast Rack, design by William A. Prindle, manufactured by Robinson Knife Co., USA, 1990.

servers – all made from the same substance and in a range of signature colours, such as cranberry, orange and dark grey, inexpensive at \$3–5 per unit or \$15–25 per set.

In addition to this merchandising strategy of continuation and seriality, the graphic design of the carton employed patriotic references and a paternalistic tone to reassure customers of the benefits of the device and the use of the microwave oven. It featured photographs of roast beef and steamed vegetables – favourite American dishes. It also confirmed that the rack met or exceeded 'all safety standards of the U.S. Food and Drug Administration for products used in food preparation', and declared that the cookware was 'crafted with pride in the USA'. The stress on substantial, wholesome meals and safety was intended to tackle head on the widespread and enduring belief that microwaves were toxic and lowered the nutritional value of food. These doubts stemmed from radiation tests published by the U.S. Department of Health, Education and Welfare on 4 January 1970, which revealed that microwave ovens sold before that year gave off radiation at a level that might be harmful to human health. Although the federal government set new standards to ensure safer microwave ovens, concerns about radiation and nutrition loss have persisted.[2]

Despite its eye-catching design, pioneering material and solid marketing logic, the Ultratemp roasting rack failed in its effort to realize a future of more efficient and nourishing cooking. For all the improvements the rack made to the cooking process, microwave ovens could not replicate the browning and texturing of roasting, nor its smells; this still required the extended cooking time of conventional ovens. Moreover, such rapid food production lacked the ritualistic connotations of roasting and its invocation of families gathered around the dining table. On 16 July 1992 the *Chicago Tribune* observed that such appliances competed needlessly with the durability, attractiveness and efficiency of traditional glass or ceramic equipment, and concluded that 'some things just aren't done as successfully in a microwave oven as they are on the stovetop.' The rack was discontinued soon thereafter, and the manufacturing division of Robinson Knife Co. was sold.

Even though the main failure of the rack was to gain acceptance among users, it was also undeniably the victim of broader social and economic forces, being introduced at a time when the market for microwave products was already saturated and business was dwindling. Although 80 per cent of American homes had microwaves by 1990, the purchasing of cookware offered by related industries since the 1970s was in decline. The demise of the rack also coincided with the economic downturn that affected the USA in 1990 and 1991. Following the loss of business confidence after the Iraqi invasion of Kuwait and the resulting shock to oil prices, the dramatic rise of unemployment forced the larger population to alter their eating habits, encouraging the consumption of canned and frozen food. Consequently, rather than being used for cooking, the microwave oven was mainly used for heating and reheating, predominantly in workplaces and by empty-nesters and older generations. The rack and its related accoutrements became obsolete, at least in terms of the way they had originally been designed. Instead of functioning as the manufacturers envisioned – helping time-pressed consumers to roast meals quickly from scratch – they were used as passive containers. Yet all was not lost for the microwave, and products that played more to its strengths, such as steamer baskets, egg boilers and micro-kettles, thrived.[3] The future of microwave cookery would be steamed, not roasted.

# UV-radiated Artificial Beach
## Maarten Liefooghe

Of all the defunct building types of the late nineteenth and early twentieth centuries, children's colonies constructed on the Belgian coast are among the most striking. They were built to prevent tuberculosis for city children through a collective stay at the seaside. Their names, including Hôpital Maritime de Grimberghe, Sanatorium Astrid and Home RTT (Régie des Télégraphes et des Téléphones), evoke the diversity of the institutions that sponsored them.[1]

With few of these complexes remaining, postcards now do a far better job of evoking them. Some show the colony buildings in the dunes, occasionally with neighbouring complexes in the picture. The architecture varies from neo-Gothic to joyous modernism, depending on the decade and the ideological outlook of the institution. Other postcards depict the strictly controlled activities that were aimed at 'strengthening' inland children physically and morally: eating in the refectory, lessons in the classroom, sports activities, sunbathing in the dunes or on a sundeck, as well as medical examinations and treatments. Finally, a somewhat strange genre of postcards portrays the modern equipment in its own right – photographs of shiny installations from industrial kitchens, laundries, bathing facilities and sports halls, and varied medical equipment.

An 'artificial beach' at the De Haan Sea Preventorium, featured on a postcard from the late 1930s, is a remarkable 'extinct object' that combines both activity and equipment. A group of white children, each wearing protective glasses, play in a sandpit lit by UV lamps on the ceiling. A nurse – nurses are ubiquitous in photos of light therapy – is supervising. Beach scenes are painted on the walls, but the real coastal landscape remains behind the bubbled window glass. Many things click together here: the institutional project of preventive and remedial holiday colonies, the medical practice of heliotherapy and phototherapy for tuberculosis, and the modern technology of UV lamps. The fact that this is more than a one-off is evident from photographs of similar artificial beaches for children in the 1930s in France.[2] Yet, within a few years this object was to be extinct, mainly because a new antibiotic treatment for tuberculosis (streptomycin) made preventive cures obsolete.

Maritime children's hospitals and colonies date back to the mid-nineteenth century with the early *ospizi marini* in Italy, soon afterwards appearing throughout Europe, fuelled by international medical congresses

Artificial beach at the Zeepreventorium/ Préventorium Marin, De Haan/Coq-sur-Mer, Belgium, postcard, late 1930s.

and publications exchanging approaches to cure a series of diseases. Much of this was focused on tuberculosis, which affected children from the urban industrial workforce. In 1885 the French doctor Henri Cazin listed seaside children's colonies in England, Italy, Austria, Germany, Holland, Belgium, France, Switzerland, Denmark, Spain, Russia, the United States and Uruguay. The Sea Preventorium in De Haan was one of the preventoria founded by the Belgian Association for the Control of Tuberculosis in the 1920s.

After sea bathing (seawater) and natural or artificial aerotherapy (healthy air by the sea or in the mountains), solar cures became the focus in the 'environmental therapy' applied by children's colonies. From the second half of the 1920s, international radiation research allowed the development of electric ultraviolet radiators that could give precise doses of UV. This made it possible to supply shortwave radiation for tuberculosis and other treatments, even in less sunny climates and seasons, and to improve the health of the population of sun-starved nations. A postcard photo from the early 1930s shows 'Les Rayons Ultra-violets' in the preventorium Gai Séjour, right next to the Sea Preventorium in De Haan. In a room in the roof, almost naked children step on a circular walking line marked on the floor, passing side-mounted UV radiators at regular intervals, tracing a path reminiscent of the disciplined figure gymnastics that were such a common part of the regime of children's colonies.

The novel and uncanny feature of the UV therapy room in the Sea Preventorium is that the simulation of sunlight is combined with the simulation of a natural beach setting and associated beach play. It is undeniably strange to see children playing on a fake indoor beach while looking out on to a real one. Here we see on a smaller scale what an unrealized Berlin project from 1928 had envisioned. The hygienist engineer J. Goldmerstein and architect Karl Stodieck had published a proposal for a new type of urban swimming pool for all advanced nations: a gigantic 'thermal palace' in which 17,800 adults and 15,000 children would spend several hours every day engaging in sports and sunbathing.[3] The architect Hans Poelzig then set out a more concrete design, envisioning a domed hall spanning 150 m (492 ft), at its centre a terraced, artificial, heated beach surrounded by a ring-shaped swimming pool and enclosed by a gigantic painted panorama of friendly, sunny landscapes – all irradiated by UV lamps. The designers even described how artificial clouds would be created to complete the climatic illusion. In the beach room in De Haan, the illusionism is less complete, but it is similarly intended to naturalize the institutional and technological of the healthy modern child.

Today, retrospectively, we read in the postcard of this artificial beach-cum-UV bath the culmination of half a century of medical progress and

Hans Poelzig, *Interior View of Thermenpalast, Berlin*, 1928, charcoal on tracing paper.

reformist experiment, and an announcement of a future in which tuberculosis prevention and treatment would be dissociated from children's colonies. Collective architecture and its advanced technological equipment were supplanted by individual pharmacological treatment. In Belgium, as in other European countries, after a proliferation of holiday colonies in the interwar period, the institutional fabric of civil and governmental organizations that set up holiday colonies was gradually taken apart after the Second World War. In those colonies that remained, recreation and ideological education rather than a medical project came to the fore. Family holidays for a broadening middle class became the new norm in the 1960s, demanding a different type of seaside accommodation. Most colony buildings were converted, redeveloped or demolished; a few became photogenic ruins.

Lastly, UV lamps are still used for tanning, but since the turn of the millennium they have become suspect, as awareness of the carcinogenic effects of natural and artificial UV light has grown, and legislation and campaigns in the UK and elsewhere have attempted to curb 'tanorexia'. Yet light therapy for mental well-being remains popular, as does the general belief in the health benefits of beaches – natural or man-made. Artificial indoor beaches continue to be built around the world, from Dubai to Singapore, their ever more grandiose simulations promising a more complete 'natural' experience than even a modernist utopian would have dared to imagine.

# Vertical Filing Cabinet
Zeynep Çelik Alexander

When Gregor Samsa, the protagonist of Franz Kafka's story *The Metamorphosis* (1915), wakes up one morning to find himself transformed into a 'monstrous insect' and therefore incapable of going to work, his first thought is practical:

> What if he called in sick? But that would be mortifying and also suspicious, since Gregor had never once been ill in all his five years of service. No doubt his boss would come calling with the company doctor, reproach Gregor's parents for their son's laziness, silencing all objections by referring them to this doctor, in whose opinion there existed only healthy individuals unwilling to work.[1]

As a high-ranking officer at the Workmen's Accident Insurance Institute in Prague at the time, Kafka must have been thoroughly familiar with the Kingdom of Bohemia's insurance programme.[2] The Austro-Hungarian Empire had passed its first laws regarding sickness, disability and old-age insurance in 1888, only a few years after the initiation of the German Chancellor Otto von Bismarck's compulsory social insurance programme, the first of its kind in Europe. After 1918, however, such sick leave in Prague would have to be filed with the Central Social Insurance Company of the newly founded Republic of Czechoslovakia. While the German model relied on stamps acquired by the insured at post offices and affixed to a receipt card, which the insured were responsible for managing themselves, in the Czechoslovakian Social Insurance System, where the premiums were collected centrally, the managerial burden fell entirely on its central agency.[3] As a result, by 1926 Samsa's file would have been one of 13,960,000 held in the company's central headquarters, in the Smíchov district of Prague. According to one report, so overwhelming was the volume of these records that, if placed on top of each other, they would have formed a pile 2 km (1¼ mi.) high. In another decade it would have fallen to one of the company's 737 clerks to locate Samsa's record from this immense mass and determine whether or not his request for sick leave for having turned into a bug was, in fact, warranted.[4]

Like countless other institutions faced with similarly impossible mandates for expansion, the Central Social Insurance Company turned to a new managerial technology of the time to solve this problem: the filing

Central Social Insurance Institution, Prague, interior, photographed by Josef Ehm, 1936.

cabinet. It is easy to forget how this humble piece of equipment – marginal to the functioning office today – revolutionized office work. Before the advent of the filing cabinet, incoming and outgoing correspondence would be folded, its content summarized on the outside, and placed in pigeonhole cabinets, the primary organizational equipment of the nineteenth-century office. With the introduction of horizontal filing in the 1880s by the American inventor Henry Brown, some of these clerical steps could be skipped; the development of vertical filing, made popular after the Chicago World's Fair in 1893, exponentially increased the capacity for storage and retrieval simply by placing documents vertically on their edges.

The filing system at the Central Social Insurance Company took the technology of the vertical filing cabinet to an architectural extreme rarely seen. In an addition built to the north of the main company building in 1931, the filing cabinets lined the longer sides of the room from floor to ceiling. The two walls accommodated eight blocks of cabinets, each of which was 3 m (10 ft) deep and contained 340 drawers. Access to this mountain of filing cabinets was provided by an elaborate mechanism of pulleys, cables and counterweights. There were no conventional desks in this office; instead, the clerks sat at workstations that moved horizontally and vertically thanks to a mobile steel structure running on tracks in the floor and the ceiling. Trapezoidal counterweights attached to the backs of these floating desks allowed clerks to pull these unusually deep drawers out when needed. A clerk could fly around in his section of the filing cabinets, push a row of records into the slide attached to his workstation, and access the file he needed even if it were at the very back. The same filing mechanism was incorporated into subsequent additions to the building, and it remained operational as late as 2002, when the restructured company transferred its records to an electronic database.

Compared to the database that replaced it, the filing system of the Czechoslovakian Central Social Insurance Company seems heavy, slow and outlandish – like a scene from a twentieth-century film about a future that never came to pass. The comic effect of such extinct technology, however, is ideological; we laugh because we think the binary code of today's computers has completely superseded the pulls, slides and rails of these mechanical contraptions. The underlying assumption is that as our world has become more complex, the hardware of such unabashedly physical systems is transcended by elegant software. And, since the cables and wires that make the transition of these electronic signals possible are increasingly being pushed out of our sight, the immaterial appears to have replaced the material and the efficient to have taken the place of the inefficient, towards a historical horizon where electronic computing is the endgame.

Yet, we forget that speed and efficiency were not the only problems that technologies such as the filing cabinet purported to solve. Consider the hapless Samsa. For the company doctor adjudicating his case, locating Samsa's record among hundreds of others would have been easy compared to the challenge of figuring out what to do with it. The filing cabinet was built on the assumption that judgement could be standardized: in the case of the Czechoslovakian Central Social Insurance Company, that cross-referencing similar cases and relevant provisions in the regulations would facilitate the insurance agent's judgement. Yet, as Samsa's unusual ailment made clear, this was more fantasy than reality. However uniform the architecture of the filing cabinet might have been, illness, accident and other contingencies in life continued to come in all forms. In this sense, Samsa's case was not as bizarre as it might seem. Every insurance case, after all, raised different dilemmas regarding judgement: What illness to file under? Which provision to apply? What extenuating circumstances to consider? Such questions require answers that are more political than technical, problems that, even after electronic computing, humans have been reluctant to leave to machines. This, then, might be the reason why the filing cabinet is not entirely extinct today, even though it has been technologically surpassed. It survives here and there as a 'contingency plan' – in case someone needs to go back and determine whether that strange case of a worker turning into a 'monstrous insect' might merit medical leave or not.

# Water Bag
## Sarah Bell

In rural Australia, if an old-timer tells you it is '110 in the water bag', they mean it's hot: 110 is degrees Fahrenheit (43° Celsius), and the water bag is a canvas bag hanging from the veranda or from a 'roo bar' attached to the front grille of a vehicle in case it collides with a kangaroo. A temperature of 110 in the water bag is especially high, because the water bag is cool. If the water bag temperature is 110, the air temperature is beyond comprehension. Air temperatures in Australia have been hotter in recent years than at any time since colonial records began. Now, as the Australian Bureau of Meteorology has added the colour purple to show a new category of heat on weather maps, 52–4°C (126–9°F), the water bag has ceased to serve as a reference point. Widespread outside Australia's cities until the middle of the twentieth century, the bags have now almost entirely vanished.

Canvas water bags were used to store and transport drinking water. With a capacity of around 7 litres (just over 12 pints), the bags were particularly useful for workers and travellers. The canvas was semi-permeable, holding the main volume of water inside but seeping water slowly so that the outside of the bag remained wet. Evaporation from the outside of the damp bag kept the water inside cool. A new water bag had to be soaked overnight before use, so that the canvas fibres would swell and seal sufficiently for the bag not to drip or leak. A water bag in use had to be topped up each night to prevent it from drying out and leaking. Bags were stitched in a rectangular shape, with a screw-top lid in one corner at the top, and heavy wire handles for hanging on the front of a vehicle or from a rafter or wall peg. The cool water was typically poured into an enamel mug for drinking, and the taste of canvas diminished with the age of the bag. The bags were manufactured in Australia, mostly from canvas imported from Scotland. Purchased from farm suppliers, they were branded with the name of the bag retailer or manufacturer. Wool growers marked their water bags with the same stencil they used to mark wool bales sent for sale. Workers might stencil their personal bags with their initials, to indicate ownership.

The 1970s and '80s brought new ways of making and keeping water cool. Improved connection to electricity grids and more efficient refrigerators meant that farms could produce their own ice and cold water. New insulating materials were incorporated into moulded plastic water coolers. Ice and water stored in a 5- or 10-litre (roughly 9- or 18-pint) flask provided cold water on the road or at work throughout the day. The new containers

Canvas water bag, Australia, c. 1950.

'The Waterbag' or 'Halt for Refreshment', wagon driver in New South Wales, photographed by George Bell, 1890–1900.

weren't damp on the outside, sealed completely, didn't flavour the water and required less attention and care. Water coolers were fitted with push-taps that poured water with more control than the open tops of the slippery water bags. The mass production of products for insulated food and water storage reduced prices and sped up the decline of the water bag. Simply freezing water in reused plastic bottles to thaw during the day became another popular alternative to the water bag, enabled by more reliable electricity, more powerful household freezers and the ubiquity of plastic containers.

The decline of the water bag also reflected the transformation of Australia's rural economy during this period. Poorer terms of trade for commodities, increased mechanization and a reduction in state support for agriculture contributed to a dramatic decline in the rural labour force and family-owned farms.[1] Markets for products specifically designed for rural workers, such as the water bag, diminished. The new water containers were equally suitable for rural and urban markets and were more consumer-friendly, allowing a range of choice in size, shape and colour. They were also easier to carry; wet and heavy, the water bag required the use of a motor vehicle for portability and couldn't be thrown into a city commuter's backpack or handbag.

In the face of these factors, the water bag could not compete. Today it persists only as a nostalgia object, a symbol of a lost economy and social order, and is effectively extinct in use. Yet it has much to offer present-day workers and travellers anxious to reduce their energy and materials foot-prints. It represents an alternative to the resource- and energy-intensive products and practices that contribute to climate change and the purple patches of extreme heat that are now a regular feature of Australian weather maps. The water bag is a reusable container, attractive in an era of water bottled in single-use plastic. Moreover, it is a passive technology. Modern cold water is extremely energy-intensive by comparison; the evaporative cooling of the water bag requires no energy input, while freezing or refrigerating water is energy-hungry. The canvas water bag was made from natural fibres. By contrast, modern water bottles and water coolers are made from plastic, glass and aluminium, and may include industrial insulating materials made from petrochemicals or mineral fibres. Even though the water bag is now long gone, the principles of passive cooling, reusable containers and the use of natural fibres may yet be revived to inform the design of low-impact products to provide workers and travellers with cool, clean water wherever they need it in a warming world.

# Writing Case
## Barry Curtis

In the early 1960s the writing case was a suitable, even coveted, present for a bookish child who might, one day, aspire to engage in 'correspondence'. The one illustrated was bought at a stationer's in south London and given to me as a fourteenth birthday present. It was an aspirational object with historic affinities to older portfolios: the despatch cases and portable writing desks associated with diplomatic or military use. It also drew on the allure of a mobile sense of self-sufficiency and alluded to enviable intimacies.

The writing case was a 'networked' object in an extensive realm of stationery. A typical case contained a small address book, a 'perpetual calendar' protected by cellophane, pockets for envelopes and a retaining leather strap for a pad, which extended into a pen loop. 'Stationers' derived their name from non-itinerant providers of everything bookish; they were originally shops licensed by universities and anchored by immovable 'presses'. By the 1960s specialist high-street stationers sold all the items contained in my case, as well as bottles of ink, blotting paper and more specialized objects such as paper knives and sealing wax.

When it came into my hands, this attractive leather object was a portal into a realm of skill, etiquette and obligation. The fountain pen, invented in the nineteenth century, was effectively the key that unlocked and activated it. In a world of class distinctions, the fountain pen played a significant role as a congratulatory present, promising initiation into white-collar work and status. The Conway Stewart that occupied the loop in my writing case was an invention of the Edwardian period, when two employees of the venerable De La Rue Company – manufacturer of paper for banknotes and passports – borrowed the imposing name from a music-hall comedy act. By the 1920s they were manufacturing affordable pens with lever-action filling.

In the 1930s the pens became ornate, marbled plastic devices with screw tops and clips for top pockets. A 'Gentleman's Set' with a propelling pencil was advertised in the late 1950s for 25 shillings, and the 'Ladies' Model', without a clip, for 21 shillings. The part-gold nib, shaped like the tip of an archaic carved quill, was a key feature. Gold resisted corrosive ink, increased 'wettability' and flow, and was soft enough to wear unevenly in response to the user's distinctive pressure.

Managing the flow of ink required considerable training and became a performance of skill and personality. Unlike typing, handwriting is single-handed, and attention focuses on forming the letters and the contact

Letter-writing case, UK, early 1960s, leather.

between pen and paper. At infant school I was taught italic handwriting with thick and thin strokes, flourishes and strange protocols of joining letters. Desks incorporated inkwells and grooves for steel-nibbed dip pens. The interest in calligraphy aroused by the Arts and Crafts movement and influential designers such as Edward Johnston was still a vague but potent ethos. The slanted pen 'hand' required a compromise between legibility and 'character'.

The next most important element was paper. The ubiquitous Basildon Bond found in my writing case was advertised as 'fountain-pen friendly'. It was created in the same Edwardian milieu, when the Tottenham-based firm Millington & Sons devised a low-priced, high-quality writing paper. The directors named it after Basildon Park, a stately home in Berkshire. Press advertising featured celebrities, such as the television announcer Leslie Mitchell and the ballerina Alicia Markova, stressing the part played in their careers by a timely letter. Qualities of taste, refinement and discrimination were advertised as present both in the analysis of their handwriting and in the paper itself. A guide sheet was included with each pad to assure steady, even lines, and each sheet was watermarked with a crest.

To become an effective correspondent, it was necessary to know about stamps, their values and the relative speed of delivery, the situation of post-boxes and the times of collection. Stamp collecting was an approved hobby, particularly for boys. A Ladybird 'Easy Reading' book of 1965 sets out the correct format for addressing the envelope. On the letter itself, the address was justified to the left, above the date. There were protocols of spacing and indenting, and it was necessary to know how to locate the letter's receiver socially as well as geographically, and how to sign off. The signature was particularly important, and, like a handshake, was considered a revealing sign of character.

Everything about my writing case spoke of permanence, and of a regulated social world. But this was the era of social shocks, and that stable order was changing rapidly; the case was already redolent of energy radiated by obsolescence. My father, a signwriter and calligrapher, used a coral-red Olivetti Lettera 22 to type his invoices, and I learned to touch-type on it. Typing required both hands and involved the distinctive experience of watching the words appear on white space without being formed by hand. I began to type letters, and then to type letters electronically. The writing case became a storage facility. After the electronic transcendence of everything 'stationery', it was redundant.

Stationery today is not quite dead. There has been a metaphorical resurrection of it in the skeuomorphic world of digital desktops, blank documents, folders and waste bins. And, despite the decline of personal correspondence, elaborate handwritten letters continue to be used to announce major life events – births, weddings and the like – their specialness reasserted by modern stationers such as Smythson and Paperchase. Written communication is now a 'sensory' and a 'giving' experience. Numerous websites promote the mystique and accoutrements of penmanship, and similar writing cases to mine are still purchasable as high-status accessories, now branded as 'business portfolios' or 'conference folders'.

It's in the nature of extinction that living fossils persist and 'Lazarus taxa' return. Filofaxes defer to electronic organizers, Moleskine evokes the hypomnema of Ernest Hemingway and Paul Bowles, electronic devices are sheathed in tactile cases. Certificates and documents retain nebulous links to calligraphy. What has been lost is their real promise of social mobility, which is now associated with an entirely different set of objects. The leather goods purveyor Stow now advertises a deluxe leather case in soft coral and dove grey, 'as seen on Meghan Markle'. Outwardly, it resembles my case, but open it up and you see it holds not pen and paper, but a mobile phone, charger and memory stick.

# Zeppelin
Jeremy Myerson

Some objects slip slowly and almost imperceptibly from view over a long period of time, their obsolescence a gentle and rueful process. By contrast, the Zeppelin became obsolete literally overnight, when the German airship *Hindenburg* abruptly fell from the sky in an explosion of smoke and fire over Lakehurst, New Jersey, on 6 May 1937. It burned to ashes in less than a minute, although 62 of the 97 people on board miraculously survived by jumping dozens of feet to safety.

The *Hindenburg* was the largest and most sophisticated commercial airship ever built, at around 244 m (800 ft) in length and more than three times the size of a modern Boeing 747 jetliner. Before the explosion, passengers on the trip from Frankfurt to the United States would have been cruising at a leisurely 130 kph (80 mph; faster than any ocean-going liner at the time), enjoying fine dining with excellent wines, watching panoramic views from 180 m (600 ft) up and listening to a pianist in an elegant lounge decorated with silk wallpaper (the piano was lightweight, to meet the airship's strict weight standards, made mostly of aluminium alloy and covered in yellow pigskin). It took about four days to cross the Atlantic, so this Zeppelin had guest bedrooms and even featured its own smoking room for passengers to sample Cuban cigars, protected by a double-door airlock.

But beneath the smooth veneer of upmarket air travel, the *Hindenburg* LZ-129 was a ticking time bomb. German airships were filled with highly flammable hydrogen gas, rather than helium, since in the 1930s the United States held a monopoly on the world supply of helium and prohibited its export to countries that might use the gas for military advantage. The *Hindenburg*'s designer, Hugo Eckener, had recognized the danger and wanted to use helium as a lifting gas, but U.S. law held firm.

The scale and impact of the horrendous accident in 1937 was enough to consign the Zeppelin to the history books, and a much-heralded new age of airship travel failed to materialize. The Zeppelin instantly became a pariah of international travel, even though the *Hindenburg* wasn't the deadliest airship disaster; seven years earlier, in 1930, a British-built R101 had crashed in France, killing 48 men. But political circumstances meant the *Hindenburg* crash was a bigger story, and in 1937, when Nazi Germany appeared technologically menacing and all-powerful, and with Europe bracing for conflict, it was expedient to call attention to such a spectacular technical failure. The *Hindenburg* had even been used for Nazi propaganda. Just a year before the

The Zeppelin Hindenburg over Manhattan, 1 April 1936.

crash, the Nazi propaganda minister Joseph Goebbels had ordered the airship to make a 6,500 km (4,000 mi.) aerial tour of Germany blaring out patriotic tunes and dropping propaganda leaflets. It was also significant that this technological disaster was one of the first to be captured on film, and that speculation about the cause of the crash was intense in the aftermath.

At the dawn of the twentieth century, the future for the Zeppelin had looked a lot brighter. The airship was named after Count Ferdinand von Zeppelin, a German military officer who had studied the use of hot-air balloons in the American Civil War. Zeppelin built his first rigid-framed airship, the LZ-1, in 1899, and it made its maiden flight in 1900 – three years before the Wright Brothers took to the skies. The Zeppelin quickly became a popular form of air travel, and by 1914 some 10,000 fare-paying passengers had travelled on 1,500 commercial flights of the airline company Deutsche Luftschiffahrts-Aktiengesellschaft (DELAG for short). Among the family of airships (defined as lighter-than-air craft), Zeppelins were considered the most functional on account of their rigid metal framework covered with fabric and containing a number of individual gasbags. They were massive in scale and relatively cost-effective to build, and could travel over long distances, propelled by several engines; they could also be adapted for military use, and were used to bomb the enemy in the First World War.

The Treaty of Versailles in 1919 prohibited Germany from building large airships, and it was forced to surrender the Zeppelins used for DELAG flights. However, in 1926 these restrictions were lifted and the German Zeppelin company founded by Zeppelin, Luftschiffbau Zeppelin GmbH, turned its attention to civilian air travel, beginning work on the LZ-127 *Graf* Zeppelin. This made its maiden flight in 1928, circumnavigated the globe in a daring stunt the following year and went on to make 590 flights covering more than 1.6 million km (1 million mi.) before it was scrapped in June 1937 following the *Hindenburg* crash, and turned into a museum.

The design of the Zeppelin had evolved rapidly, outstripping the blimp (an airship with inflatable air compartments) in terms of technical advance. Its grander scale also dwarfed the flying boat, which was a largish rival for long-distance air travel but lacked the Zeppelin's majesty and the ability for passengers to stroll about and suggest songs to the pianist. By the mid-1930s affluent passengers were being whisked across the Atlantic in luxury and comfort from Germany to North America and Brazil, first by the *Graf* Zeppelin and then by the *Hindenburg*, which made its maiden flight in 1936. A vast infrastructure was built to accommodate these floating monsters of the skies, from hangars to mooring towers. The sole Zeppelin hangar to survive today can be found an hour's drive from Rio de Janeiro: this immense structure – 58 m (190 ft) tall and the size of several football pitches – is now used by the Brazilian air force.

Perhaps it could be argued that the Zeppelin was always destined for obsolescence, even without the *Hindenburg* disaster. Three months before the airship made its final trip, a Pan American Airways M-130 China Clipper completed its first scheduled flight across the Pacific, demonstrating the potential of fixed-wing commercial airlines to revolutionize travel. The outbreak of the Second World War in 1939 then accelerated the development and refinement of aircraft technology, to the point at which modern aeroplanes in the post-war era were able to offer a faster and more efficient alternative to the slow, supersized and anachronistic airship.

In the decades following the *Hindenburg* disaster, the operation of the flying machines that Zeppelin inspired has become a forgotten and fringe activity. Even so, despite its impractical nature, there is something romantic and special about the Zeppelin that refuses to leave the public consciousness. My own interest in the subject was kindled by one unique encounter. In 1993 I met Victor Papanek, designer, social activist and environment campaigner, and the author of the influential and controversial book *Design for the Real World: Human Ecology and Social Change* (1971). Papanek was speaking at the International Design Congress in Glasgow, and told his audience that his first childhood memory was travelling from Vienna to Berlin on a Zeppelin with his father and cycling the entire length of the structure on a tricycle. It was a vivid image from a forgotten world, and it stuck with me. Papanek then speculated about whether airships could be revived as an eco-friendly way to transport cargo over very long distances, reducing aeroplane-borne freight. It was a promise of rebirth for an obsolete form of transport that has yet to materialize. The rising global cost of helium hasn't helped to make services viable, and old drawbacks remain; for example, the weight of water prohibits tourists from taking the type of high-pressure shower on board that we expect today, so a luxury sky cruise lasting several days might require passengers to go without washing.

Even so, the glorious sight of a giant Zeppelin appearing slowly and serenely through a break in the clouds evokes a slower, gentler era of air travel. As we queue through security in hot and congested airports to cram into packed aluminium tubes for several hours of bone-numbing boredom, we might just recall diners on the *Hindenburg* sampling the red wine as they floated through the sky, or an infant Papanek cycling along the superstructure for hours at a time with his dad in tow. Is the Zeppelin overdue a comeback?

# Biographies

**Daniel M. Abramson** is professor of architectural history in the department of History of Art and Architecture at Boston University. His most recent book is *Obsolescence: An Architectural History* (2016).

**Zeynep Çelik Alexander** is associate professor in the department of Art History and Archaeology at Columbia University. She is the author of *Kinaesthetic Knowing: Aesthetics, Epistemology, Modern Design* (2017).

**Pedro Ignacio Alonso** is an architect and MSc from the Pontificia Universidad Católica de Chile, and holds a PhD in architecture from the Architectural Association. With Hugo Palmarola, he was awarded a Silver Lion at the 14th Venice Architecture Biennial (2014). They are the authors of *Panel* (2014) and *Monolith Controversies* (2014), and are the curators of the exhibition 'Flying Panels: How Concrete Panels Changed the World' at ArkDes, Stockholm (2019–20).

**Tim Ainsworth Anstey** is professor at the Institute of Form, Theory and History and head of the PhD programme at the Oslo School of Architecture and Design.

**Daniel Barber** is associate professor of architecture at the University of Pennsylvania. His book, *Modern Architecture and Climate: Design before Air Conditioning*, was published by Princeton University Press in 2020.

**Maude Bass-Krueger** is professor of art history at Ghent University in Belgium. Her research focuses on French fashion historiography, the culture of architecture in the nineteenth century and the history of museums and collections.

**Sarah Bell** is City of Melbourne Chair in Urban Resilience and Innovation at the University of Melbourne. She is a Fellow of the Institution of Civil Engineers and the Chartered Institution of Water and Environmental Management.

**Barry Bergdoll** is Meyer Schapiro Professor of Art History at Columbia University and the former chief curator in the department of Architecture and Design at the Museum of Modern Art, New York.

**Eirik A. G. Bøhn** is an art historian working on early modern print culture. He is the editor and curator, alongside Mari Lending and Tim Anstey, of the book and exhibition *Images of Egypt* (Historical Museum, Oslo, 2018).

**Tim Boon**, Head of Research and Public History for the Science Museum Group in London, is also a historian and curator of the public culture of science.

**Paul Bouet** is PhD candidate and lecturer at the École nationale supérieure d'architecture Paris-Est (Université Gustave Eiffel). His PhD investigates the alternative future shaped by solar architecture in post-war France and Africa.

**Mario Carpo** is professor of architectural history and theory at the Bartlett School of Architecture, UCL, and at the School of Applied Arts (Die

Angewandte) of the University of Vienna. He is the author of *Architecture in the Age of Printing* (2001) and other books.

**Pippo Ciorra** is an architect, critic and professor at SAAD_UNICAM (Scuola di Ateneo Architettura e Design, Università di Camerino) and IUAV in Venice (PhD programme). He is the author of books and essays on Italian architecture, urbanism and museums. He has curated exhibitions in Italy and elsewhere, and since 2009 he has been senior curator at MAXXI in Rome.

**Steven Connor** is Grace 2 Professor and Director of the Centre for Research in Arts, Social Sciences and Humanities (CRASSH) at the University of Cambridge.

**Miranda Critchley** is completing her PhD at the Bartlett School of Architecture, UCL, on the subject of railways and colonial narratives of progress. She has taught in the History department at Queen Mary University, the History of Art department at UCL and the Architecture School at Greenwich University.

**Barry Curtis** graduated in English and architecture and fine arts from the University of Cambridge, and studied film at the British Film Institute and the University of Westminster. He has taught at Middlesex University, the Open University, the London Consortium and the Royal College of Art, and is currently at the University of the Arts, London. He has written on cultural history and theory, and the history of design, film and popular culture, and has a long-standing interest in the life and afterlife of things.

**Gillian Darley** is a writer, broadcaster and biographer with a particular interest in visionary projects. Among her books are *Villages of Vision*

(1975; revised 2007), *Ian Nairn: Words in Place* (with David McKie; 2013) and *Excellent Essex* (2019).

**Carlotta Darò** is an art and architectural historian at the École nationale supérieure d'architecture Paris-Malaquais and a member of the Laboratoire Infrastructure, Architecture, Territoire. Her research explores the impact of sound technology, telecommunications infrastructure and media on architectural and urban culture in the nineteenth and twentieth centuries. She is the author of *Avant-gardes sonores en architecture* (2013) and *Les murs du son: Le Poème électronique au Pavillon Philips* (2015).

**Tacita Dean** was born in Canterbury, England, in 1965, and lives and works in Berlin and Los Angeles. She uses a variety of mediums in her practice, but is primarily known for her works on photochemical film.

**Edward Denison and Guang Yu Ren** combine academic work at the Bartlett School of Architecture, UCL, with professional practice. Their award-winning research focuses on challenging canonical histories and has been published in more than twenty books.

**Paul Dobraszczyk** is a writer and researcher and teaching fellow at the Bartlett School of Architecture, London. His books include *Future Cities: Architecture and the Imagination* (2019) and *The Dead City: Urban Ruins and the Spectacle of Decay* (2017).

**David Edgerton** is the author of *Shock of the Old: Technology and Global History since 1900* (2006; 2019) and *The Rise and Fall of the British Nation: A Twentieth-century History* (2019).

**Kjetil Fallan** is professor of design history at the University of Oslo. His most recent books are *The Culture of Nature in the History of Design* (2019) and *Designing Modern Norway* (2017).

**Adrian Forty** is professor emeritus of architectural history at the Bartlett School of Architecture, UCL. He is the author of *Objects of Desire: Design and Society Since 1750* (1986), and *Concrete and Culture: A Material History* (2012). From 2010 to 2014 he was president of the European Architectural History Network.

**Hal Foster** is Townsend Martin Class of 1917 Professor of Art and Archaeology at Princeton University. *What Comes after Farce?*, a collection of essays on art and politics, and his 2018 Mellon lectures, *Brutal Aesthetics*, were published in 2020.

**Tony Fretton** is a principal of Tony Fretton Architects and Emeritus Professor, Chair of Interiors Buildings and Cities at TU Delft the Netherlands. Buildings designed and realized by the practice include the Lisson Gallery and the Red House, Chelsea, both in London; Fuglsang Kunstmuseum in Denmark, which was shortlisted for the Stirling Prize; the new British Embassy in Warsaw; Solid 11, a multi-purpose building in Amsterdam; and most recently two apartment towers in Antwerp Harbour and the City Hall of Deinze, Belgium. Fretton's sketchbooks are in the Archive of Art and Design at the Victoria and Albert Museum, London, and project models and drawings by the practice are in the Drawing Matter Trust collection. See also www.tonyfretton.com.

**Kristen Gallerneaux**, PhD, is a sonic researcher, museum curator and artist. Her monograph *High Static, Dead Lines: Sonic Spectres and the Object Hereafter* (2018) is available via Strange Attractor and MIT Press.

**Lucy Garrett** is an LPC student at BPP University in the UK. Previously she worked as an antique map and books specialist at Daniel Crouch Rare Books in London.

**Elizabeth Guffey** is professor of art and design history and head of the MA programme in art history at the State University of New York, Purchase College. She is the author of *Retro: The Culture of Revival* (2006) and *Posters: A Global History* (2015). She is also co-editor of *Making Disability Modern* (with Bess Williamson; 2020) and author of *Designing Disability* (2018).

**Gökçe Günel**'s first book, *Spaceship in the Desert: Energy, Climate Change, and Urban Design in Abu Dhabi* (2019), examines the construction of Masdar City.

**Harriet Harriss** is a UK-registered architect and dean of the Pratt School of Architecture in Brooklyn, New York. Her publications include *Radical Pedagogies: Architectural Education and the British Tradition* (2015), *A Gendered Profession* (2016), *Interior Futures* (2018) and *Architects After Architecture* (2020).

**Tanya Harrod** is a design historian who also writes about craft and art. She is co-editor of the *Journal of Modern Craft*. Her book *The Last Sane Man: Michael Cardew: Modern Pots, Colonialism, and the Counterculture* (2012) won the 2012 James Tait Black Memorial Prize for biography.

**Lucinda Hawksley** is an author, broadcaster and lecturer. Her books include biographies of the artists Lizzie Siddal, Kate Perugini (née Dickens) and Princess Louise, as well as *Dickens and Christmas* (2017); *March, Women, March* (about the suffrage movement; 2013); *Charles Dickens and His Circle* (2016); *The Writer Abroad: Literary Travels from Austria to Uzbekistan* (about historic travel writing;

2017); and *Bitten by Witch Fever: Wallpaper and Arsenic in the Victorian Home* (2016). Her upcoming book is *Dickens and Travel* (2021).

**Amy de la Haye** is professor of dress history and curatorship and joint director of the Centre for Fashion Curation at London College of Fashion, University of the Arts London.

**Edwin Heathcote** is the architecture and design critic of the *Financial Times*. He is an architect and designer, and the author of around a dozen books. He is also the founder and editor of the online archive www.readingdesign.org.

**Carola Hein** is professor of history of architecture and urban planning at TU Delft, the Netherlands. Books she has authored or edited include *The Routledge Handbook of Planning History* (2017); *Port Cities: Dynamic Landscapes and Global Networks* (2011); *Cities, Autonomy, and Decentralization in Japan* (with Philippe Pelletier; 2006); and she contributed to Nishiyama Uzo's *Reflections on Urban, Regional and National Space* (2019).

**Olivia Horsfall Turner** is a historian of architecture and design. She is senior curator of designs at the Victoria and Albert Museum, and the V&A's lead curator for the V&A + RIBA Architecture Partnership.

**Mari Hvattum** is professor of architectural history at the Oslo School of Architecture and Design.

**Thomas Kador** has a background combining chemical engineering and archaeology. He currently leads a number of UCL's flagship undergraduate object-based learning modules. His interests include material culture and the use of objects in learning and well-being.

**Lydia Kallipoliti** is an architect, engineer and scholar whose research focuses on the intersections of architecture, technology and environmental politics. She is an assistant professor of architecture at the Cooper Union, New York, and the author of *The Architecture of Closed Worlds, Or, What Is the Power of Shit?* (2018).

**Priya Khanchandani** is the head of curatorial at the Design Museum and former editor of *Icon* magazine. A graduate with distinction from the Royal College of Art, London, she previously worked at the Victoria and Albert Museum on design acquisitions and at the British Council on cultural programmes.

**Robin Kinross** is a typographer and editor who runs Hyphen Press in London. His own books include *Modern Typography* (1992; 2004) and *Unjustified Texts* (2002).

**Hannah le Roux** is an associate professor in the School of Architecture and Planning at University of the Witwatersrand, South Africa, and is an architect, educator and theorist. Her work in all these areas revisits the modernist project in architecture in Africa, and considers how its transformation through the agency of Africa presents a conceptual model for contemporary design.

**Maarten Liefooghe** is assistant professor in architectural history and theory at Ghent University, Belgium.

**Thandi Loewenson** is an architectural designer-researcher who operates through design, fiction and performance to interrogate our perceived and lived realms and to speculate on the possible worlds in our midst.

**Nikos Magouliotis** is an architectural historian and PhD candidate in the Institute for the

History and Theory of Architecture at ETH Zurich. His work focuses on the history and historiography of architecture from the mid-eighteenth to early twentieth century, particularly on the notions of the 'primitive', the 'vernacular' and the 'anonymous' as historical entities and as theoretical constructs.

**Niall McLaughlin** is an architect. He was born in Geneva in 1962. He was educated in Dublin and his practice is in London. He teaches at the Bartlett School of Architecture, UCL.

**Thomas McQuillan** is professor of architecture at the Oslo School of Architecture and Design, where he is also head of the Institute of Architecture.

**Iris Moon** is an assistant curator at the Metropolitan Museum of Art in New York, and has lectured and written broadly on eighteenth- and nineteenth-century European architecture and the decorative arts.

**Mark Morris** is an architect, historian and curator, and head of teaching and learning at the Architectural Association in London. He is the author of *Models: Architecture and the Miniature* (2006).

**Anders V. Munch** is a design historian, educated in the history of ideas and art history at the University of Aarhus, Denmark. He is a professor in the Department of Design and Communication at the University of Southern Denmark. His most recent book is *The Gesamtkunstwerk in Design and Architecture. From Bayreuth to Bauhaus* (2021).

**Jeremy Myerson** is Helen Hamlyn Professor of Design at the Royal College of Art, London, and director of the global online knowledge network WORKTECH Academy.

**Bob Nicholson** is the curator of the Old Joke Archive and a historian of Victorian popular culture, based at Edge Hill University, Lancashire. He tweets @DigiVictorian.

**Emily M. Orr** is the assistant curator of modern and contemporary American design at Cooper Hewitt, Smithsonian Design Museum, New York. She holds a PhD in the history of design from the Royal College of Art/Victoria and Albert Museum in London.

**Hugo Palmarola** is a designer from the Pontificia Universidad Católica de Chile (2004) and PhD in Latin American Studies at UNAM, Mexico (2018). He won the Student Essay Prize from the Design History Society (2018). With Pedro Ignacio Alonso, he was awarded a Silver Lion at the 14th Venice Architecture Biennial (2014). They are the authors of *Panel* (2014) and *Monolith Controversies* (2014), and are the curators of the exhibition 'Flying Panels: How Concrete Panels Changed the World' at ArkDes, Stockholm (2019–20).

**Christian Parreno** is assistant professor of history and theory of architecture at Universidad San Francisco de Quito, Ecuador. He is the author of *Boredom, Architecture, and Spatial Experience* (2021).

**Jacob Paskins** teaches architectural history in the department of History of Art, UCL. He is the author of *Paris Under Construction: Building Sites and Urban Transformation in the 1960s* (2016).

**Angus Patterson** is a senior curator at the Victoria and Albert Museum in London, responsible for collections of European metalwork and arms and armour from 1450 to 1900. With Alistair Grant, Angus co-authored *The Museum and the Factory: The V&A, Elkington and the Electrical Revolution* (2018).

**Barbara Penner** is professor in architectural humanities at the Bartlett School of Architecture, UCL. She is the author of *Bathroom* (2013) and has co-edited numerous books on gender, space and architecture. She is a contributing editor of *Places Journal*.

**James Purdon** is a lecturer in modern and contemporary literature at the University of St Andrews, and the author of *Modernist Informatics: Literature, Information, and the State* (2016).

**Bryony Quinn** is a writer, editor and lecturer on contextual studies and design research at the University of East London, London College of Communication, and the Royal College of Art, London. Her research focuses on figurative and spatial obliquity: things that lean, slopes, diagonals, digression, and so on.

**Charles Rice** is professor of architecture at the University of Technology Sydney. He is the author of *The Emergence of the Interior: Architecture, Modernity, Domesticity* (2007) and *Interior Urbanism: Architecture, John Portman and Downtown America* (2016).

**David Rooney** is a writer and curator. He was formerly keeper of technologies and engineering at the Science Museum, London, and curator of timekeeping at the Royal Observatory, Greenwich.

**Johanna Agerman Ross** is the curator of twentieth-century and contemporary furniture and product design at the Victoria and Albert Museum in London. She is also the founder and director of the leading quarterly design journal *Disegno*, founded in 2011.

**Simon Sadler** is a professor in the department of Design at the University of California, Davis. His

publications, which examine how design makes change, include *Archigram: Architecture without Architecture* (2005); *Non-Plan: Essays on Freedom, Participation and Change in Modern Architecture and Urbanism* (co-edited with Jonathan Hughes; 2000); and *The Situationist City* (1998).

**Shahed Saleem** is a London-based architect, a design studio leader at the University of Westminster School of Architecture, and an honorary research fellow at the Bartlett School of Architecture, UCL. His particular interest is in user-led design and how this leads to new forms of architecture and urbanism, particularly among migrant and diaspora communities. He is the author of *The British Mosque: An Architectural and Social History* (2018).

**Catherine Slessor** is an architecture critic, editor and writer. She originally trained as an architect before defecting to architectural journalism. She is a former editor of the *Architectural Review* and has written for numerous publications, including *The Observer, Icon, Dezeen* and the *Architects' Journal*. She lived with a militant smoker for nineteen years and has a world-class collection of purloined ashtrays.

**Laurent Stalder** is an architectural historian at ETH Zurich. The main focus of his research and publications is the history and theory of architecture where it intersects with the history of technology.

**Eszter Steierhoffer** is senior curator at the Design Museum in London. Previously she worked as curator of contemporary architecture at the Canadian Centre for Architecture in Montreal. She has a PhD in critical and historical studies from the Royal College of Art in London, and her research focuses on the modern and contemporary histories of architecture exhibitions.

**Deyan Sudjic** is professor of architecture and design studies at the University of Lancaster and was director of the Design Museum in London from 2006 to 2020. He has taught at the Royal College of Art, London, and the University of the Applied Arts in Vienna, and has curated exhibitions in Istanbul, Glasgow and Copenhagen. He was the founding editor of *Blueprint* magazine, and the editor of *Domus*. He was the director of the Venice Architecture Biennale in 2002. His most recent book, *The Language of Cities* (2017), was published by Penguin.

**Richard Taws** is reader in the history of art at UCL. He is currently completing a book about art and telegraphy in nineteenth-century France.

**Carsten Timmermann** is based at the University of Manchester's Centre for the History of Science, Technology and Medicine. He has published widely on the history of cancer and other non-communicable diseases, and teaches course units in the history of psychiatry, biology and medicine. He is the academic lead for the University of Manchester's Museum of Medicine and Health.

**David Trotter** is an emeritus professor of English literature at the University of Cambridge. He has written widely about nineteenth- and twentieth-century literature and culture, and about the history of media.

**Rachel Siobhán Tyler** is an artist and cultural researcher. She is currently undertaking doctoral research into East London's fashion industry, funded by Techne AHRC, within the Centre for Geohumanities at Royal Holloway, University of London.

**Ben Vandenput** is an art historian and urbanist. He is working on his doctoral thesis, entitled *Victor Hugo's Elephants*, in the department of Architecture and Urban Planning of Ghent University, Belgium.

**Richard Wentworth** is an artist and educator. He has played a leading role in New British Sculpture since the end of the 1970s. He documents the everyday, paying attention to objects, occasional and involuntary geometries, and uncanny situations that often go unnoticed. He lives and works in London.

**Tom Wilkinson** is a lecturer in the history of art at the Courtauld Institute, before which he was a Leverhulme Early Career Fellow at the Warburg Institute. He is history editor of the *Architectural Review*.

**Danielle S. Willkens** is an assistant professor at Georgia Institute of Technology's School of Architecture in Atlanta. She was the Society of Architectural Historians' H. Allen Brooks Travelling Fellow in 2015.

# References and Further Reading

## Introduction

1 Amartya Sen, 'On the Darwinian View of Progress,' *London Review of Books*, XIV/21 (November 1992), www.lrb.co.uk.

2 Beatriz Colomina and Mark Wigley, *Are We Human? Notes on an Archaeology of Design* (Zurich, 2016), p. 10.

3 Lewis Mumford, *Technics and Civilization* (Chicago, IL, 2010), p. 187.

4 All quotations from Charles C. Gillispie, 'Introduction', in *A Diderot Pictorial Encyclopedia of Trades and Industry: 485 Plates Selected from 'L'Encyclopédie' of Denis Diderot*, vol. I (New York, 1959), pp. 9, 24.

5 All quotations from Sigfried Giedion, *Mechanization Takes Command: A Contribution to Anonymous History* (Oxford, 1948), p. 692.

6 Adrian Forty, *Objects of Desire* (London, 1986), p. 8.

7 See, for instance, the contributions to Irene Cheng, Charles L. Davis II and Mabel O. Wilson, eds, *Race and Modern Architecture: A Critical History from the Enlightenment to the Present* (Pittsburgh, PA, 2020).

8 Jill Lepore, 'The Robot Caravan: Automation, AI, and the Coming Invasion', *New Yorker*, 4 March 2019, p. 22.

9 All quotations from David Edgerton, *The Shock of the Old: Technology and Global History Since 1900* (London, 2006), pp. ix, xii.

10 David Farrier, *Footprints: In Search of Future Fossils* (London, 2020), p. 9.

11 Rupert Neate, 'UK Firm Flying High as Eco-friendly Airship Project Gathers Pace: Bedford's HAV Crowdfunds a Further £1.6m to Bring Low-emissions "Zeppelin" Back to the Skies', www.theguardian.com, 29 August 2020.

12 See Emily Orr's contribution, 'ConvAirCar', in this volume, p. 85.

Further reading: Beatriz Colomina and Mark Wigley, *Are We Human? Notes on an Archaeology of Design* (Zurich, 2016); Charles Darwin, *On the Origin of Species by Means of Natural Selection or the Preservation of Favoured Races in the Struggle for Life* (London, 2009); *A Diderot Pictorial Encyclopedia of Trades and Industry: 485 Plates Selected from 'L'Encyclopédie' of Denis Diderot*, vol. I (New York, 1959); David Edgerton, *The Shock of the Old: Technology and Global History Since 1900* (London, 2006); Adrian Forty, *Objects of Desire* (London, 1986); Sigfried Giedion, *Mechanization Takes Command: A Contribution to Anonymous History* (Oxford, 1948); Lewis Mumford, *Technics and Civilization* (Chicago, IL, 2010)

## Acoustic Location Device

Further reading: Sir Alfred Rawlinson, *The Defence of London, 1915–1918* (London, 1923); Richard N. Scarth, *Echoes from the Sky: A Story of Acoustic Defence* (Hythe Civic Society, 1999)

## Action Office Acoustic Area Conditioner: The 'Maskitball'

1 Conversation between the author and Jack Kelley, May 2019.

2 Quoted in Kristen Gallerneaux, 'Robert Propst', *Henry Ford Magazine* (Winter 2019), pp. 30–31.

3 Robert Propst, 'Letter to Lyman Blackwell, 25 March 1963', Additive acoustic component for AO2, Box 42, Accession 2010.83, Robert Propst Papers, Benson Ford Research Center, The Henry Ford, Dearborn, Michigan.

4 Robert Propst and Michael Wodka, *The Action Office Acoustic Handbook* (Ann Arbor, MI, 1972), p. 15.

5 Ibid., p. 7.

6 Conversation between the author and Jack Kelley, May 2019.

7 Herman Miller, Inc., *Ideas from Herman Miller* (Zeeland, MI, 1976), p. 7.

Further reading: John Biguenet, *Silence* (London, 2015); Mack Hagood, *Hush: Media and Sonic Self-control* (Durham, NC, 2019); Robert Propst, *The Office: A Facility Based on Change* (Elmhurst, IL, 1968)

**Air-curtain Roof**

1 Werner Ruhnau, 'Von der monumentalen Steinstadt zur flexiblen Struktur in der klimatisierten Stadtlandschaft', *Bauen und Wohnen*, 19 (1965), pp. v, 20, 22.

2 B. Etkin and P.L.E. Goering, 'Air-curtain Walls and Roofs – "Dynamic" Structures', *Philosophical Transactions of the Royal Society of London, Series A: Mathematical and Physical Sciences*, CCLXIX/1199: A Discussion on Architectural Aerodynamics (13 May 1971), section V: 'Future Possibilities and Challenges', pp. 541–2; Peter L. E. Goering, 'Intermittent Environments: The Air-curtain as an Adaptive Enclosure', *IL 14: Anpassungsfähig bauen/Adaptable Architecture – 10 Jahre IL, 1964–1974, Mitteilungen des Instituts für leichte Flächentragwerke (IL) Universität Stuttgart*, 14 (1975), p. 272.

3 See for instance A. A. Haasz and S. Raimondo, 'Performance of Adjacent Dual-jet Air-curtain Roofs', *Journal of Wind Engineering and Industrial Aerodynamics*, 10 (1982), pp. 79–87; A. A. Haasz and B. Kamen, 'Annular Air-curtain Domes for Sports Stadia', *Journal of Wind Engineering and Industrial Aerodynamics*, 26 (1986), pp. 75–92.

Further reading: Valéry Didelon, 'Aire conditionnée – Utopies domestiques: Yves Klein et Werner Ruhnau', *Faces*, 63 (Autumn 2006), pp. 8–11; Kim Förster, 'Air Curtain', *Arch+*, ed. Laurent Stalder, Elke Beyer, Anke Hagemann and Kim Förster, 191/192, special issue of *Schwellenatlas* (2009), pp. 27–8; Cyrille Simonnet, *Brève histoire de l'air* (Versailles, 2014)

**All-plastic House**

1 'President's Report', *Massachusetts Institute of Technology Bulletin (1956)*, p. 89.

2 Quoted ibid., p. 129.

3 Meg Neal, 'A Map of the Last Remaining Flying Saucer Homes', 13 September 2016, www.atlasobscura.com; see also www.thefuturohouse.com.

4 All quotations in this paragraph from the Plastics Institute, *Plastics in Building Structures: Proceedings of a Conference Held in London, 14–16 June 1965* (London, 1966), pp. 1–2.

Further reading: Elke Genzel and Pamela Voigt, *Kunststoffbauten* (Weimar, 2005); Marko Home and Mika Taanila, eds, *Futuro: Tomorrow's House from Yesterday* (Eastbourne, 2003); Jeffrey L. Meikle, *American Plastic: A Cultural History* (New Brunswick, NJ, 1995); Mika Taanila, *Futuro: A New Stance for Tomorrow* (film), 1998, https://mikataanila.com

## Arundel Print

1 John Ruskin, *The Works*, ed. E. T. Cook and A. Wedderburn (London, 1903–12), vol. XI, p. 241.
2 Richard Offner, 'An Outline of the Theory of Method', in *Studies in Florentine Painting: The Fourteenth Century* (New York, 1927), p. 136.
3 Ernest William Tristram, *English Medieval Wall Painting: The Twelfth Century* (Oxford, 1944), pp. vi–vii.
4 Christian Barman, 'Printed Pictures', *Penrose Annual*, XLIII (1949), p. 56.

Further reading: Robyn Cooper, 'The Popularisation of Renaissance Art in Victorian England', *Art History*, I/3 (1978), pp. 263–92; Tanya Ledger [Harrod], 'A Study of the Arundel Society 1848–1897', doctoral thesis, University of Oxford, 1978; Michael Twyman, *A History of Chromolithography: Printed Colour for All* (London, 2013)

## Asbestos-cement Rondavel

1 John Comaroff and Jean Comaroff, *Of Revelation and Revolution*, vol. II: *The Dialectics of Modernity on a South African Frontier* (Chicago, IL, 1997), p. 313.
2 Karel A. Bakker, R. C. De Jong and A. Matlou, 'The "Mamelodi Rondavels" as Place in the Formative Period of Bantu Education and in Vlakfontein (Mamelodi West)', *South African Journal of Cultural History*, XVII/2 (2003), pp. 1–21.
3 Laurie Kazan-Allen, 'Chronology of Asbestos Bans and Restrictions', International Ban Asbestos Secretariat, updated 12 August 2020, www.ibasecretariat.org; and J. C. Wagner, C. A. Sleggs and Paul Marchand, 'Diffuse Pleural Mesothelioma and Asbestos Exposure in the North Western Cape Province', *British Journal of Industrial Medicine*, XVII/4 (1960), pp. 260–71.

Further reading: Everite Limited, *Asbestos-cement Rondavel Roofs: Brochure with Comprehensive Erection Instructions* (Johannesburg, 1974), and —, *Everite Rondavel/Everite Rondawel* (Johannesburg, 1974); Brian Gibson, 'Asbestos and Health at Everite: A Review of Critical Milestones' (Johannesburg, 2002); Mauritz Naude, 'A Legacy of Rondavels and Rondavel Houses in the Northern Interior of South Africa', *South African Journal of Art History*, XXII/2 (2007)

## Cab-fare Map

1 Henry Moore, *Omnibuses and Cabs: Their Origin and History* (London, 1902), pp. 237–8; 'The Cabman's Shelter. Enter Mrs Giacometti Prodgers. Tableau!', *Punch*, 68 (6 March 1875), p. 106.
2 M. Roberts, 'Cabs and Cabmen', *Murray's Magazine*, VII/39 (1890), p. 385.

Further reading: Paul Dobraszczyk, 'Useful Reading? Designing Information for London's Victorian Cab Passengers', *Journal of Design History*, XXI/2 (2008), pp. 121–41; Ralph Hyde, 'Maps that Made London's Cabmen Honest', *The Ephemerist*, I/26 (1980), pp. 138–9; Thomas May, *Gondolas and Growlers: The History of the London Horse Cab* (Stroud, 1995)

## Central Heating

1 Le Corbusier, *Précisions sur un état présent de l'architecture et de l'urbanisme* (Paris, 1930), pp. 64–6. See also Reyner Banham, *The Architecture of the Well-tempered Environment* [1969] (Chicago, IL, 1984), pp. 159–60.
2 Robert Bean, Bjarne W. Olesen and Kwang Woo Kim, 'History of Radiant Heating and Cooling Systems, Part II', *Journal of the American Society of Heating, Refrigerating, and*

*Air-conditioning Engineers*, LIV/2 (February 2010), pp. 51–5.

3 Chamberlin, Powell and Bon, 'Report to the Court of Common Council of the Corporation and City of London on Residential Developments within the Barbican Area', May 1956, Appendix VI, p. ix; and ibid., April 1959, Section 3, pp. 5, 8–9.

4 Ibid., p. 9.

5 Emmanuelle Gallo, 'Jean Simon Bonnemain (1743–1830) and the Origins of Hot Water Central Heating', paper presented at the Second International Congress on Construction History, Queens' College, Cambridge, 29 March – 2 April 2006.

### Chaparral 2J: The 'Sucker Car'

1 Dick Rutherford, 'The Incredible Chaparral 2J: An Entirely New Concept of Downforce Promises to Make this Can-Am Car Stick Like Glue in the Corners', *Corvette News* (August/September 1970), pp. 4–9.

2 Ibid.

3 Quoted in Marshall Pruett, 'Racer Redux: The Chaparral 2J', www.racer.com, 8 April 2016.

4 Brian Redman, interviewed under free practice at the Targa Florio, Sicily, May 1970, in untitled documentary footage. Collezione Targapedia.

Further reading: R. M. Clarke, *Can-Am Racing Cars, 1966–1974* (Little Chalfont, Buckinghamshire, 2001); Blake Z. Rong, 'Jim Hall and the Chaparral 2J: The Story of America's Most Extreme Race Car', *Road & Track*, 23 January 2017; Dick Rutherford, 'The Incredible Chaparral 2J: An Entirely New Concept of Downforce Promises to Make This Can-Am Car Stick Like Glue in the Corners', *Corvette News* (August/September 1970), pp. 4–9

### Chatelaine

1 Genevieve E. Cummins and Nerylla D. Taunton, *Chatelaines: Utility to Glorious Extravagance* (Woodbridge, 1994), p. 49.

2 See curatorial note by Clare Vincent, curatorial files, European Sculpture and Decorative Arts, Metropolitan Museum of Art, New York.

Further reading: Barbara Burman and Ariane Fennetaux, *The Pocket: A Hidden History of Women's Lives, 1660–1900* (New Haven, CT, and London, 2019); Peter McNeil, *Pretty Gentlemen: Macaroni Men and the Eighteenth-century Fashion World* (New Haven, CT, and London, 2018)

### The Clapper

1 Sidney A. Boguss, *United States Patent no. USD299127S*, 27 December 1988, http://patents.google.com.

2 Abstract, Joe Pedott Oral History Interview, 20 September 2004. Joseph Pedott Papers, Archives Centre, National Museum of American History, Smithsonian Institution, Washington, DC, NMAH.AC.0898.

3 Carlile R. Stevens and Dale E. Reamer, *United States Patent no. US5493618A*, 20 February 1996, http://patents.google.com.

4 Brian M. King et al., *United States Patent no. US20140164562A1*, pending (published 12 June 2014), http://patents.google.com.

Further reading: Sigfried Giedion, *Mechanization Takes Command: A Contribution to Anonymous History* (Oxford, 1948)

### Close-constraint Key

1 Iain Boyd Whyte and David Frisby, eds, *Metropolis Berlin: 1880–1940* (Berkeley and Los Angeles, CA, and London, 2012), pp. 134–5.

2 Werner Sombart, 'Domesticity', ibid., pp. 151–2. First published in Werner Sombart, *Das Proletariat* (Frankfurt am Main, 1906).

3 Hans Kurella, *Wohnungsnot und Wohnungsjammer, ihr Einfluß auf die Sittlichkeit, ihr Ursprung aus dem Bodenwucher und ihre Bekämpfung durch demokratische Städteverwaltung* (Frankfurt am Main, 1900), quoted in Whyte and Frisby, eds, *Metropolis Berlin*, p. 151.

4 Victor Noack, 'Housing and Morality', in Whyte and Frisby, eds, *Metropolis Berlin*, pp. 167, 170. First published as 'Wohnungen und Sittlichkeit', *Die Aktion*, 19/20 (1912), pp. 584–6, 618–20.

Further reading: Jens Sethmann, 'Der Siegeszug des Doppelschlüssels. Wider die Säumigen und Vergesslichen', www.berliner-mieterverein.de, 28 November 2005, and —, 'Berlins Stille Portiers. Ein hölzerner Diener', www.berliner-mieterverein.de, 28 May 2012; Petra Ullmann, 'Vor fast 90 Jahren erfand ein Weddinger Handwerker den Durchsteckschlüssel', www.tagesspiegel.de, 8 March 2000; Iain Boyd Whyte and David Frisby, eds, *Metropolis Berlin: 1880–1940* (Berkeley and Los Angeles, CA, and London, 2012)

### Concorde

1 *Flight International*, 10 April 1976, p. 874.

2 Ministère de l'Équipement des Transports et du Logement-bureau d'Enquêtes et d'Analyses pour la Sécurité de l'Aviation Civile France, 'Accident survenu le 25 juillet 2000 au lieu-dit La Patte d'Oie de Gonesse au Concorde immatriculé F-BTSC exploité par Air France' (Paris, 2002).

Further reading: Lawrence Azerrad, *Supersonic: The Design and Lifestyle of Concorde* (Munich, 2018); Frédéric Beniada and Michel Fraile, *Concorde* (London, 2006); Geoffrey Knight, *Concorde: The Inside Story* (London, 1976); Christopher Orlebar, *The Concorde Story* (Oxford, 2011)

### ConvAirCar

1 Consolidated Vultee Aircraft Corporation, *Preliminary Design Report No. ZP 47 118 001-A* (San Diego, CA, 3 July 1947), n.p., Henry Dreyfuss Archive, Gift of Henry Dreyfuss, Cooper Hewitt, Smithsonian Design Museum, New York.

2 Ibid.

3 Sam McDonald, 'Your Air Taxi Could Be Hovering Just Beyond the Horizon, Industry Leader Says', 15 June 2018, www.nasa.gov.

Further reading: Henry Dreyfuss Archive, Gift of Henry Dreyfuss, Cooper Hewitt, Smithsonian Design Museum, New York; Russell Flinchum and Henry Dreyfuss, *Henry Dreyfuss, Industrial Designer: The Man in the Brown Suit* (New York, 1997)

### Cybernetic Anthropomorphic Machines

1 Heinrich Wölfflin, *Renaissance and Baroque*, trans. Kathrin Simon (London, 1964), p. 77.

2 General Electric Company, Corporate Research and Development Report, 'Research and Development Prototype for Machine Augmentation of Human Strength and Endurance', 1 May 1971. In the archives of the Museum of Innovation and Science (MiSci), Schenectady, New York.

3 Ralph S. Mosher, 'Handyman to Hardiman', SAE Technical Paper 670088, 1967, doi: 10.4271/670088. Research and Development Center, General Electric Co., Schenectady, New York.

4 Leo Marx, *The Machine in the Garden: Technology and the Pastoral Ideal in America*

(Oxford, 1964), pp. 7–8. Quoting José Ortega y Gasset, *The Revolt of the Masses* [1930] (New York, 1950).

Further reading: Beatriz Colomina and Mark Wigley, *Are We Human? Notes on an Archaeology of Design* (Zurich, 2016); Donna Haraway, 'A Cyborg Manifesto: Science, Technology and Socialist-Feminism in the Late Twentieth Century', in *Simians, Cyborgs and Women: The Reinvention of Nature* (New York, 1991); Leo Marx, *The Machine in the Garden: Technology and the Pastoral Ideal in America* (Oxford, 1964)

## Cybersyn

1 Salvador Allende, 'Presentación de la Sala de Operaciones', in speech to be recorded by the President of the Republic Salvador Allende (Cybersyn's Operations Room, Santiago; possibly written by Stafford Beer, 1972 or 1973).
2 Stafford Beer, 'Proyecto SYNCO. Práctica cibernética en el Gobierno', Dirección Informática CORFO (Santiago, 1973), p. 66. Original: Stafford Beer, 'Fanfare for Effective Freedom', Richard Goodman's Third Commemorative Conference, 14 February 1973, Brighton Polytechnic, Moulsecoomb, East Sussex.
3 Eden Medina, 'The Politics of Networking a Nation', *Diseña*, 11 (2017), p. 49.

Further reading: 'Diseño de una sala de operaciones', INTEC, 4 (June 1973), pp. 19–29; Eden Medina, *Cybernetic Revolutionaries: Technology and Politics in Allende's Chile* (Cambridge, MA, 2011); Hugo Palmarola, 'Productos y Socialismo: Diseño Industrial Estatal en Chile', in *1973: La vida cotidiana de un año crucial*, ed. Claudio Rolle (Santiago, 2003), pp. 225–95

## Cyclegraph

1 Sigfried Giedion, *Mechanization Takes Command: A Contribution to Anonymous History* (New York, 1948), pp. 100–108.
2 Quoted in Suren Lalvani, 'Photography and the Industrialization of the Body', *Journal of Communication Inquiry*, XIV/2 (July 1990), p. 99.
3 'Ironing Procedures', in *America's Housekeeping Book*, ed. New York Herald Tribune Home Institute (New York, 1941), pp. 272–80.
4 'Easier Housekeeping: Scientific Analysis Simplifies a Housewife's Work', *Life*, 9 September 1946, p. 97.

Further reading: Ellen Lupton and J. Abbott Miller, *The Bathroom, the Kitchen and the Aesthetics of Waste* (New York, 1992); Allan Sekula, 'Photography Between Labour and Capital', in *Mining Photographs and Other Pictures 1948–1968: A Selection from the Negative Archives of Shedden Studio, Glace Bay, Cape Breton*, ed. Benjamin H. D. Buchloh and Robert Wilkie (Cape Breton, Nova Scotia, 1983), pp. 193–268

## *Cyclops 1*

1 'Dr' Edward Festus Mukuka Nkoloso, 'The Moon and I', *Abercornucopia*, 10 January 1964.
2 Edward Festus Mukuka Nkoloso, 'We're Going to Mars!' (1964), reproduced in Louis de Gouyon Matignon, 'Edward Makuka Nkoloso, The Afronauts and the Zambian Space Program', 4 March 2019, www.spacelegalissues.com; Dennis Lee Royle, 'Zambia Countdown – 10, 9, 3, 8, 1, 5, 0', *Miami News*, 28 October 1964.
3 Ibid.
4 Arthur Hoppe, 'Our Man Hoppe', *Journal Herald*, 3 December 1964.

5  Kabinda Lemba, 'Mukuka Nkoloso the Afronaut', https://vimeo.com, accessed 10 October 2018.

6  Namwali Serpell, 'The Zambian "Afronaut" Who Wanted to Join the Space Race', *New Yorker*, 11 March 2017, www.newyorker.com.

7  'Zambia's Forgotten Space Program', *Lusaka Times*, 28 January 2011, www.lusakatimes.com.

Further reading: Clarence Chongo, 'Decolonising Southern Africa: A History of Zambia's Role in Zimbabwe's Liberation Struggle, 1964–1979', doctoral thesis, University of Pretoria, June 2015, p. 12; Kabinda Lemba, 'Mukuka Nkoloso the Afronaut', https://vimeo.com, accessed 10 October 2018; Namwali Serpell, 'The Zambian "Afronaut" Who Wanted to Join the Space Race', *New Yorker*, 11 March 2017, www.newyorker.com

**Dougong**

1  Chen Zhanxiang, 'Recent Architecture in China', *Architectural Review*, special China issue (July 1947), p. 28.

2  Tong Jun, 'Architecture Chronicle', *T'ien Hsia*, October 1937, p. 308.

Further reading: Edward Denison, *Architecture and the Landscape of Modernity in China before 1949* (London, 2017); Liu Dunzhen, *History of Ancient Chinese Architecture* (Beijing, 1980); Liang Sicheng, *A Pictorial History of Chinese Architecture* (Cambridge, MA, 1984)

**Dymaxion House**

Further reading: Barry Bergdoll and Peter Christensen, *Home Delivery: Fabricating the Modern Dwelling* (New York, 2008); Joachim Krauss, ed., *Your Private Sky: Buckminster Fuller* (Zurich, 1999)

**Edison's Anti-gravitation Under-clothing**

1  George Du Maurier, 'Edison's Anti-gravitation Under-clothing', in *Punch's Almanack for 1879* (London, 1878), p. 2.

2  Ivy Roberts, '"Edison's Telephonoscope": The Visual Telephone and the Satire of Electric Light Mania', *Early Popular Visual Culture*, xv/1 (2017), pp. 1–25.

3  For examples of the Victorian press coverage of Edison, see 'American Affairs', *Birmingham Daily Post*, 6 November 1878, p. 7; 'Edison the Inventor', *[Dundee] Evening Telegraph*, 15 October 1878, p. 4; 'The Inventor of the Phonograph', *Herts & Cambs Reporter*, 2 August 1878, p. 6; 'Mr Edison at Home', *Freeman's Journal*, 11 October 1878, p. 2; 'An Interview with Edison', *Funny Folks*, 30 June 1888, p. 202; 'Our Extra-special and Mr Edison', *Fun*, 8 January 1879, p. 17; and 'An Evening with Edison', *Funny Folks*, 23 November 1878.

Further reading: Ivy Roberts, '"Edison's Telephonoscope": The Visual Telephone and the Satire of Electric Light Mania', *Early Popular Visual Culture*, xv/1 (2017), pp. 1–25; Randall Stross, *The Wizard of Menlo Park* (New York, 2007)

**Electrotype Pattern**

1  'Stayed nearly two hours. Went overall into Photographic & Electrotyping Rooms', *Henry Cole's Diary*, 17 May 1854, National Art Library, 55.AA.17.

2  *Memorandum: P. Flood to Charles Oman*, 23 May 1947, V&A Archive, London, Board of Survey SF710, Disposal of Electrotypes 47/1507.

3  Robert Hunt, 'On the Applications of Science to the Fine and Useful Arts: The Electrotype', *Art Journal*, X (1848), pp. 102–3.

4 Harry Howells Horton, *Birmingham: A Poem – In Two Parts with Appendix* (Birmingham, 1853), Appendix note IV, pp. 124–37. Source supplied by Alistair Grant, who also read and commented on this essay.

Further reading: Julius Bryant, ed., *Art and Design for All: The Victoria and Albert Museum*, exh. cat., Kunst- und Ausstellungshalle der Bundesrepublik Deutschland, Bonn, 18 November 2011–15 April 2012 (London, 2011); Moncure Daniel Conway, *Travels in South Kensington with Notes on Decorative Art and Architecture in England* (London, 1882); Alistair Grant and Angus Patterson, *The Museum and the Factory: The V&A, Elkington and the Electrical Revolution* (London, 2018)

### Fisher-Price Peg Figures

1 Steven G., 'Thank Goodness for Edward M. Swartz, Esq.!!!', review of Edward Swartz, *Toys That Kill*, Amazon, 21 December 2018, www.amazon.com.

### Flashcube

1 Susan Sontag, *On Photography* (New York, 1977), p. 18.
2 Mimi Sheller, *Aluminum Dreams: The Making of Light Modernity* (Cambridge, MA, 2014).
3 Sherry Turkle, ed., *Evocative Objects: Things We Think With* (Cambridge, MA, 2007), p. 4.

### Flying Boat

1 Robin Higham, *Speedbird: The Complete History of BOAC* (London, 2013), p. 77.
2 Historic England, *Nine Thousand Miles of Concrete: A Review of Second World War Temporary Airfields in England* (2016), p. 7.

Further reading: Robin Higham, *Speedbird: The Complete History of BOAC* (London, 2013);

Richard K. Smith, 'The Intercontinental Airliner and the Essence of Airplane Performance, 1929–1939', *Technology and Culture*, 24 (1983), pp. 428–49

### Glass Lantern Slide

1 Theaster Gates, *Theaster Gates*, with contributions by Lisa Lee, Carol Becker and Achim Borchardt-Hume (London and New York, 2015).

Further reading: Frederick N. Bohrer, 'Photographic Perspectives: Photography and the Institutional Formation of Art History', in *Art History and Its Institutions: Foundations of a Discipline*, ed. E. Mansfield (London, 2002), pp. 246–59; Steve Humphries, *Victorian Britain Through the Magic Lantern* (London, 1989); Howard B. Leighton, 'The Lantern Slide and Art History', *History of Photography*, VIII/2 (1984), pp. 107–18

### Globe of Mars

1 A phalanstery was a building for a utopian socialist working community, conceived by the French thinker François-Marie-Charles Fourier (1772–1837).
2 George Basalla, *Civilized Life in the Universe: Scientists on Intelligent Extraterrestrials* (Oxford, 2006), p. 82.
3 Emmy Ingeborg Brun, *Smaa Notitser til Darwinismens Religion* (Copenhagen, 1911), p. 16.

Further reading: George Basalla, *Civilized Life in the Universe: Scientists on Intelligent Extraterrestrials* (Oxford, 2006); Maria D. Lane, 'Mapping the Mars Canal Mania: Cartographic Projection and the Creation of a Popular Icon', *Imago Mundi*, LVIII/2 (2006), pp. 198–211; Robert Markley, *Dying*

*Planet: Mars in Science and the Imagination* (Durham, NC, 2005)

### High-pressure Water Mains
1 T. Smith, 'Hydraulic Power in the Port of London', *Industrial Archaeology Review*, XIV/1 (1991), pp. 64–88.
2 E. B. Ellington, 'The Distribution of Hydraulic Power in London', *Proceedings of the Institution of Civil Engineers*, XCIV (1888), pp. 1–31.

Further reading: London Metropolitan Archive, records of the General Hydraulic Power Company (of which the London Hydraulic Power Company was a wholly owned subsidiary); Ian McNeil, *Hydraulic Power* (London, 1972); Roger Morgan, 'The Watery Death of Electricity's Rival', *New Scientist*, LXXV/1062 (28 July 1977), pp. 221–3

### House Environment
1 Emilio Ambasz, *Italy: The New Domestic Landscape. Achievements and Problems of Italian Design* (New York, 1972).
2 Quoted in Ettore Sottsass, 'Could Anything Be More Ridiculous?', *Design*, 262 (1970), p. 30.
3 Milco Carboni, 'Ettore Sottsass e l'esperienza con Poltronova 1957–1972', in Ettore Sottsass, *Catalogo ragionato dell'archivio 1922–1978*, ed. Francesca Zanella (Parma, Italy, 2018), p. 111.

Further reading: Emilio Ambasz, *Italy: The New Domestic Landscape. Achievements and Problems of Italian Design* (New York, 1972); Ettore Sottsass, *Catalogo ragionato dell'archivio 1922–1978*, ed. Francesca Zanella (Parma, Italy, 2018); Deyan Sudjic, *Ettore Sottsass and the Poetry of Things* (London, 2015)

### Incandescent Light Bulb
1 Edvard Munch, sketchbook notes 1891–2. The Munch Museum, Oslo, MM T 2760.
2 Alfred Dolge, 'Testimonial', 18 February 1882, in *The Edison Light: The Edison System of Incandescent Electric Lighting as Applied in Mills, Steamships, Hotels, Theatres, Residences &c.* (New York, 1883), n.p.
3 Tom Kertscher, 'Donald Trump's Complaints about Light Bulbs, Fact-checked', *Politifact*, 16 September 2019, www.politifact.com.

Further reading: Brian Bowers, *Lengthening the Day: A History of Lighting Technology* (Oxford, 1998); Sandy Isenstadt, *Electric Light: An Architectural History* (Cambridge, MA, 2018); David Nye, *Electrifying America* (Cambridge, MA, 1990); Wolfgang Schivelbusch, *Disenchanted Night: The Industrialization of Light in the Nineteenth Century* (Berkeley, CA, 1995)

### Integrated Radio/TV Cabinet
1 Per Arnoldi and Torben Schmidt, 'The Decision-makers: An Interview with Jens Bang', *Mobilia*, 180 (1970), n.p.
2 Jens Bang, *Bang & Olufsen: From Spark to Icon* (n.p., 2005).
3 Raimonda Riccini, 'The Appliance-shaped Home', in *Italy: Contemporary Domestic Landscape 1945–2000*, ed. Giampiero Bosoni (Milan, 2001).

Further reading: Jens Bang, *Bang & Olufsen: From Spark to Icon* (n.p., 2005); Raimonda Riccini, 'The Appliance-shaped Home', in *Italy: Contemporary Domestic Landscape 1945–2000*, ed. Giampiero Bosoni (Milan, 2001)

## Invacar: The 'Invalid Carriage'

1 HC Deb, 11 June 1959, vol. DCVI, cc1333-6, see https://hansard.parliament.uk.
2 Anonymous, 'Break the Law . . . or Neglect My Children', *Daily Mail*, 3 August 1966, p. 7. Further reading: Stuart Cyphus, *An Introduction to the British Invalid Carriage: 1850–1978* (Buffalo, NY, 2012); Colin Sparrow, *Greeves: The Complete Story* (Ramsgate, 2014)

## Letraset

1 Tony Brook and Adrian Shaughnessy, *Letraset: The DIY Typography Revolution* (London, n.d.). I have mostly followed this book in the dating of developments, but there are points at which it is still unclear what happened when.

Further reading: Tony Brook and Adrian Shaughnessy, *Letraset: The DIY Typography Revolution* (London, n.d.)

## Leucotome

1 António Egas Moniz, *Tentatives Opératoires dans le Traitement de Certaines Psychoses* (Paris, 1936).
2 David Crossley, 'The Introduction of Leucotomy: A British Case History', *History of Psychiatry*, IV/16 (1 December 1993), pp. 553–64.
3 James S. McGregor and John R. Crumbie, 'Prefrontal Leucotomy', *Journal of Mental Science*, LXXXVIII/373 (October 1942), pp. 534–40; Peter Mohr and Julie Mohr, 'The Warlingham Park Hospital Leucotome', *Historical Medical Equipment Society Bulletin*, 17 (February 2007), pp. 2–4.

Further reading: Jack El-Hai, *The Lobotomist: A Maverick Medical Genius and His Tragic Quest to Rid the World of Mental Illness* (Hoboken, NJ, 2005); Jack David Pressman, *Last Resort: Psychosurgery and the Limits of Medicine* (Cambridge, 1998); Elliot S. Valenstein, *Great and Desperate Cures: The Rise and Decline of Psychosurgery and Other Radical Treatments for Mental Illness* (New York, 1986)

## Manchester Pail System: 'Dolly Vardens'

1 Quoted in Harold L. Platt, *Shock Cities: The Environmental Transformation and Reform of Manchester and Chicago* (Chicago, IL, 2005), p. 7.
2 For the use of the moniker 'Dolly Vardens', see Frank Davy, letter to the editor, *Manchester Courier and Lancashire General Advertiser*, 18 September 1879, p. 7; and William W. Hulse, letter to the editor, *Manchester Courier and Lancashire General Advertiser*, 18 November 1884, p. 3.
3 W. A. Power, 'The Pail Closet System: Progress at Manchester', *Manchester Selected Pamphlets* (1877), p. 2.
4 Platt, *Shock Cities*, pp. 70–74. See also Shena D. Simon, *A Century of City Government: Manchester, 1838–1938* (London, 1938), pp. 182–3.
5 'Gordon and the Amalgamation Scheme', *Manchester Weekly Times*, 25 January 1890, p. 2.
6 Simon, *A Century of City Government*, p. 183.

Further reading: Stephen Graham and Simon Marvin, *Splintering Urbanism: Networked Infrastructures, Technological Mobilities and the Urban Condition* (London, 2001); Stephen Halliday, *The Great Stink of London: Sir Joseph Bazalgette and the Cleansing of the Victorian Metropolis* (London, 1999); Harold L. Platt, *Shock Cities: The Environmental Transformation and Reform of Manchester and Chicago* (Chicago, IL, 2005)

## Mechanical Polygraph

1 Letter to Maria Cosway, 30 July 1788, and letter to John Adams, 11 January 1817, www.founders.archives.gov.
2 Letter to John Page, 21 February 1779, ibid.
3 Letter to Thomas Jefferson, 2 October 1803, ibid.

Further reading: Silvio A. Bedini, *Thomas Jefferson and His Copying Machines* (Charlottesville, VA, 1984); Sydney Hart, '"To Increase the Comforts of Life": Charles Willson Peale and the Mechanical Arts', *Pennsylvania Magazine of History and Biography*, CX/3 (1986), pp. 323–57; David Philip Miller, '"Men of Letters" and "Men of Press Copies": The Cultures of James Watt's Copying Machine', *Archimedes*, 52 (2017), pp. 65–80

## Medical Wax Model

Further reading: Robin A. Cooke, 'A Moulage Museum Is Not Just a Museum: Wax Models as Teaching Instruments', *Virchows Archiv*, 457 (2010), pp. 513–20; Elena Giulia Milano et al., 'Current and Future Applications of 3D Printing in Congenital Cardiology and Cardiac Surgery', *British Journal of Radiology*, 92 (2019), p. 1094; World Health Organization, 'Smallpox', www.who.int, accessed 22 February 2021; Fabio Zampieri, Alberto Zanatta and Maurizio Rippa Bonati, 'Iconography and Wax Models in Italian Early Smallpox Vaccination', *Medicine Studies*, 2 (2011), pp. 213–27

## Memo

1 Du Pont Company, High Explosive Operating Division, Efficiency Division, report on letter-writing, 1913, quoted in JoAnne Yates, *Control Through Communication: The Rise of System in American Management* (Baltimore, MD, and London, 1989), pp. 252–3.

Further reading: John Guillory, 'The Memo and Modernity', *Critical Inquiry*, XXXI/1 (2004), pp. 108–32; JoAnne Yates, *Control Through Communication: The Rise of System in American Management* (Baltimore, MD, and London, 1989), and —, 'The Emergence of the Memo as a Managerial Genre', *Management Communication Quarterly*, 11/4 (May 1989), pp. 485–510

## Milk Spoon

1 Jorge Rojas, *Historia de la infancia en el Chile republicano, 1810–2010* (Santiago, 2010), p. 636.
2 *Programa nacional de leche. Instructivo para personas que participan en labores educativas* (Santiago, 1972), pp. 11–13.
3 'Diseño de envases', INTEC, 4 (June 1973), pp. 41, 42.

Further reading: 'Diseño de envases', INTEC, 4 (June 1973), pp. 41–7; Hugo Palmarola, 'Productos y Socialismo: Diseño Industrial Estatal en Chile', in *1973: La vida cotidiana de un año crucial*, ed. Claudio Rolle (Santiago, 2003), pp. 225–95

## MiniDisc

Further reading: Om Malik, 'Back to the Future: The MiniDisc', www.forbes.com, 18 April 1998; Brian Mooar, 'MiniDisc's Second Coming?', *Washington Post*, 26 September 1997, www.washingtonpost.com; Sony Archive, www.sony.net, accessed 30 April 2019

## Minitel

1 Julien Mailland and Kevin Driscoll, *Minitel: Welcome to the Internet* (Cambridge, MA, 2017), pp. 9, 28.

2 Angelique Chrisafis, 'France Says Farewell to the Minitel, the Little Box that Connected a Country', www.theguardian.com, 28 June 2012.

Further reading: Peter Large, 'Computers in Society – Plus ça change', www.theguardian.com, 12 June 1980; Julien Mailland and Kevin Driscoll, *Minitel: Welcome to the Internet* (Cambridge, MA, 2017); Simon Nora and Alain Minc, *The Computerisation of Society: A Report to the President of France* (Cambridge, MA, 1980), originally published as *L'information de la societe* (Paris, 1978)

## Moon Towers

Further reading: Ernest Freeberg, *The Age of Edison: Electric Light and the Invention of Modern America* (London, 2014); John A. Jakle, *City Lights: Illuminating the American Night* (Baltimore, MD, 2001)

## Nikini

1 *New Tampax Compak*, UK television advertising campaign, 1994, www.youtube.com, accessed September 2020.
2 Robinson & Sons, *Achievement . . . The Story of Robinsons of Chesterfield* (Chesterfield, 1963), published on the occasion of the visit of HRH Princess Margaret and the Earl of Snowdon.
3 Glenda Lewin Hufnagel, *A History of Women's Menstruation from Ancient Greece to the Twenty-first Century: Psychological, Social, Medical, Religious, and Educational Issues* (New York, 2012), pp. 67–8.

Further reading: Lara Freidenfelds, *The Modern Period: Menstruation in Twentieth-century America* (Baltimore, MD, 2009); Chris Bobel et al., eds, *The Palgrave Handbook of Critical Menstruation Studies* (Springer Nature, 2020); Sara Read, *Menstruation and the Female Body in Early-Modern England* (London, 2013)

## 'No Nonsense' Fountain Pen

1 See Francesco Moschini, ed., *Segno, disegno e progetto nell'architettura italiana del dopoguerra*, exh. cat., Hangang Gallery, Seoul (Seoul, 2002).
2 Glenn Adamson and Jane Pavitt, eds, *Postmodernism: Style and Subversion, 1970–1990*, exh. cat., Victoria and Albert Museum, London (London, 2011), p. 23.
3 See Jean-Louis Cohen, *La Coupure entre architectes et intellectuels: ou les enseignements de l'italophilie* (Brussels, 2015).
4 Francesco Moschini, 'Architettura Disegnata', in *Annisettanta: Il decennio lungo del secolo breve*, ed. M. Belpoliti, S. Chiodi and G. Canova, exh. cat., Milan Triennale (Milan, 2007).

Further reading: M. Belpoliti, S. Chiodi and G. Canova, eds, *Annisettanta: Il decennio lungo del secolo breve*, exh. cat., Milan Triennale (Milan, 2007); Pippo Ciorra, 'A Territory Without Conflicts', in *Looking at European Architecture: A Critical View*, ed. Sylvie Lemaire et al. (Brussels, 2008), pp. 269–82; Paolo Portoghesi, *I nuovi architetti Italiani* (Rome and Bari, Italy, 1985); Franco Purini, *Una lezione sul disegno* (Rome, 2008)

## North Bucks Monorail City

1 Ian Nairn, 'The Best in Britain', *The Observer*, 22 November 1964.
2 Jane Jacobs, *The Death and Life of Great American Cities: The Failure of Town Planning* (Harmondsworth, 1965), pp. 16–17.
3 See Fred Pooley, *North Bucks New City* (Aylesbury, 1966).

4 Guy Ortolano, *Thatcher's Progress: From Social Democracy to Market Liberalism Through an English New Town* (Cambridge, 2019), p. 64.

Further reading: www.billberrett.info (Berrett died in 2014, but this website includes interesting material on the monorail city); Guy Ortolano, *Thatcher's Progress: From Social Democracy to Market Liberalism Through an English New Town* (Cambridge, 2019); Fred Pooley, papers at the Centre for Buckinghamshire Studies, Aylesbury

### *Notgeld*

1 Eric Rowley, *Hyperinflation in Germany: Perceptions of a Process* (Aldershot, 1994), p. 117.
2 Kurt Biging, *Geldscheine in Halle an der Saale 1916–1922* (Halle, Germany, 2003), p. 11.
3 Claire Zimmerman, 'Promissory Notes of Architecture', talk given at the Warburg Institute, London, on 13 March 2019.

### Oil from Coal

Further reading: David Edgerton, *Britain's War Machine: Weapons, Experts and Resources in the Second World War* (London, 2011); Anthony Stranges, 'Germany's Synthetic Fuel Industry, 1927–1945', in *The German Chemical Industry in the Twentieth Century*, ed. John Lesch (Dordrecht, the Netherlands, 2000), pp. 147–216

### Optical Telegraph

1 Thomas Carlyle, *The French Revolution: A History*, ed. K. J. Fielding and David Sorensen (Oxford, 1989), vol. II, pp. 372–3.
2 Anon., 'Theatres', *The Satirist, or the Censor of the Times*, 752 (13 September 1846), p. 294.
3 Charles Dickens, *The Pickwick Papers*, ed. Robert L. Patten (Harmondsworth, 1972), p. 700.

Further reading: Anon., 'Tales of the Telegraph', *Punch*, 29 May 1847, p. 220; Gerard J. Holzmann and Björn Pehrson, *The Early History of Data Networks* (Hoboken, NJ, 2003), pp. 47–96

### Paper Aeroplane Ticket

1 Quoted in IATA, 'Industry Bids Farewell to Paper Ticket', press release, 31 May 2008, www.iata.org.

Further reading: Chandra D. Bhimull, *Empire in the Air: Airline Travel and the African Diaspora* (New York, 2017); Lisa Gitelman, *Paper Knowledge: Toward a Media History of Documents* (Durham, NC, 2014)

### Paper Dress

1 Alexandra Palmer, 'Paper Clothes: Not Just a Fad', in *Dress and Popular Culture*, ed. Patricia Anne Cunningham and Susan Voso Lab (Bowling Green, OH, 1991), p. 90.
2 Marylin Bender, *The Beautiful People* (New York, 1967), p. 16.
3 Quoted in Helen Carlton, 'Answer to Laundry in Space', *Life*, 25 November 1966, p. 136.

Further reading: Beate Schmuck and Stadtmuseum Nordhorn, *Oneway Runway: Paper Dresses zwischen Marketing und Mode* (Münster, Germany, 2018); Vasilēs Zēdianakēs, Mouseio Benakē and Mouseion Benakē, eds, *Rrripp!! Paper Fashion* (Athens, 2007)

### Pasilalinic-sympathetic Compass

1 H. D. Justesse, *Histoire de la Commune de Paris* (Zurich, 1879), p. 242; 'Jules Allix', *La Petite Presse*, 21 September 1871.
2 M. Clairville, *Les Escargots sympathiques; à-propos, mélé de couplets* (Paris, 1850), pp. 4–5.

Further reading: Jules Allix, 'Communication universelle de la pensée, à quelque distance que ce soit, à l'aide d'un appareil portatif appelé Boussole pasilalinique sympathique', *La Presse*, 25 and 26 October 1850; Yves Couturier, ed., *Le Patrimoine des télécommunications françaises* (Paris, 2002); Justin E. H. Smith, 'The Internet of Snails', *Cabinet*, 58 (Summer 2015), www.cabinetmagazine.org

### Phase-change Chemical Heat-storage Barrel

1 Mária Telkes, 'Solar House Heating: A Problem of Heat Storage', *Heating and Ventilating*, XLIV/5 (1947), pp. 12–17.

Further reading: Daniel A. Barber, *A House in the Sun: Modern Architecture and Solar Energy in the Cold War* (New York, 2016); Mária Telkes and Eleanor Raymond, 'Storing Solar Heat in Chemicals: A Report on the Dover House', *Heating and Ventilating News*, November 1949, pp. 79–86; United Nations Department of Economic and Social Affairs, *New Sources of Energy and Economic Development: Solar Energy, Wind Energy, Tidal Energy, Geothermic Energy, and Thermal Energy of the Seas* (New York, 1957)

### Player Piano

1 William Gaddis, 'Agapē Agape: The Secret History of the Player Piano', in *The Rush for Second Place*, ed. Joseph Tabbi (New York, 2002), p. 142.
2 William Gaddis, *Agapē Agape* (New York, 2002), pp. 1–2; unless otherwise specified, all other quotations are from this work, or from the earlier proposals, in *The Rush for Second Place*, pp. 7–13, 142–4.
3 Quoted in *The Rush for Second Place*, p. 147.

### Pneumatic Postal System

Further reading: Anne-Laure Cermak and Elisa Le Briand, *Le Réseau avant l'heure: la poste pneumatique à Paris* (Paris, 2006); Thierry Poujol, 'Des Égouts au musée, splendeur et déclin de la poste atmosphérique', *Culture & Technique*, 19 (1989), pp. 143–9; Molly W. Steenson, 'Interfacing with the Subterranean', *Cabinet: A Quarterly of Art and Culture*, 41 (2011), pp. 82–6

### Polaroid SX-70

1 Phil Patton, 'The Polaroid SX-70', *American Heritage*, XLIII/7 (November 1992).

Further reading: Victor K. McElheny, *Insisting on the Impossible: The Life of Edwin Land. An Inventor Is Born* (New York, 1999)

### Public Standards of Length

1 *Report of the Commissioners Appointed to Consider the Steps to Be Taken for Restoration of the Standards of Weight & Measure* (London, 1841), pp. 16–17.
2 Simon Schaffer, 'Accurate Measurement Is an English Science', in *The Values of Precision*, ed. M. Norton Wise (Princeton, NJ, 1995), p. 135.
3 Memorandum, 'Results of Re-verification of Public and Mural Standards of Length at Trafalgar Square', 14 June 1887, National Archives BT 101/165; Memorandum, 'Report on Visit to Edinburgh and Inspection of Public Standards of Length', 10 October 1916, National Archives BT 101/818.
4 Kenneth Hume, *Engineering Metrology*, 2nd edn (London, 1963), p. 56.

Further reading: M. Norton Wise, ed., *The Values of Precision* (Princeton, NJ, 1995); Michael Wright, 'Length, Measurement Of', in *Instruments of Science: An Historical*

*Encyclopedia*, ed. Robert Bud and Deborah Jean Warner (New York and London, 1998), pp. 97–8

**Pyrophone**

1 Henri Dunant, 'Description of M. Kastner's New Musical Instrument, the Pyrophone', *Journal of the Society of Arts*, XXIII/1161 (1875), pp. 293–7.
2 Anon., 'The Pyrophone: A Scientific and Musical Curiosity', undated but likely 1953, typescript article in Science Museum, London, file T/1876-590.

Further reading: Robert Bud, 'Responding to Stories: The 1876 Loan Collection of Scientific Apparatus and the Science Museum', *Science Museum Group Journal*, 1/1 (2016), doi: 10.15180/140104; Colin Harding, 'Singing Flames: The Story of the Pyrophone', *Theatrephile Magazine*, 11/5 (1984), pp. 51–3

**Realistic Wax Mannequin**

Further reading: Maude Bass-Krueger, 'Fashion Collections, Collectors, and Exhibitions in France, 1874–1900: Historical Imagination, the Spectacular Past, and the Practice of Restoration', *Fashion Theory*, XXII/4–5 (2018), pp. 405–33; Caroline Evans, *The Mechanical Smile: Modernism and the First Fashion Shows in France* (New Haven, CT, 2013); Jane Munro, *Silent Partners, Artist and Mannequin from Function to Fetish* (New Haven, CT, 2014)

**Rotring, Letratone, MiniCAD**

I would like to thank my friend Paul Notley for his reminiscence of Arup Associates, where we worked and drew together, and Dr Inge Andritz of TU Wien for her astute insights.

Further reading: Tony Fretton, *AEIOU: Articles Essays Interviews and Out-takes* (Prinsenbeek, the Netherlands, 2017)

**Scaphander: 'Man-boat'**

1 Jean-Baptiste de La Chapelle, *Traité de la construction théorique et pratique du scaphandre, ou du bateau de l'homme* (Paris, 1775), p. x (my translation).
2 Ibid., p. xxxiv.
3 Pierre Denys de Montfort, *Conchyliologie systématique et classification méthodique des coquilles* (Paris, 1808–10), vol. II, p. 335.

Further reading: Paolo Bertucci, *Artisanal Enlightenment: Science and the Mechanical Arts in Old Regime France* (New Haven, CT, 2018); Jean-Baptiste de La Chapelle, *Le Ventriloque, ou l'engastrimythe* (Paris, 1772), and *Traité de la construction théorique et pratique du scaphandre, ou du bateau de l'homme* (Paris, 1775); Lloyd Mallan, *Suiting Up for Space: The Evolution of the Space Suit* (New York, 1971); Philippe Poulet, *L'odyssée des scaphandres* (Paris, 2010)

**Scarificator**

Further reading: Gerry Greenstone, 'The History of Bloodletting', *British Columbia Medical Journal*, LII/1 (2010), pp. 12–14; Lawrence Hill, *Blood: A Biography of the Stuff of Life* (London, 2014); Liakat Ali Parapia, 'History of Bloodletting by Phlebotomy', *British Journal of Haematology*, 143 (2008), pp. 490–95; Kuriyama Shigehisa, 'Interpreting the History of Bloodletting', *Journal of the History of Medicine and Allied Sciences*, 50 (1995), pp. 11–46

**Serving Hatch**

1 Randal Phillips, *The Book of Bungalows* (London, 1920), pp. 54–5, reviewed and

advertised in *Country Life*, XLIX/1252 (1 January 1921), p. 28. Pont (Graham Laidler), 'At Home: The Epicure', in *The British at Home* (London, 1939), p. 17.

2 On the experimentation behind, and ubiquitous nature of, the eye-level grill by 1965, see 'Survey: Fifteen Years of Cooker Design', *Design*, CIC (July 1965), pp. 52–9. Available from the Online Resource for the Visual Arts, www.vads.ac.uk

3 For examples of the change to the recirculating cooker hood, see J. A. Saltmarsh, 'Design Management: Investigating the Consumer', *Design*, CCIII (November 1965), pp. 52–7, and Joan S. Ward, 'Ergonomics in the Kitchen', *Design*, CCLIIIVI (October 1972), pp. 58–9. Both available from the Online Resource for the Visual Arts, www.vads.ac.uk.

Further reading: Michelle Corodi, 'On the Kitchen and Vulgar Odors', in *The Kitchen: Lifeworld, Usage, Perspectives*, ed. Klaus Spechtenhauser (Berlin, 2006), pp. 30–33; Martina Heßler, 'The Frankfurt Kitchen: The Model of Modernity and the "Madness" of Traditional Users, 1926 to 1933', in *Cold War Kitchen: Americanization, Technology, and European Users*, ed. Ruth Oldenziel and Karin Zachmann (Cambridge, MA, 2009), pp. 163–84; Andrew Higgott, *Mediating Modernism: Architectural Cultures in Britain* (London, 2006)

### Sinclair C5

1 Ian Adamson and Richard Kennedy, *Sinclair and the 'Sunrise' Technology: The Deconstruction of a Myth* (Harmondsworth, 1986), p. 330.

2 'Sinclair C5 – A New Power in Personal Transport', quoted ibid., p. 332.

3 Quoted in Rob Gray, *Great Brand Blunders* (Bath, 2014), p. 93.

4 A. Klarenberg in *Marketing Magazine*, 31 January 1985, p. 8, quoted in Andrew P. Marks, 'The Sinclair L5: An Investigation into Its Development, Launch, and Subsequent Failure', *European Journal of Marketing*, 1 (1989), p. 62.

5 Stephen Pile, 'It's Not Kid's Stuff, This Formula One Bath-chair', *Sunday Times*, 13 January 1985, p. 5.

6 Quoted in Anon., 'Miles Ahead of Its Time', *The Scotsman*, 10 January 2005.

Further reading: Ian Adamson and Richard Kennedy, *Sinclair and the 'Sunrise' Technology: The Deconstruction of a Myth* (Harmondsworth, 1986); Rodney Dale, *The Sinclair Story* (London, 1985); Andrew P. Marks, 'The Sinclair C5: An Investigation into Its Development, Launch, and Subsequent Failure', *European Journal of Marketing*, 1 (1989), pp. 61–71

### Skirt Grip

1 Quoted in Mary Sawdon, *A History of Victorian Skirt Grips* (n.p., 1995), p. 77.

Further reading: Anon. [Eliza Cheadle], *Manners of Modern Society: Being a Book of Etiquette* (London, Paris and New York, 1877); Barbara Kotzin, *The Art of the Skirt Lifter: A Practical and Passionate Guide* (self-published, 2015); Mary Sawdon, *A History of Victorian Skirt Grips* (n.p., 1995)

### Slide Rule

1 William Oughtred, *To the English Gentrie, and all Other Studious of the Mathematicks which Shall Bee Readers Hereof. The Just Apologie of Wil: Oughtred, against the Slaunderous Insimulations of Richard Delamain, in a Pamphlet Called Grammelogia, or The mathematicall Ring,*

or *Mirifica logarithmorum projectio circularis* (n.p., 1633).

Further reading: Florian Cajori, *A History of the Logarithmic Slide Rule and Allied Instruments* (London, 1909); Cliff Stoll, 'When Slide Rules Ruled', *Scientific American*, CCXCIV/5 (May 2006), pp. 80–87; Frances Willmoth, 'William Oughtred', *Oxford Dictionary of National Biography*, published online 23 September 2004, revised 3 January 2008

### Slotted Screwdriver
Further reading: Witold Rybczynski, *One Good Turn: A Natural History of the Screwdriver and the Screw* (London, 2000)

### Space Frame
1 Konrad Wachsmann, 'Vom bauen in unserer Zeit', *Baukunst und Werkform*, January 1957, pp. 26–31. Translated as 'On Building in Our Time', now in Joan Ockman, ed., *Architecture Culture 1943–1968: A Documentary Anthology* (New York, 1993), p. 267.

Further reading: Yona Friedman and Manuel Orazi, *The Dilution of Architecture*, ed. Nader Seraj (Chicago, IL, 2015); Joan Ockman, ed., *Architecture Culture 1943–1968: A Documentary Anthology* (New York, 1993); Wendel R. Wendel, 'The Geometry of Space', *World Architecture*, I/4 (1989), pp. 76–83

### Stanley 55 Combination Plane
1 *Stanley Tools Catalogue no. 34* (Harrisburg, PA, 1926), pp. 99–100.
2 Frank Lloyd Wright, *Modern Architecture, Being the Kahn Lectures for 1930* (Princeton, NJ, 1987), pp. 50–59.

Further reading: Josef M. Greber, *Die Geschichte des Hobels – Von der Steinzeit bis zum Entstehen der Holzwerkzeugfabriken im frühen 19. Jahrhundert* (Zurich, 1956); David R. Russel, *Antique Woodworking Tools: Their Craftsmanship from the Earliest Times to the Twentieth Century* (Cambridge, 2010); Raphael A. Salaman, *Dictionary of Tools Used in the Working and Allied Trades, c. 1700–1970* (London, 1975)

### Telephone Table
1 Robin Evans, 'Figures, Doors and Passages', in *Translations from Drawing to Building and Other Essays* (London, 1997), pp. 55–92.

### *Théâtrophone*
1 Brian Kane, *Sound Unseen: Acousmatic Sound in Theory and Practice* (New York, 2014).
2 See Jonathan Crary, *Techniques of the Observer: On Vision and Modernity in the Nineteenth Century* (Cambridge, MA, 1990).
3 Marcel Proust, *À un ami, correspondance de Marcel Proust avec Georges de Lauris* (Paris, 1948), p. 243.

Further reading: Julien Lefèvre, *L'électricité au théâtre* (Paris, 1894), pp. 322–37; Eugène Testavin, 'L'organisation actuelle du théâtrophone', *Annales des Postes, Télégraphes et Téléphones*, XIX/1 (1930), pp. 1–24; Melissa van Drie, 'Hearing Through the Théâtrophone: Sonically Constructed Spaces and Embodied Listening in Late Nineteenth-century French Theatre', *Sound Effects: An Interdisciplinary Journal of Sound and Sound Experience*, V/1 (2015), pp. 73–90

### Think City Electric Vehicle
1 Jim Motavalli, 'Inside the Think City Electric Car', www.forbes.com, 30 March 2010.
2 Arne Asphjell, Øystein Asphjell and Hans Håvard Kvisle, *Elbil på norsk* (Oslo, 2013), pp. 119–31.

3 Katinka von der Lippe, interviewed in Tony Lewin and Ryan Borroff, *How to Design Cars Like a Pro* (Minneapolis, MN, 2010), p. 92.

Further reading: Arne Asphjell, Øystein Asphjell and Hans Håvard Kvisle, *Elbil på norsk* (Oslo, 2013), pp. 119–31; D. Sperling, 'Electric Vehicles: Approaching the Tipping Point', in *Three Revolutions: Steering Automated, Shared, and Electric Vehicles to a Better Future*, ed. D. Sperling (Washington, DC, 2018), pp. 21–54; L. Tillemann, *The Great Race: The Global Quest for the Car of the Future* (New York, 2015)

## Trombe Wall

1 Douglas Kelbaugh, 'Solar Home in New Jersey', *Architectural Design*, 11 (1976), pp. 653–6.

Further reading: Jacques Michel, 'Utilisation de l'énergie solaire', *L'Architecture d'Aujourd'hui*, 167 (May–June 1973), pp. 88–96; Félix Trombe, 'L'utilisation de l'énergie solaire: État actuel et perspectives d'avenir', *Journal des Recherches du CNRS*, 25 (1953), pp. 193–215; Mirko Zardini and Giovanna Borasi, eds, *Sorry, Out of Gas: Architecture's Response to the 1973 Oil Crisis* (Montreal, Québec, and Mantua, Italy, 2007)

## T-shirt Plastic Bag

1 Börje Gavér, '50 år sedan plastpåsen uppfanns', www.nt.se, 22 July 2015.
2 Mateo Kries, ed., *Atlas of Furniture Design* (Weil am Rhein, Germany, 2019), p. 970.
3 Susan Freinkel, *Plastic: A Toxic Love Story* (Boston, MA, and New York, 2011), p. 144.
4 United Nations Environment Programme, 'Single-use Plastics: A Roadmap for Sustainability,' 5 June 2018, p. viii, www.unep.org.
5 United Kingdom Department for Environment, Food and Rural Affairs, 'Single-use Plastic Carrier Bags Charge: Data in England for 2019 to 2020', 30 July 2020, www.gov.uk.

## Ultratemp® Roasting Rack

1 The patent application was filed on 13 March 1990, being approved on 26 February 1991 with the number 4,996,404. See http://patents.google.com and www.uspto.gov.
2 Andrew F. Smith, *Eating History: 30 Turning Points in the Making of American Cuisine* (New York, 2011), p. 207; Anahad O'Connor, 'The Claim: Microwave Ovens Kill Nutrients in Food', *New York Times*, 17 October 2006, www.nytimes.com; Bob Barnett, 'Does Microwaving Food Remove Its Nutritional Value?', https://edition.cnn.com, 7 February 2014.
3 'Campbell Microwave Institute Studies Reveal Microwave Usage Trend', *Pittsburgh Post-Gazette*, 7 March 1990, p. 122.

Further reading: Andrew F. Smith, *Eating History: 30 Turning Points in the Making of American Cuisine* (New York, 2011)

## UV-radiated Artificial Beach

1 Martine Vermandere, *We zijn goed aangekomen! Vakantiekolonies aan de Belgische kust [1887–1980]* (Brussels, 2010).
2 Beatriz Colomina, *X-Ray Architecture* (Zurich, 2019), pp. 74–8.
3 J. Goldmerstein and Karl Stodieck, ed., *Thermenpalast: Kur-, Erholungs-, Sport-, Schwimm- und Badeanlage* (Berlin, 1928).

Further reading: Valter Balducci and Smaranda Bica, eds, *Architecture and Society of the Holiday Camps: History and Perspectives* (Timișoara, Romania, 2007);

Niklaus Ingold, *Lichtduschen: Geschichte einer Gesundheitstechnik, 1890–1975* (Zurich, 2015); Tania Anne Woloshyn, *Soaking Up the Rays: Light Therapy and Visual Culture in Britain, c. 1890–1940* (Manchester, 2017)

### Vertical Filing Cabinet
I thank Peter Sealy and Christy Anderson for bringing this filing system to my attention.
1  Franz Kafka, *The Metamorphosis* [1915], trans Susan Bernofsky (New York, 2014), p. 25.
2  See Franz Kafka, *The Office Writings*, ed. Stanley Corngold, Jack Greenberg and Benno Wagner; trans. Eric Patton with Ruth Hein (Princeton, NJ, 2009).
3  Evžen Erban, *Czechoslovak National Insurance: A Contribution to the Pattern of Social Security* (Prague, 1948), and Jan Gallas and Václav Heral, *Social Security in Czechoslovakia* (Prague, 1952), as well as the company periodical *Věstník Ústřední sociální pojišťovny*.
4  *Ústřední sociální pojišťovna 1926–1936* (Prague, 1936).

Further reading: Markus Krajewski, *Paper Machines: About Cards and Catalogs, 1548–1929* [2002] (Cambridge, MA, 2011); Cornelia Vismann, *Files: Law and Media Technology*, trans. Geoffrey Winthrop-Young (Stanford, CA, 2008); JoAnne Yates, *Control Through Communication: The Rise of System in American Management* (Baltimore, MD, and London, 1989)

### Water Bag
1  Matthew Tonts, Neil Argent and Paul Plummer, 'Evolutionary Perspectives on Rural Australia', *Geographical Research*, L/3 (2012), pp. 291–303.

Further reading: Lenore Layman and Criena Fitzgerald, *110 Degrees in the Waterbag* (Perth, Australia, 2011)

### Writing Case
Further reading: Donald Jackson, *The Story of Writing* (London, 1981)

### Zeppelin
Further reading: Dan Grossman, *Zeppelin Hindenburg: An Illustrated History of LZ-129* (Cheltenham, 2017); Anna von der Goltz, *Hindenburg: Power, Myth, and the Rise of the Nazis* (Oxford, 2009)

# Acknowledgements

This book is the product of a Humanities in the European Research Area (HERA) project, *Printing the Past: Architecture, Print Culture, and Uses of the Past in Modern Europe* (PriArc) led by Professor Mari Hvattum at the Oslo School of Architecture and Design. It additionally involved researchers from Leiden University, Ghent University and UCL, curators from the Victoria and Albert Museum, London, the Musée d'Orsay, Paris, and the digital media lab Factum Arte. PriArc examined the relationship between architecture, print culture and uses of the past in modern Europe. As part of this project, the UCL team, consisting of the four editors of *Extinct*, explored the theme of 'Projecting the Past', which specifically sought to understand and challenge the operation of a smooth narrative of progress through evolutionary design histories and institutional collecting strategies. It was from these discussions that the idea for *Extinct* emerged.

We are most grateful to HERA and to the Arts and Humanities Research Council for providing the framework within which this book could develop, and for funding it so generously. We are extremely thankful to our colleagues on the PriArc project for the stimulating and always congenial intellectual exchanges that occurred over a three-year period between 2016 and 2019. It was a privilege. Many PriArc colleagues who participated in our discussions and events are represented through essays in this volume. We sincerely thank them all: Tim Ainsworth Anstey, Mari Hvattum, Mari Lending, Eirik Arff Gulseth Bøhn, Caroline van Eck, Maude Bass-Krueger, Maarten Delbeke, Maarten Liefooghe, Nikos Magouliotis, Ben Vandenput and Alice Thomine. We also thank the 'honorary' PriArc members who were regular contributors to events: Anne Hultzsch, Alina Payne, Victor Plahte Tschudi and Richard Wittman. Lastly, we are grateful to our research assistant, Rachel Tyler, for so capably shepherding this complex project through to its submission, and to Vivian Constantinopoulos and Alex Ciobanu at Reaktion Books for their constant guidance and support.

# Photo Acknowledgements

The editors and publishers wish to express their thanks to the below sources of illustrative material and/or permission to reproduce it. Every effort has been made to contact copyright holders; should there be any we have been unable to reach or to whom inaccurate acknowledgements have been made please contact the publishers, and full adjustments will be made to any subsequent printings. Some locations of artworks are also given below, in the interest of brevity:

Collection of Daniel M. Abramson (photo Daniel M. Abramson): p. 132; © ADAGP, Paris and DACS, London 2021: pp. 144 (photo The Museum of Modern Art, New York), 146 (photo © The Museum of Modern Art, New York/Scala, Florence); akg-images: p. 224; Albany Institute of History & Art, NY: p. 71; photo Max Anstey Hayes: p. 284; courtesy Architekturmuseum der Technischen Universität Berlin: p. 343; courtesy Austin History Center, Austin Public Library, TX: p. 204; courtesy Bang & Olufsen, Struer: p. 156; photo Bettmann/Getty Images: p. 304; Bibliothèque nationale de France, Paris: p. 186; courtesy Kara K. Bigda: p. 123; © Gui Bonsiepe, 1973: p. 92; photo Brown Eyed Rose: p. 120; courtesy Buckinghamshire Archives, Aylesbury: p. 216; Michael Burrell/Alamy Stock Photo: p. 232; photo Bernard Cahier/Hulton Archive via Getty Images: p. 64; from Chamberlin, Powell and Bon, *Report to the Court of Common Council of the Corporation of the City of London on Residential Development within the Barbican Area Prepared on the Instructions of the Barbican Committee* (London, 1959): p. 60; collection of Pippo Ciorra (photo Stefano Gobbi): p. 212; collection of Marc Constandt, Middelkerke: p. 340; Cooper Hewitt, Smithsonian Design Museum, New York (photos Matt Flynn © Smithsonian Institution): pp. 63 (gift of Honeywell Inc., 1994-37-1), 84, 256 (Art Resource, NY/Scala, Florence); photo Ralph Crane/The LIFE Picture Collection via Getty Images: p. 32; courtesy Culture NL, Coatbridge: pp. 208, 210; collection of Barry Curtis (photo Barry Curtis): p. 352; © Salvador Dalí, Fundació Gala-Salvador Dalí, DACS 2021/photo National Gallery of Victoria, Melbourne (gift of Lady Potter, 2009.170): p. 48; courtesy Daniel Crouch Rare Books, London: p. 136; Open Access Image from the Davison Art Center, Wesleyan University, Middletown, CT (photo M. Johnston): p. 240; collection of Tacita Dean (photo Tacita Dean): p. 164; DeGolyer Library, Southern Methodist University, Dallas, TX: p. 298; photos Edward Denison: pp. 104, 107; courtesy Drawing Matter and Tony Fretton Architects: p. 272; photo Ewald Ehtreiber (CC BY-SA 4.0): p. 126; © Estate of Josef Ehm/photo Canadian Centre for Architecture, Montreal: p. 344; Martyn Evans/Alamy Stock Photo: p. 163; courtesy Everite Building Products Collection: pp. 44, 47; © Fashion Museum Bath/Bridgeman Images: p. 268; collection of Adrian Forty (photos Adrian Forty): pp. 140, 188, 296; from Peter Goering, 'Energy Structures', in *Canadian Architect*, 16/11 (November 1971): p. 28; from Garrett Hack, *The Handplane Book* (Newtown, CT, 1999), reproduced with permission of the Taunton Press (photo John S. Sheldon): p. 308; collection of Harriet Harriss (photo Harriet Harriss): p. 124; collection of Amy de La Haye (photo Amy de La Haye): p. 292; from the Collections of The Henry

Ford (2010.83.834/THF140432): p. 24; from the *Illustrated Exhibitor and Magazine of Art*, vol. I (London, 1852), photo Getty Research Institute, Los Angeles, CA: p. 119; from the *Illustrated London News*, vol. IV/no. 88 (6 January 1844): p. 52; INTERFOTO/Alamy Stock Photo: p. 159; courtesy International Air Transport Association (IATA): p. 234; © IWM (Q 69051): p. 20; courtesy Joseph Enterprises, Inc.: p. 74; © Douglas Kelbaugh, reproduced with permission/photo Douglas Kelbaugh fonds, Canadian Centre for Architecture, Montreal (gift of Douglas Kelbaugh): p. 328; collection of Robin Kinross (photo Robin Kinross): p. 168; © Kodak Collection/National Science & Media Museum/Science & Society Picture Library: p. 248; from Jean-Baptiste de La Chapelle, *Traité de la construction théorique et pratique du scaphandre, ou du bateau de l'homme* (Paris, 1775), photos courtesy ETH-Bibliothek Zürich: pp. 276, 278; Library of Congress, Prints and Photographs Division, Washington, DC: pp. 128 (Harris & Ewing Collection), 307 (Gilbert H. Grosvenor Collection of Photographs of the Alexander Graham Bell Family); London Metropolitan Archives: pp. 56 (photo Paul Dobraszczyk), 143 (B/GH/LH/08/001); © Manchester Daily Express/Science & Society Picture Library: p. 80; MARKA/Alamy Stock Photo: p. 252; The Metropolitan Museum of Art, New York: p. 68; photo Gjon Mili/The LIFE Picture Collection via Getty Images: p. 96; from *Les Modes*, no. 128 (August 1911), photo Bibliothèque nationale de France, Paris: p. 271; courtesy Monticello and the Thomas Jefferson Foundation, Charlottesville, VA: p. 180; Motoring Picture Library/Alamy Stock Photo: p. 160; courtesy Museo Morgagni di Anatomia Patologica, Padua (photo © Giovanni Magno): p. 184; collection of Museum of Applied Arts and Sciences, Ultimo, Sydney (gift of Australian Consolidated Press under the Taxation Incentives for the Arts Scheme, 1985): p. 350; courtesy the archives of the Museum of Innovation and Science (MISCI), Schenectady, NY: p. 88; courtesy Museum of Medicine and Health, The University of Manchester: p. 172; © Museum of Science & Industry/Science & Society Picture Library: p. 176; Patrick Nairne/Alamy Stock Photo: p. 55; National Museum of American History, Division of Work and Industry, Smithsonian Institution, Washington, DC: p. 152; NY Daily News Archive/Getty Images: p. 356; courtesy Orange/DGCI: p. 320; collection of Hugo Palmarola, photo © Hugo Palmarola 2020 (based on a photograph by Gui Bonsiepe, 1973): p. 192; from *Le Petit Journal. Supplément illustré*, XII/576 (1 December 1901), photo Bibliothèque nationale de France, Paris: p. 228; photo Popular Science via Getty Images: p. 244; private collection: p. 179; Punch Cartoon Library/TopFoto: p. 287; from *Punch, or The London Charivari*, vol. LXVIII (January – June 1875), photo Robarts Library, University of Toronto: p. 58; from *Punch's Almanack for 1879* (9 December 1878), photo Universitätsbibliothek Heidelberg (CC BY-SA 4.0): p. 112; photo Wendy Ribadeneira: p. 336; collection of Charles Rice (photo Charles Rice): p. 72; collection of David Rooney (photos David Rooney): pp. 260, 263; © Science Museum/Science & Society Picture Library: pp. 148, 196, 200, 264, 316; courtesy Systembolaget Archive at Centrum för Näringslivshistoria, Stockholm: pp. 332, 335; courtesy Think City: p. 324; © Thomson Reuters/ScreenOcean: pp. 100, 102; Marc Tielemans/Alamy Stock Photo: p. 312; courtesy UCL Culture (photo Thomas Kador): p. 280; UPI/Alamy Stock Photo: p. 319; collection of Ben Vandenput (photo Ben Vandenput): p. 76; © Victoria and Albert Museum, London: pp. 36 (E.1243-1937), 40 (E.104-1995, photo Adrian Forty), 116 (REPRO.1854A-18), 236 (T.30-1992); collection of Richard Wentworth (photo Richard Wentworth): p. 300; © Western Australian Museum, Welshpool, Perth: p. 348; Wichita-Sedgwick

County Historical Museum, KS: p. 108;
collection of Tom Wilkinson (photo Tom
Wilkinson): p. 220; World Image Archive/Alamy
Stock Photo: p. 288.